Interfacial Electrochemistry
An Experimental Approach

E. Gileadi (Eliezer)
E. Kirowa-Eisner
J. Penciner
Tel-Aviv University

Interfacial Electrochemistry
An Experimental Approach

1975
ADDISON-WESLEY PUBLISHING COMPANY, INC.
ADVANCED BOOK PROGRAM
Reading, Massachusetts
London · Amsterdam · Don Mills, Ontario · Sydney · Tokyo

Library of Congress Cataloging in Publication Data

Gileadi, Eliezer.
  Interfacial electrochemistry.

  Bibliography: p.
  1. Electrodes. 2. Electric double layer.
I. Kirowa-Eisner, E., joint author. II. Penciner, J.,
joint author. III. Title.
QD571.G54   541'.3724   75-2451
ISBN 0-201-02398-9
ISBN 0-201-02399-7 pbk.

Copyright © 1975 by Addison-Wesley Publishing Company, Inc.
Published simultaneously in Canada

All rights reserved. No part of this publication may be reproduced, stored in a retrieval system, or transmitted, in any form or by any means, electronic, mechanical, photocopying, recording, or otherwise, without the prior written permission of the publisher, Addison-Wesley Publishing Company, Inc., Advanced Book Program, Reading, Massachusetts 01867, U. S. A.

Manufactured in the United States of America

This book is dedicated to the memory of Yaki Zemer, our student and friend, whose devotion to his people, to his friends, and to science will always be cherished.

CONTENTS

I. PREFACE xiii
   1. Definition and Importance of the Field xiii
   2. Scope of the Book xv
   3. Acknowledgments xviii

II. SOME THEORETICAL CONSIDERATIONS 1
   1. The Structure of the Ionic Double Layer 1
      1.1 The Helmholtz Model
      1.2 The Gouy-Chapman Model
      1.3 The Stern Model
      1.4 The Effect of Solvent
      1.5 The Dielectric Constant in the Double Layer
   2. Potential Differences Across Interphases 16
      2.1 General
      2.2 The Potential Difference Between Two Identical Phases
      2.3 The Potential Difference Between Two Phases at Equilibrium
   3. Polarizable and Nonpolarizable Interphases 21
   4. Electrocapillary Thermodynamics 27
      4.1 The Equations of Electrocapillary Thermodynamics
      4.2 Experimental Methods for the Measurement of γ
      4.3 The Shape of the γ - E and C - E Curves in Different Systems

5. Electrode Kinetics 43

   5.1 Introduction

   5.2 Equations for a One-Step Reaction

   5.3 The Symmetry Factor, β

   5.4 Equations for Multistep Processes

   5.5 Double-Layer Effects: The Frumkin Correction

   5.6 Types of Overpotential

6. Evaluation of Kinetic Parameters 60

   6.1 The Overall Reaction

   6.2 The Tafel Slope b

   6.3 The Exchange-Current Density i_o

   6.4 The Stoichiometric Number ν

   6.5 Reaction Orders in Electrode Kinetics

   6.6 The Effect of pH

   6.7 Isotope Effects

   6.8 Apparent Energy of Activation

7. Adsorption of Intermediates 75

   7.1 The Quasi-Equilibrium Assumption

   7.2 Calculation of Tafel Slopes With the Use of the Quasi-Equilibrium Assumption

   7.3 The Effect of the Adsorption Isotherm on the Kinetic Parameters

8. The Adsorption Pseudocapacity 86

   8.1 Physical Significance of the Adsorption Pseudocapacity

   8.2 The Equivalent Circuit and Frequency Response

   8.3 Equations for Adsorption Pseudocapacity

   8.4 Open-Circuit Decay

CONTENTS

9. Adsorption of Neutral Species — 96
   - 9.1 General
   - 9.2 Adsorption Isotherms for the Electrosorption of Neutral Species
   - 9.3 The Combined Adsorption Isotherm

10. Analysis of Reactions Occurring Under Mixed Activation and Diffusion Control — 105
    - 10.1 Statement of the Problem
    - 10.2 The Variation of Current with Time at Constant Potential
    - 10.3 The Significance of the αn_a Term in Irreversible Polarography, Chronopotentiometry and Linear Potential Sweep

11. Fundamentals of Corrosion — 111
    - 11.1 Corrosion Rates and Mixed Potentials
    - 11.2 Passivity
    - 11.3 Potential/pH Diagrams
    - 11.4 Corrosion Prevention

12. Batteries and Fuel Cells — 127
    - 12.1 Introduction
    - 12.2 Batteries
    - 12.3 Fuel Cells

13. Fundamentals of Electroplating — 138
    - 13.1 The Cathodic Discharge of Metal Ions
    - 13.2 Electrocrystallization and Electroplating
    - 13.3 The Properties of a Plating Bath
    - 13.4 Plating of Active Metals

III. EXPERIMENTS

1. Experiments in Electronics Related to Electrochemistry 151

   1.1 The Operational Amplifier

   1.2 Experimental

   1.3 Basic Circuits in Electrochemical Measurements

2. Reference Electrodes 206

   2.1 The Dynamic Hydrogen Electrode

   2.2 The Palladium/Hydrogen Electrode

   2.3 The $Ag/AgClO_4$ Electrode in Propylene Carbonate

3. Double Layer Measurements 233

   3.1 Systems without Specific Adsorption

   3.2 Systems with Specific Adsorption

   3.3 Determination of the Relative Surface Excess

   3.4 Strong Adsorption: Potential Dependence of Charge

4. Structure of the Ionic Double Layer and Electrode Processes 262

   4.1 The Reduction of Iodate Ion on Mercury

   4.2 The Reduction of Persulfate Ion on Mercury

   4.3 The Reduction of Nickel Ion on Mercury

5. Kinetics of the Hydrogen-Evolution Reaction on Solid Electrodes 293

   5.1 The Hydrogen-Evolution Reaction on Bright Platinum

   5.2 The Hydrogen-Evolution Reaction on Copper and Lead

CONTENTS

6. The Rotating-Disc Electrode 308

   6.1 The Diffusion Coefficient of Hydroquinone

   6.2 Determination of the Rate Constant and Transfer Coefficient for the Fe^{2+}/Fe^{3+} System

   6.3 Determination of the Rate Constant and Transfer Coefficient for $[Fe(CN)_6]^{4-}/[Fe(CN)_6]^{3-}$ System

7. Reaction Order Studies 328

   7.1 The Variation of i_o with E_r and with Concentration

   7.2 Measurements at High Overpotentials

8. Determination of Kinetic Parameters by Galvanostatic and Potentiostatic Transients 347

   8.1 Galvanostatic Transients

   8.2 Potentiostatic Transients

9. Cyclic Voltammetry 368

   9.1 The Effect of Medium and of Electrode Material

   9.2 Kinetic Parameters from Linear Potential Sweep

   9.3 Mechanism of Oxidation of p-phenylenediamine (i) Determination of the Number of Electrons Transferred by Constant-Current Electrolysis

   9.4 Mechanism of Oxidation of p-phenylenediamine (ii) Determination of the Number of Electrons Transferred by Constant-Potential Electrolysis

   9.5 Mechanism of Oxidation of p-phenylenediamine (iii) Steps in the Overall Reaction

10. Voltammetry at Controlled Current and at Controlled Potential 397

    10.1 Chronopotentiometry (constant-current voltammetry)

    10.2 Chronopotentiometry with Current Reversal

- 10.3 Voltammetry at Constant Potential
- 10.4 Two-Step Voltammetry at Controlled Potential

11. **Adsorption of Oxygen and Hydrogen on Noble-Metal Electrodes** — 431
    - 11.1 Fast Linear Potential Sweep
    - 11.2 Potential-Step Transients

12. **Electrosorption** — 444
    - 12.1 Linear Potential Sweep Transients on Platinum
    - 12.2 Determination of the Surface Concentration of Benzene as a Function of Sweep Rate and of Potential
    - 12.3 Kinetics of Electrosorption of Benzene on Platinum
    - 12.4 The Isotherm for Electrosorption of Benzene

13. **Absorption of Hydrogen in Palladium and Diffusion Through It** — 471
    - 13.1 The Solubility and Diffusion Coefficient of Hydrogen in Pd
    - 13.2 Equilibrium Between Adsorbed and Absorbed Hydrogen in Pd

14. **Kinetic Studies of the $Ni(OH)_2/NiOOH$ System** — 494
    - 14.1 The Charge Capacity of NiO_x Electrodes
    - 14.2 Open-Circuit-Decay Measurements on NiO_x Electrodes

INDEX — 523

I.   PREFACE

   1.   Definition and Importance of the Field

   The field of electrochemistry is commonly divided into two branches: "Ionics" and "Electrodics". The former deals with the properties and behavior of charged particles (e.g. ions, colloidal particles, etc.) in the bulk of the solution, while the latter deals with equilibria established and processes taking place <u>at</u> the electrode. A more precise definition may be called for to specify the boundary between these two branches of electrochemistry; in other words, just how close to an electrode does an ion have to be to be considered as being <u>at</u> the electrode. The definition is linked to that of an <u>interphase</u> - a phase between two phases. When two phases (e.g. a metal electrode and an aqueous solution of a salt) are brought into contact with each other, a transition region usually forms in which the properties of each component deviate markedly from its properties in the bulk of either phase. This region is called the <u>interphase.</u> Electrodics deals with the structure and properties of the interphase between an electrode and a solution and the processes taking place across such interphases. In practice, the thickness of the interphase is rarely more than 100 Å, and the critical part of it is a very thin layer extending 4–6 Å out into the solution from the electrode surface. The understanding of the structure of this very narrow region, and particularly of the processes taking place across it, is the essence of electrodics.

Ionics, often referred to as the physical chemistry of ionic solutions, is given fair representation in general textbooks and treatises on physical chemistry, in addition to many monographs dealing with specific aspects of the field. Electrodics has not been similarly favored by the authors of textbooks on physical chemistry. In most available textbooks on physical chemistry and even on electrochemistry, the subject is either omitted altogether or is given a brief, out-of-date and superficial accounting, which in no way reflects either its importance in the understanding of physical chemistry or its enormous technological implications. Although several advanced monographs and series of review articles have been published in the last decade (see bibliography at the end of the introduction section), a textbook of electrochemistry dealing with electrodics from a modern point of view at the undergraduate level, was not available until very recently.

In searching through texts of "Experiments in Physical Chemistry", the authors discovered an even more extreme situation. The sections entitled "Electrochemistry" deal exclusively with the physical chemistry of ionic solutions (Ionics) and experiments in electrodics are almost entirely absent from

SCOPE OF THE BOOK   xv

these texts.

It was with this situation in mind that we set out to produce a book which will contain some 40-50 experiments in various aspects of electrodics. We have restricted ourselves to experiments in electrodics only, because this book is intended to complement the available literature in electrochemistry rather than to compete with it. We wish to close a gap in the wall rather than to attempt strengthening sections already standing.

The importance of electrodics in both science and technology can hardly be overemphasized. An understanding of the mechanism of charge transfer, which is a central problem in electrodics, will lead to better understanding of organic and inorganic redox processes, of electrocrystallization of metals, etc. Adsorption, another important aspect of electrodics, affects the clotting of blood in the vicinity of metallic objects inserted into the body (e.g. artificial heart valves) or the rate of corrosion of a pipeline carrying oil or water. In technology, the understanding of such processes as corrosion, electroplating, electroforming and fields such as batteries, fuel cells and electroorganic synthesis is greatly dependent on a fundamental and detailed understanding of the respective electrodic processes taking place. Thus, while electrodics has its proper place among the many branches of fundamental physical chemistry, it can be linked almost directly to problems of great technological importance.

2.   Scope of the Book

This book is divided into two parts. In the first part, a

short account is given of the fundamentals of electrode kinetics and double-layer theory. This is meant to serve as a refresher and a quick reference rather than as a source for studying electrodics. Accordingly, only some of the basic assumptions and descriptions of the physical situation are presented and these are followed by the final equations. The long and winding road in between is left for other, more complete texts in the field. A bibliography of major books and series of reviews in the field has been added at the end of this section, but detailed references have been left out because this part deals mostly with well established fundamentals.

The second part of the book is devoted to a description of the experiments. These are set in groups of 2-5 experiments, each group corresponding to one particular aspect of electrodics. Each group is preceded by an introduction giving some of the theory relevant to it. The student is encouraged to begin by reading the first part of the book and is referred to specific chapters in other books and reviews for a proper understanding of the theory underlying each experiment. The notes supplied with each experiment can then be used for quick reference to the relevant equations and as a reminder of the theory.

The experiments are designed so that each group can be performed in one or two days. This poses a major difficulty, since in most properly conducted experiments in electrodics, the procedures for cleaning the cell and the electrodes and purification of the solution take several days As far as possible, experiments were chosen which are least sensitive to the impurity level in solution. The results obtained may nevertheless be expected to deviate numerically from the best values reported in the literature; however, the underlying mechanisms will be the same and it is hoped that the student will acquire a feeling for the way in which the results are used to arrive at the physical picture. It cannot be overemphasized, however, that proper experiments in a research laboratory in electrode kinetics and double-layer studies must always be performed with utmost attention paid to the purity of all parts of the system.

Most instruments used in research in electrodics can be purchased "off the shelf" from a number of companies. Nevertheless, some familiarity with the way instruments operate is very useful in electrodics, as it is in many other fields of physical chemistry. This allows full use to be made of the capabilities of available instruments and is helpful in avoiding artificious results which may sometimes arise from the use of instruments outside the limits of their capabilities. A number of experiments in electronics is accordingly included. These will help to familiarize students with some basic properties of operational amplifiers which are the central building blocks for instruments in electrodics, as well as in many other fields of applied science.

This book is intended primarily for graduate students in physical chemistry, particularly those wishing to specialize in electrochemistry. It will also be useful, we hope, to the scientist who wishes to acquire a basic competence in electrochemical techniques or do research in the field. Some prior knowledge in the field of interfacial electrochemistry will be necessary for full appreciation of the experiments performed, although part II on "Some Theoretical Considerations" may serve as an introductory text for the understanding of the physical basis of the phenomena encountered in the following experiments.

3. Acknowledgments

Thanks are due to many of our colleagues and students who helped in various ways in the preparation of this book. Mr. N. Tshernikovski and N. Lavie provided invaluable assistance in the design and construction of electronic circuits and mechanical parts, respectively. Mr. S. Efrima tried out the experiments and made many useful suggestions for improving them. Dr. S. Gottesfeld read the manuscript and made many useful comments. Mr. D. Lankovitzki provided skillful assistance in the preparation of the experiments. Experiment No. 1 on electronics related to electrochemistry was written by Dr. Ch. Yarnitzky of the Technion, Israel Institute of Technology. Finally, we wish to thank the secretarial staff for bearing with us in typing the various drafts of the book and Mrs. R. Magen for preparing the drawings.

II. SOME THEORETICAL CONSIDERATIONS

1.   The Structure of the Ionic Double Layer.

1.1.   The Helmholtz Model.

The interphase between a metallic electrode and an aqueous solution of an electrolyte behaves like an electric capacitor in that it is capable of storing electric charge. This fact was realized by Helmholtz almost a century ago. To account for this phenomenon, Helmholtz proposed a model of the interphase in which all the excess charge on the metal is located at its surface and there exists in the solution a rigidly held layer of oppositely charged ions in a plane parallel to the surface of the electrode and very close to it. This is the Helmholtz parallel-plate-capacitor model of the ionic double layer.

The modern reader will hardly be surprised to find that the actual physical situation is more complicated than that envisaged by Helmholtz a century ago. Accurate measurements of the numerical value of the double-layer capacity show that the interphase can never be represented simply by a parallel-plate capacitor. First, all double-layer capacitors are "leaky", i.e., they can be self-discharged by electrochemical reactions taking place across the interphase. This can be represented formally by considering the total impedance of the double layer as that of a capacitor and resistor in parallel[*]. Second, the charge held by the capacitor is not a linear function of the potential across it; in other words, the numerical value of the capacity depends on potential. One therefore has to define an integral capacitance K and a differential capacitance C, given by

$$K = \frac{q}{\Delta\phi} \quad ; \quad C = \frac{dq}{d(\Delta\phi)} \qquad [\text{II. 1}]$$

where q is the charge held by the capacitor and $\Delta\phi$ is the potential drop across its terminals, in this case, across the interphase.

Third, the ionic-double-layer capacity is found to depend not only on the composition of the solution but also on the concentration of the electrolyte in a given system.

The double-layer capacitance per unit area, according to Helmholtz, is given by

$$C_H = \frac{\varepsilon}{4\pi d} \qquad [\text{II. 2}]$$

where ε is the dielectric constant and d is the thickness of the

---

[*] In certain cases, notably that of the interphase between mercury and a pure solution of an aqueous electrolyte, the rate of the electrochemical reaction is negligibly low over a wide potential range, so that the capacitor can be regarded as essentially non-leaky.

1.2 GOUY-CHAPMAN MODEL   3

double layer, i.e. the effective distance between the plates of the capacitor. Thus, there is nothing in this model to account for the dependence of the measured capacity on potential or solution concentration, although the different values obtained in the presence of different electrolytes could be attributed to a variation of d with the ionic radius.

1.2 The Gouy-Chapman Model

A different model, which predicts a dependence of the measured capacity on both potential and electrolyte concentration, was proposed by Gouy and by Chapman independently. This model came to be known later as the diffuse-double-layer model.

Associated with an excess charge on the surface of the metal there is always an electric field in the interphase. A positive charge on the electrode will attract the negative ions in the solution and repel the positive ions. In the view of Gouy and Chapman, a Helmholtz-type double layer does not form because the electrostatic interaction between the field and the charges on the ions is counteracted by random thermal motion which tends to level all concentration differences in the solution. An equilibrium is reached which can be represented by the Boltzmann-type distribution equation

$$n_i(x) = n_i^o \exp\left(-\frac{z_i e_o \phi_x}{kT}\right) \qquad [\text{II}.3]$$

where $n_i(x)$ and n_i^o are the numbers of ions in unit volume at a distance x from the electrode surface and in the bulk of the solution, respectively, ϕ_x is the potential at x with respect to the potential in solution taken as zero, e_o is the electronic charge and z_i is the valency of the ion, including sign. Equation [II.3] is one of the two fundamental equations in diffuse-double-layer theory, relating the charge density to potential. The other is Poisson's equation, which, for the one-dimensional case considered here, can be written as

$$\frac{d^2\phi_x}{dx^2} = -\frac{4\pi\rho_x}{\varepsilon} \qquad [\text{II.4}]$$

where ρ_x is the volume charge density, given by

$$\rho_x = \Sigma z_i n_i(x) = \Sigma z_i n_i^o \exp\left(-\frac{z_i e_o \phi_x}{kT}\right) \qquad [\text{II.5}]$$

Solving these equations for the case of a symmetrical electrolyte, Gouy and Chapman were able to derive the following two relationships between the excess surface charge density q_M on the metal, the differential capacity C_G of the diffuse double layer and the potential ϕ_o at the limit of the diffuse double layer

$$q_M = \left(\frac{2kT\varepsilon}{\pi}\right)^{1/2} c^{1/2} \sinh\left(\frac{ze_o \phi_o}{2kT}\right) \qquad [\text{II.6}]$$

$$C_G = \left(\frac{z^2 e_o^2 \varepsilon}{2\pi kT}\right)^{1/2} c^{1/2} \cosh\left(\frac{ze_o \phi_o}{2kT}\right) \qquad [\text{II.7}]$$

Gouy and Chapman identified the potential ϕ_o with the potential ϕ_M on the surface of the metal. This is equivalent to the assumption that the ions can be regarded as point charges. We shall see in the next section that an improvement on the Gouy-Chapman model can be made by taking into consideration the finite size of the ions, which defines their distance of closest

1.2 GOUY-CHAPMAN MODEL

approach to the electrode.

The general expression for the variation of potential with distance in the diffuse double layer is rather complex. At some distance from the electrode this expression reduces to a simple exponential decay law for values of ϕ_o not exceeding ca. 0.1 volt.

$$\phi_x = \phi_o \exp(-\kappa r) \qquad [\text{II.8}]$$

where κ is the well known Debye reciprocal length defined as

$$\kappa = \left(\frac{4\pi e_o^2}{\varepsilon kT} \Sigma n_i^o z_i^2 \right)^{1/2} = \left(\frac{8\pi N e_o^2}{1000 \,\varepsilon\, kT} \right)^{1/2} I^{1/2} \qquad [\text{II.9}]$$

where I is the ionic strength in solution and N is Avogadro's number.

In the theory of Debye-Hückel the quantity κ^{-1} defines the effective radius of the ionic atmosphere around a specific ion. In much the same way it defines the effective thickness of the diffuse double layer in the Gouy-Chapman theory. Thus κ^{-1} may also be regarded essentially as the thickness of the interphase.

The Gouy-Chapman model represents a substantial improvement over the Helmholtz model, in that a dependence of the differential capacity on both potential and concentration is predicted (equation [II.7]). However, wide discrepancies between theoretical predictions and experimental results still exist. The measured capacity is much lower than that calculated from equation [II.7], except in very dilute solutions. Its variation with potential does not follow the hyperbolic cosine function except in very dilute solutions and at potentials near the potential of zero charge.*

1.3. The Stern Model

The next step in the development of the double-layer theory was a very simple one, although it may only appear simple now that the results are known. A combination of the two previous models, with some of the ions adhering to the electrode as suggested by Helmholtz and some forming a Gouy-Chapman-type diffuse layer, was suggested by Stern as a more realistic way of describing the physical situation at the interphase. This could have been deduced from the Gouy-Chapman model if the fact had been taken into account that ions have a finite size and have a distance of closest approach $\underline{a}$ where the potential is $\phi_o = \phi_a < \phi_M$. Thus, in the Stern model, two capacitors are effectively operative in determining the total differential capacity of the interphase, one, C_H, due to the Helmholtz layer and the other, C_G, due to the diffuse layer. Since the two regions are consecutive in space, the two capacities will act in series, and the total capacitance C will be given by

$$\frac{1}{C} = \frac{1}{C_H} + \frac{1}{C_G} \qquad [\text{II.10}]$$

---

* The potential of zero charge, E_z, is an important quantity which is characteristic of the electrode material and depends to some degree on the components of the solution. It is defined as the potential at which $q_M = 0$ in a given system. Its significance and methods for its determination will be discussed below. (cf. section II.4).

1.3 STERN MODEL

This result is very significant because it shows that the <u>lower</u> of the two contributions will be predominant in determining the total capacitance. On considering equations [II.2], [II.6], [II.7] and [II.10], it is realized that C_G will be predominant in very dilute solutions and at potentials in the vicinity of the potential of zero charge where q_M and ϕ_o or ϕ_a have low values.

Stern also recognized the effect of specific interactions and suggested a distinction between ions which were adsorbed on the electrode and those which merely approached it to the distance of closest approach. This view was used by Grahame to develop a model of the interphase which consists of three regions, although it is still commonly referred to as the double layer. The first region extends from the electrode to a plane passing through the centers of the specifically adsorbed ions. This is the inner Helmholtz plane (IHP) and its potential is denoted by ϕ_1. Next is the outer Helmholtz plane (OHP) which passes through the centers of the hydrated ions at their distance of closest approach to the electrode. The potential at the OHP is denoted by ϕ_2. Beyond the OHP lies the diffuse double layer. The potential ϕ_o at its boundary is identified with the potential ϕ_2 at the OHP. The models discussed so far are shown in Figs. II.1a-II.4a and the corresponding potential variations across the interphase are given schematically in Figs. II.1b-II.4b. Of particular importance is the fact that the potential changes linearly with distance up to the outer Helmholtz plane and then exponentially through the diffuse double layer. Numerical computation based on the Gouy-Chapman theory (cf. equations [II.8] and [II.9]), shows that in a 1<u>M</u> solution of a 1:1 electrolyte the thickness of the diffuse double

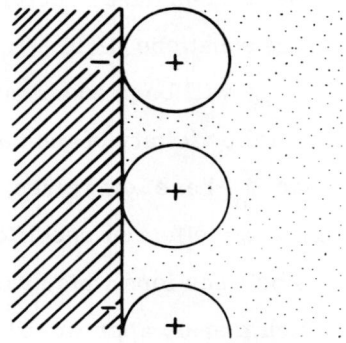

FIG. II 1a HELMHOLTZ MODEL

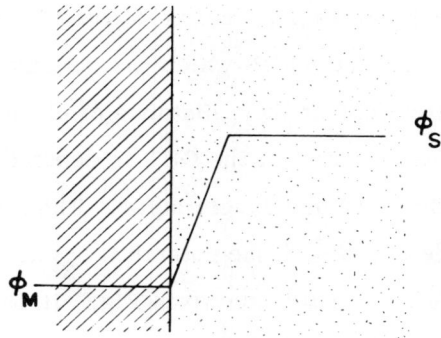

FIG. II 1b POTENTIAL PROFILE CORRESPONDING TO THE HELMHOLTZ MODEL

1.3 STERN MODEL

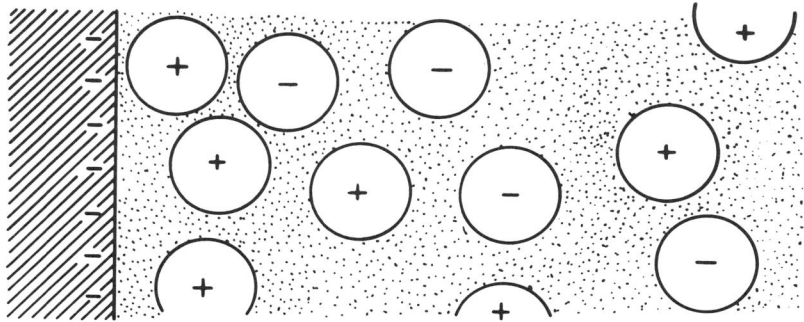

FIG. II 2a GOUY-CHAPMAN MODEL

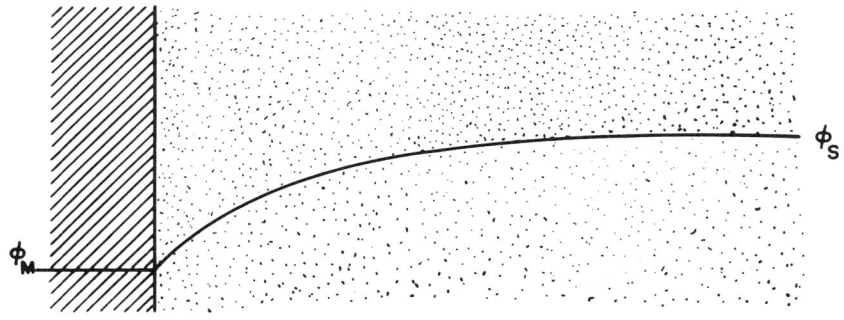

FIG. II 2b POTENTIAL PROFILE CORRESPONDING TO THE GOUY-CHAPMAN MODEL

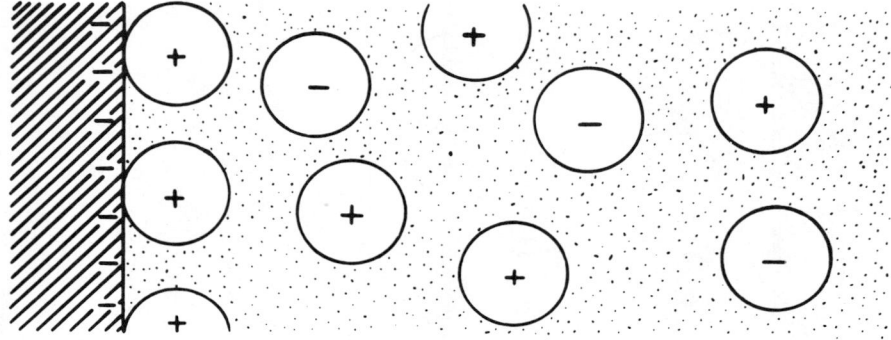

FIG. II 3a   STERN MODEL

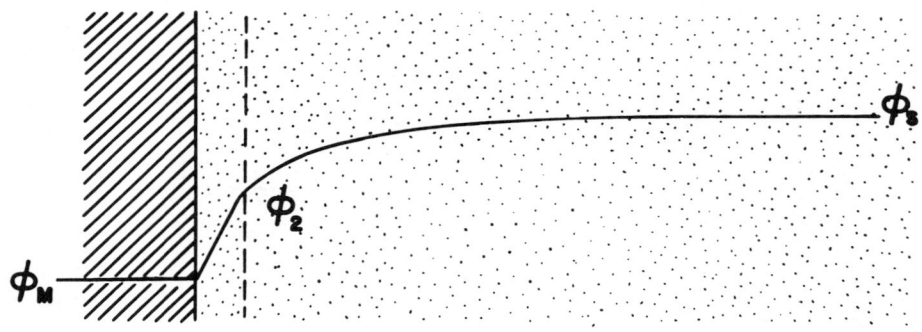

FIG. II 3b   POTENTIAL PROFILE CORRESPONDING TO THE STERN MODEL

1.3 STERN MODEL

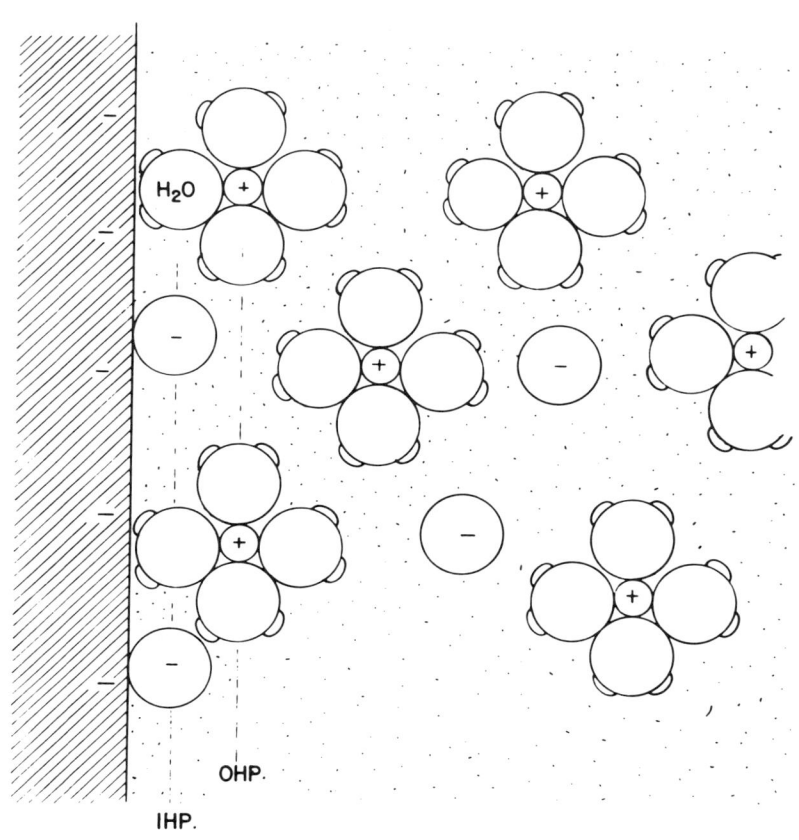

FIG. II 4a GRAHAME MODEL

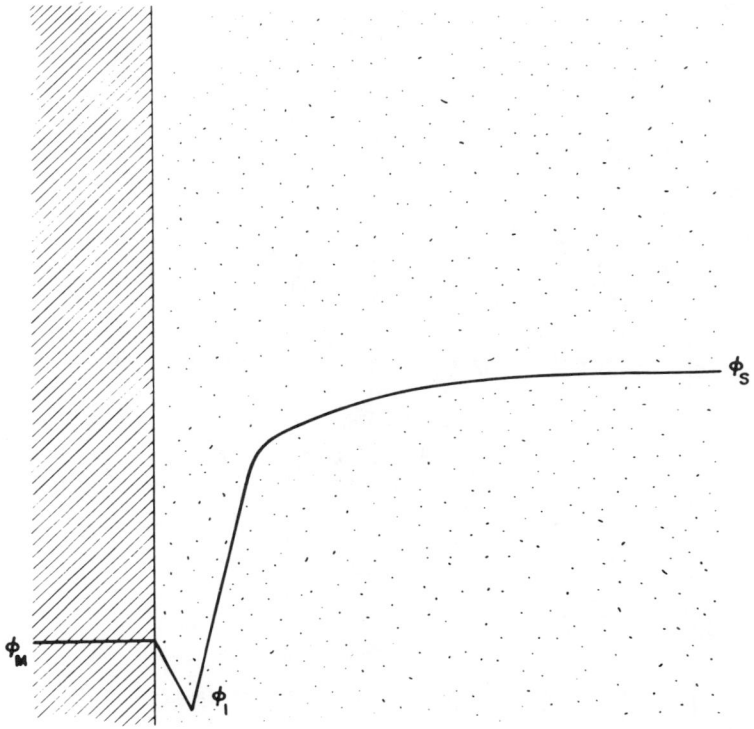

FIG. II 4b   POTENTIAL PROFILE CORRESPONDING
TO THE GRAHAME MODEL

layer would be approximately 3Å, about equal to the distance of closest approach of hydrated ions at the electrode. Thus the diffuse double layer does not, in effect, exist under these conditions. It must be admitted that the equations of the Gouy-Chapman theory do not represent a very good approximation at such high concentrations. Nevertheless, it is clear that diffuse-double-layer effects diminish with increasing concentration in solution. For this reason, electrode-kinetic studies are very often performed in concentrated solution and the variation of

1.4   EFFECT OF SOLVENT

diffuse-double-layer potential (ϕ_2 or ϕ_a) with the total metal/ solution potential difference is neglected.

1.4.   The Effect of Solvent

For some reason, the effect of the solvent was taken explicitly into account rather late in the development of double-layer theory. In the theory of Gouy-Chapman, it was seen that the diffuse part of the double layer is very similar to the ionic atmosphere around an ion. The electrode may be considered as a very large central ion, the radius of which tends to infinity.*
One would accordingly expect to find a layer of solvent molecules attached to the electrode surface in fixed orientation and a few more layers in orientation intermediate between that of the first layer and that in the bulk - an arrangement corresponding to the primary and secondary solvation shells around an ion. This is indeed found to be the best available model for the role of the solvent in the double layer. The model of the double layer proposed by Bockris, Devanathan and Muller, which explicitly takes into account the predominant existence of the solvent in the interphase, is shown in Fig. II.5 for water as the solvent. Most of the electrode is covered with an oriented layer of water molecules (the equivalent of the first hydration shell). On certain sites, the water molecules are replaced by specifically adsorbed

---

* In spite of the large radius, the electric field near the electrode is comparable to that near an ion in solution, because the electrode carries an excess of many unit charges.

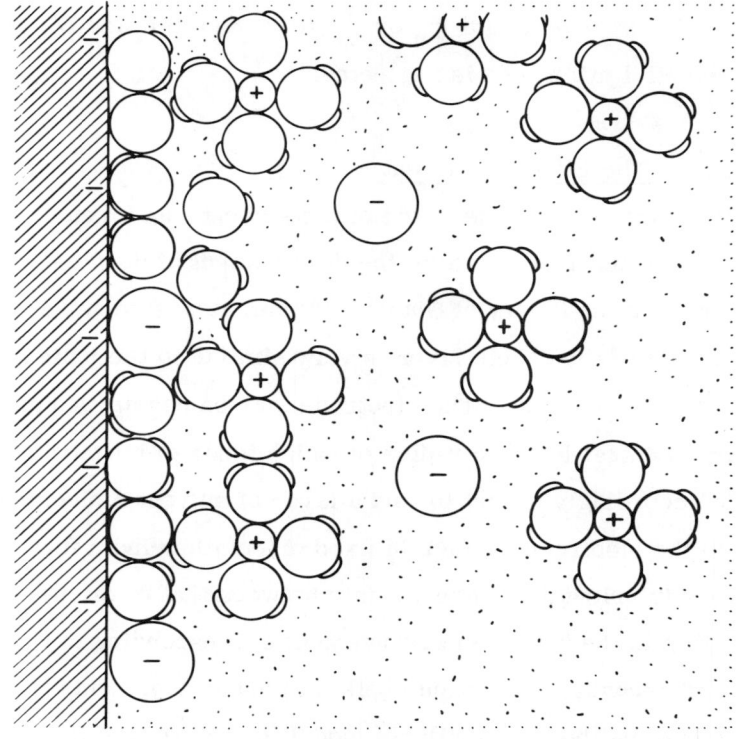

FIG. II 5   THE MODEL OF BOCKRIS DEVANATHAN AND MULLER

ions* (usually anions) which do not carry a hydration shell. The plane going through the centers of these ions is the inner Helmholtz plane. Next, one finds ions which maintain their primary hydration shell and are placed outside the first layer of water molecules adsorbed on the electrode surface. The plane through

---

* The term "contact adsorption" has been used to describe such situations since the ion may be regarded as being in direct contact with the electrode surface with no solvent molecules interposed.

the centers of these ions constitutes the outer Helmholtz plane. A fundamental difference between ions in the inner and outer Helmholtz planes is noted. In order for the ion to be contact-adsorbed, chemical work must be done to rid it of its primary hydration sheath and to remove one or more water molecules from the electrode surface. On the other hand, the final state is stabilized by the specific interactions between the ion and the surface. For ions adsorbed in the OHP, these effects are absent and the interactions are mainly electrostatic. The adsorption of neutral organic molecules is always contact adsorption and the energy required to remove the appropriate number of water molecules from the surface plays a central role in determining the energetics of the process, as we shall see below.

1.5. The Dielectric Constant in the Double Layer

The question of the proper value of the dielectric constant to be used in treating the double layer is an interesting one. If the value of approximately 80 for bulk water is used in, say, equation [II.2], the value of the Helmholtz capacity turns out to be about 200 $\mu F/cm^2$. This is an order of magnitude higher than the values found experimentally. An accurate theoretical calculation of the dielectric constant in different regions of the interphase is not possible. In the first layer of water molecules, which are tightly held to the electrode surface, a value of $\varepsilon = 6$ is most often used. This is consistent with measured values of C_H on the one hand and with measurements of ε in bulk water at very high frequencies on the other. (In the latter case, the contribution of rotation of the water molecules to the measured dielectric constant is negligible). In the second layer the water molecules are much more free to rotate, but the structure of bulk water is not yet attained. In this region, the value of ϵ has been estimated to be 30-40. Beyond the second layer of water molecules (i.e. essentially outside the OHP) the dielectric constant is assumed to be that of bulk water and this value is used in equations [II.4] to [II.9] which relate to diffuse-double-layer theory.

2. Potential Differences Across Interphases

2.1 General

The driving force for all chemical processes is the gradient of chemical potential μ along a generalized coordinate x_i

$$\text{driving force} = -\left(\frac{\partial \mu}{\partial x_i}\right)_{T,P} \qquad [\text{II}.11]$$

The coordinate x_i may represent distance, a change in concentration, gravitational field etc. Equilibrium is reached when this gradient approaches zero. This happens when the chemical potential of the system (at a fixed temperature and pressure) reaches a minimum. For most applications in chemistry, the chemical potential of a species i can be written as

$$\mu_i = \mu_i^o + RT \ln a_i \qquad [\text{II}.12]$$

where μ_i^o is a constant and a_i is the activity of the species <u>i</u> in the system. When charged particles are involved in a chemical equilibrium, it is convenient to include a term explicitly related to the coulombic interaction of the charge with the electric field. One then writes

$$\bar{\mu}_i = \mu_i + z_i F \phi \qquad [\text{II}.13]$$

2.1 GENERAL

or

$$\Delta \bar{\mu}_i = \Delta \mu_i + z_i F \Delta \phi \qquad [\text{II.14}]$$

where $\bar{\mu}_i$ is the electrochemical potential of the species $\underline{i}$, ϕ is the inner potential of a phase, $\Delta \phi$ is the potential difference across the interphase and $z_i F$ is the electric charge on the particle, per mole. In equation [II.13] there is an apparent separation between the chemical part μ and the electrical part $zF\phi$ of the electrochemical potential. It must be remembered that this is only a formal separation assumed for convenience, since all chemical forces are basically electrical in nature. The usefulness of this assumption will, however, become apparent below.

Equations [II.12] and [II.13] combine to yield

$$\bar{\mu}_i = \mu_i^o + RT \ln a_i + z_i F \phi \qquad [\text{II.15}]$$

It will be noted that the electrochemical potential $\bar{\mu}$ in equations [II.13] to [II.15] is equal to the chemical potential μ if the charge on the particle is zero, that is, where neutral species are concerned.

The quantity $\Delta \phi$ in equation [II.14] is the <u>inner potential difference</u> between two phases. Multiplied by the charge, $z_i e_o$, on a particle, it represents the electrical work involved in transferring the particle across the interphase (short-range chemical interactions of the particle with the two phases being ignored). It was previously noted that the separation between the "chemical" and "electrical" parts of the electrochemical potential is only a formal one and while $\bar{\mu}_i$ or changes in it can be measured, it is impossible to obtain experimentally either μ_i or ϕ in equation

[II.13]. <u>The potential difference $\Delta\phi$ across a single interphase cannot be measured</u>. It will be shown below that, although $\Delta\phi$ is inaccessible experimentally, the change, $\delta\Delta\phi$, can be measured quite easily.

The contributions to $\Delta\phi$ are of two kinds; those which arise as a result of excess charge and those which are due to re-orientation of surface dipoles.

$$\Delta\phi = \Delta\psi + \Delta\chi \qquad [\text{II.16}]$$

where $\Delta\psi$ is the <u>outer potential difference</u> due to excess charge and $\Delta\chi$ is the <u>surface or "chi" potential difference</u>, generated by surface dipole re-orientation. The outer potential ψ is the potential "just outside" a phase. It can be shown that a particle may be placed near a phase in such a position that it is far enough to avoid all short-range interactions, yet close enough that the potential is essentially equal to that on the surface, or differs from it in a calculable manner. The work required to bring a charged particle from infinity to such a position near a phase defines the outer potential ψ of the phase. <u>Both ψ and $\Delta\psi$ can, in principle, be measured experimentally</u>.

From equation [II.16], it is clear that since $\Delta\phi$ is not measurable and $\Delta\psi$ is, the surface potential difference $\Delta\chi$ is also a quantity which cannot be measured experimentally. In the case of a metal-solution interphase, both phases contribute to the surface potential difference

$$\Delta\chi = \chi_{(\text{metal})} - \chi_{(\text{solution})} \qquad [\text{II.17}]$$

Neither of these contributions can be measured. Several

attempts have been made to calculate $\chi_{(solution)}$ theoretically, on the basis of various models, while attempts to calculate $\chi_{(metal)}$ have so far been rare.

2.2 The Potential Difference between Two Identical Phases

A special case of equation [II.14] arises when the two phases considered are chemically identical and are <u>not</u> placed in direct contact with each other. In this case $\Delta\mu_i = 0$ and equation [II.14] yields

$$\Delta\phi = \frac{\Delta\bar{\mu}_i}{z_i F} \qquad [\text{II}.18]$$

since $\Delta\bar{\mu}_i$ is a thermodynamically defined, measurable quantity, $\Delta\phi$ can also be measured in this particular instance. <u>The inner potential difference between two chemically identical phases can be measured.</u> This fact is of great practical importance. Consider the measurement of the potential of any d.c. source (e.g. a battery, thermocouple, power supply etc.). The measurement is performed by connecting two copper wires between the terminals of the source and the terminals of a suitable voltmeter. The relevant species in this case are the electrons in the two copper wires ($z_i = -1$) and equation [II.18] becomes

$$\Delta\phi = -\frac{\Delta\bar{\mu}_e}{F} = E \qquad [\text{II}.19]$$

The potential E measured by the voltmeter equals the **inner** potential difference between the two chemically identical phases and is proportional to the difference in the electrochemical potential of the electrons in the two wires. It may be added here that the two wires need not be identical. The voltmeter measures the difference in $\bar{\mu}_e$ between its two terminals (which are chemically identical). In each terminal $\bar{\mu}_e$ is equal to that in the wire connected to it, since the electrons are at equilibrium in two pieces of metal which make contact. Thus the instrument still measures the difference in electrochemical potential of the electrons in the two wires, even if one is made of copper and the other is made of nickel. However, the inner potential difference between the copper and nickel wires will not be measurable, because an additional interphase has been formed between one of the terminals of the voltmeter and the nickel wire; the potential difference $\Delta\phi$ across this interphase is not measurable (cf. next section).

2.3 The Potential Difference between Two Phases at Equilibrium

Consider now the application of equation [II.14] to systems at equilibrium. To be specific, assume two different metal wires welded together, the electrons in one piece of metal being in equilibrium with the electrons in the other piece of metal. Equation [II.14] will take the form

$$\Delta\mu_e = F\Delta\phi \qquad [II.20]$$

since $\Delta\bar{\mu}_e = 0$ and $z_i = -1$. Thus, at the contact between the two metals a potential difference will exist, which is proportional to the difference in the chemical potential of the electrons in the two phases.

Even more interesting is the case of an ion in solution in equilibrium with the same ion in the metal lattice (the electrode), e.g. a copper electrode in equilibrium with Cu^{++} ions in solution. Here $\Delta \bar{\mu}_{Cu^{++}} = 0$ and $z_i = +2$. Thus, instead of equation [II. 20] one has

$$\Delta \mu_{Cu^{++}} = -2F\Delta \phi \qquad [II.21]$$

When the value of $\Delta \mu_{Cu^{++}}$ is substituted in this equation, it leads, after minor transformation, to the equation

$$\Delta \phi = \Delta \phi^o + \frac{RT}{2F} \ln a_{Cu^{++}} \qquad [II.22]$$

which is the equivalent of the Nernst equation, applied to a single electrode.

It cannot be overemphasized that $\Delta \phi$ as well as $\Delta \phi^o$ in equation [II. 22] are still not measurable quantities, although it is possible, in principle, to calculate these potential differences on the basis of a specific model.

3. Polarizable and Nonpolarizable Interphases

When a current is passed through an interphase or a potential difference is applied across it, two extremes of behavior may be distinguished. Fig. II 6a shows the response of a highly polarizable interphase. A small current forced across the interphase gives rise to a large change in the potential difference across it and vice-versa. In Fig. II 6b, the behavior of a nonpolarizable interphase is described. A relatively high current can be passed with little change in the potential. Whether an inter-

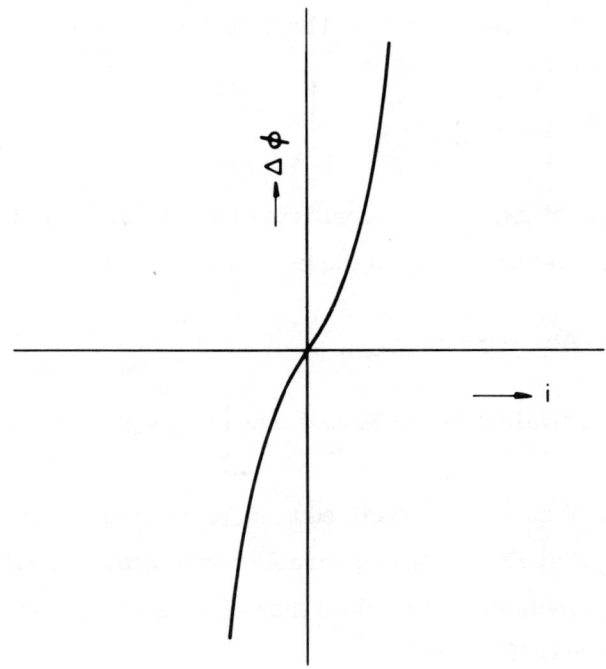

FIG. II 6a CURRENT - POTENTIAL RESPONSE OF HIGHLY POLARIZABLE ELECTRODE

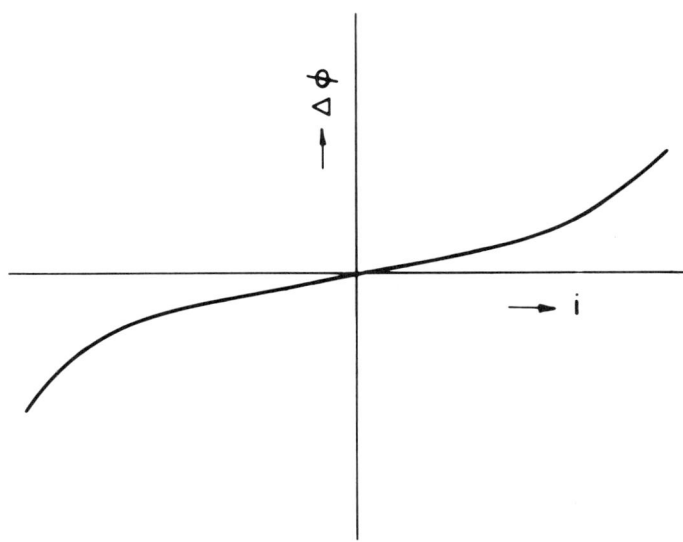

FIG. II 6b CURRENT - POTENTIAL RESPONSE OF HIGHLY NON-POLARIZABLE ELECTRODE

phase will be polarizable or not, or the degree to which it will be polarizable, depends on the specific rate constants for the electrochemical reaction taking place across it, as will be shown below. An ideal polarizable interphase is one which can allow the passage of any current without causing a change in the potential difference across it. Figs. II 7a and II 7b show the electrical analogs of the two extreme cases. Fig. II 7c represents an intermediate real case. The resistance in parallel with the double-layer capacitance is called the faradaic resistance R_F. Its value is constant near the reversible potential and decreases with potential in other regions. In all cases it is inversely proportional to the electro-

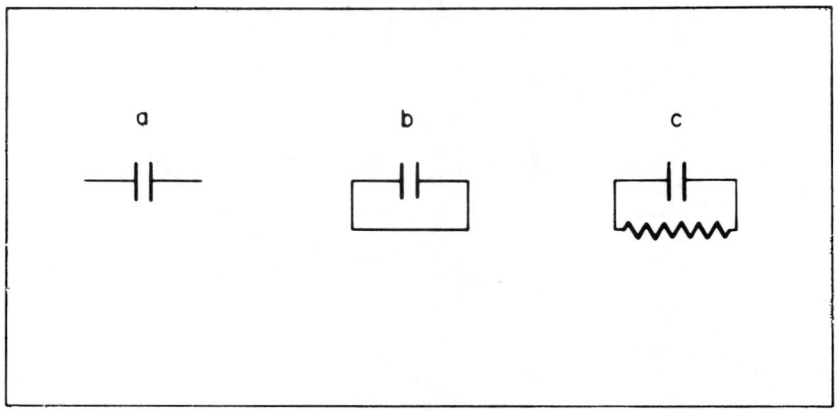

FIG. II 7   ELECTRICAL EQUIVALENT OF
(a) IDEAL POLARIZABLE ELECTRODE
(b) IDEAL NON-POLARIZABLE ELECTRODE
(c) PARTIALLY POLARIZABLE ELECTRODE

chemical specific rate constant.*

Consider now an experimental setup comprising two electrodes in an electrolyte solution, with a variable source of potential, which can be applied to the electrodes to drive a current through the cell (Fig. II 8a). The electrical equivalent of this system is shown in Fig. II 8b. The total potential E across the cell terminals when a current i is passed through the cell will be given by

---

* The equivalent circuits in Fig. II 7 are somewhat oversimplified, since they imply that the potential across all interphases and the current across real interphases are zero when the charge is zero. This is not the case in practice. Additional circuit elements may be introduced to correct for the discrepancy. However, since we shall be interested here only in changes in $\Delta\phi$ due to the current crossing the interphase, the representation in Fig. II 7 will suffice.

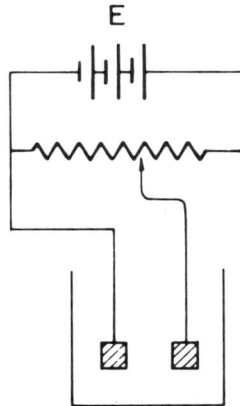

FIG. II 8a SCHEMATIC REPRESENTATION OF SIMPLE EXPERIMENTAL SETUP

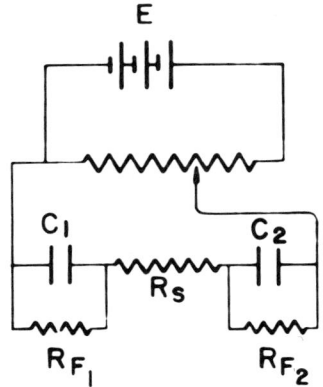

FIG. II 8b ELECTRICAL EQUIVALENT OF SETUP SHOWN IN FIG. II 8a

$$E = iR_{F_1} + iR_s + iR_{F_2} + \text{const}$$

$$= \frac{q_1}{K_1} + iR_s + \frac{q_2}{K_2} + \text{const} \qquad [\text{II. 23}]$$

where q_1, K_1 and q_2, K_2 are the charges and the integral capacities of the two interphases, and the constant depends on solution composition and electrode material, but not on the current (see footnote on p. 24). If interphase 1 is highly polarizable and interphase 2 is almost nonpolarizable, (i.e. $R_{F_1} \gg R_{F_2}$) and the solution resistance R_s is kept to a minimum, one finds from equation [II. 23] at constant composition in solution

$$\left(\frac{\partial E}{\partial i}\right)_\mu = R_{F_1} + R_s + R_{F_2} \cong R_{F_1} \qquad [\text{II. 24}]$$

Thus the total change in potential across the cell terminals is essentially equal to the change across one of the interphases. Alternatively, one could write

$$\Delta\phi_1 = iR_{F_1} + \text{const} \quad \text{and} \quad \Delta\phi_2 = iR_{F_2} + \text{const} \qquad [\text{II. 25}]$$

and hence

$$\left[\frac{\partial(\Delta\phi_1)}{\partial i}\right]_\mu = R_{F_1} \gg \left[\frac{\partial(\Delta\phi_2)}{\partial i}\right]_\mu = R_{F_2}$$

By combining a polarizable interphase with a non-polarizable one, it is possible, therefore, to measure the change in $\Delta\phi$, although for fundamental reasons, its absolute value cannot be

4.1 ELECTROCAPILLARY THERMODYNAMICS

measured.*

This approach can be extended to the measurement of changes in $\Delta\phi$ as a result of applied current even for quite nonpolarizable electrodes. A three-electrode system is used for this purpose. The current is applied between the electrode being studied (the working electrode) and a counter or auxiliary electrode. The potential of the working electrode is measured with respect to a suitable reference electrode (the nonpolarizable electrode). The current through the reference electrode is kept to a very low value and its potential is not affected by the much larger currents applied between the working and counter electrodes.

4. Electrocapillary Thermodynamics
4.1 The Equations of Electrocapillary Thermodynamics

The understanding of electrode/solution interphases owes a great deal to electrocapillary measurements on liquid-mercury electrodes. These measurements involve the determination of the metal/solution interfacial tension γ as a function of solution composition and of potential. Accurate measurements can be made by means of the electrocapillary electrometer originally developed by Lippmann in 1878. This and other methods for the

---

* To be precise, one should remember that R_{F_1} and R_{F_2} are the respective integral faradaic resistances, which may be potential dependent. Thus $(\partial E/\partial i)_\mu$ or $(\partial \Delta\phi/\partial i)_\mu$ may also depend on potential.

measurement of γ will be discussed in the next section. To date, no accurate experimental method for the determination of surface tension on solid electrodes is available.

The fundamental equation upon which electrocapillary thermodynamics is founded is the Gibbs adsorption isotherm

$$d\gamma = -\Sigma \Gamma_i d\mu_i \qquad \text{[II.27]}$$

where Γ_i is the amount, per unit area of the interphase, of the i-th component in the whole interphase, in excess of the amount which would be there if the uniform bulk concentration were maintained right up to the surface. Γ_i is the <u>surface excess</u> of the i-th component in the interphase. It is noted that according to the above definition, Γ can take on both positive and negative values.

The Gibbs adsorption isotherm is applicable to any interphase. In the special case of an ideally polarizable metal/solution interphase, it can be shown that the change in interfacial tension $d\gamma$ is given by

$$d\gamma = -q_M dE - \Sigma \Gamma_i d\mu_i \qquad \text{[II.28]}$$

where q_M is the excess charge density on the metal, E is the potential with respect to some reference electrode, and the sum is taken over all species in the solution.

A specific example will help to clarify the meaning and possible applications of Eq. [II.28]. Consider the following electrochemical cell:

$$Cu^I/Ag/AgCl/KCl, CH_3OH, H_2O/Hg/Cu^{II}$$

4.1 ELECTROCAPILLARY THERMODYNAMICS

in which a highly polarizable mercury electrode is combined with a reversible (nonpolarizable) Ag/AgCl electrode. The solution phase contains KCl dissolved in a mixture of two solvents, methanol and water. The electrometer used to measure the potential will measure the difference in electrochemical potential between the electrons in Ag and in Hg, since these two metals are in intimate contact with the two copper wires leading to the electrometer (cf. section 2.2). This potential will be denoted E_-, to imply that it is a potential measured with respect to a reference electrode reversible to the anions in solution. Equation [II. 28] applied to this system takes the form

$$-d\gamma = q_M dE_- + \Gamma'_{K^+} d\mu_{KCl} + \Gamma'_{CH_3OH} d\mu_{CH_3OH} \quad [II.29]$$

where Γ'_i is the relative surface excess of the i-th species, defined in this case as

$$\Gamma'_{K^+} = \Gamma_{K^+} - \frac{X_{K^+}}{X_{H_2O}} \Gamma_{H_2O} \;;\; \Gamma'_{CH_3OH} = \Gamma_{CH_3OH} - \frac{X_{CH_3OH}}{X_{H_2O}} \Gamma_{H_2O} \quad [II.30]$$

where X_i is the mole fraction of the i-th species in the bulk of the solution. In dilute aqueous solutions the relative surface excess Γ'_i of any species is essentially equal to the surface excess Γ_i since $X_i/X_{H_2O} \ll 1$. Nevertheless, it should be remembered that only the relative surface excess is measurable.

The second and third terms on the right-hand side of equation [II. 29] depend on the components of the cell chosen, but the first term is independent of them. From equation [II. 29] the following useful relationships can be obtained:

$$\left(\frac{\partial \gamma}{\partial E}\right)_\mu = -q_M \qquad [\text{II. 31}]$$

$$\left(\frac{\partial \gamma}{\partial \mu_{KCl}}\right)_{\mu_{CH_3OH}, E_-} = -\Gamma'_{K^+} \qquad [\text{II. 32}]$$

$$\left(\frac{\partial \gamma}{\partial \mu_{CH_3OH}}\right)_{\mu_{KCl}, E_-} = -\Gamma'_{CH_3OH}$$

Equation [II. 31] is known as the <u>Lippmann equation</u>. It allows a determination of the charge density as a function of potential from measurement of the variation of surface tension with potential of an ideally polarizable electrode at constant solution composition.* Differentiation of the Lippmann equation with respect to potential yields

$$-\left(\frac{\partial^2 \gamma}{\partial E^2}\right)_\mu = \left(\frac{\partial q_M}{\partial E}\right)_\mu = C \qquad [\text{II. 33}]$$

The differential capacity obtained in this manner is rather inaccurate because it is the result of double differentiation of the experimental $\gamma - E$ data. The main importance of equation [II. 33] is in its integral forms

---

* The potential E is used in equations [II. 31] and [II. 33] instead of E_- since these equations apply regardless of the type of reference electrode used.

4.1 ELECTROCAPILLARY THERMODYNAMICS

$$q_M = \int_{E_z}^{E} C\,dE \quad ; \quad \gamma - \gamma_{max} = \iint_{E_z}^{E} C\,dE\,dE \qquad [\text{II.34}]$$

From the Lippmann equation it is seen that $q_M = 0$ at the potential where γ reaches its maximum value. In other words, the potential of zero charge coincides with the electrocapillary maximum.* The lower limits of the double integral in equation [II. 34] which should be E_z and $E_{\gamma=\gamma_{max}}$ are therefore identical.

By the use of equation [II. 34], the charge density and surface tension at different potentials can be calculated from accurate measurements of the differential capacity as a function of potential. The first operation requires a knowledge of the potential of zero charge and the second requires in addition a knowledge of γ_{max}, both as integration constants. These quantities are obtained independently from electrocapillary measurements.

Equation [II. 32] or its equivalents in other systems is the basis for adsorption studies employing electrocapillary measurements. Adsorption is associated with a decrease in the interfacial tension γ. The great accuracy attainable in the determination of the interfacial tension γ allows accurate determinations of Γ' even though this quantity is obtained by a differentiation process. When a single salt in a single solvent is studied, one may write equation [II. 32] in the form

---

* This statement is strictly true only if the interphase is ideally polarizable.

$$\left(\frac{\partial \gamma}{\partial \mu_{salt}}\right)_{E_{\mp}} = -\Gamma'_{\pm} \qquad [\text{II. 35}]$$

where the sign on E indicates the type of reference electrode used and the corresponding (opposite) sign on Γ' refers to the ion for which Γ' is determined.

4.2 Experimental Methods for the Measurement of γ

The most accurate and reliable measurements of the metal/solution interfacial tension γ are obtained with the electrocapillary electrometer devised by Lippmann. A schematic diagram of this instrument is shown in Fig. II 9. A fine glass capillary, slightly tapered towards the lower end, is dipped into the solution containing a reference electrode which serves also as the counter electrode. Mercury is forced into the capillary from a reservoir of variable height and the pressure applied is measured on a manometer. The mercury in the capillary is brought to a fixed position very near to the end of the capillary and the pressure, ΔP, required to do this is measured accurately. According to the equation of Young and Laplace the pressure difference between two phases at equilibrium when the interphase is not planar, is given by

$$\Delta P = \gamma \left(\frac{1}{r_1} + \frac{1}{r_2}\right) \qquad [\text{II. 36}]$$

where r_1 and r_2 are the two radii of curvature of the interphase. Thus, if the mercury in the capillary is always brought to the same position, r_1 and r_2 are constant and the pressure applied is

4.2 MEASUREMENT OF γ 33

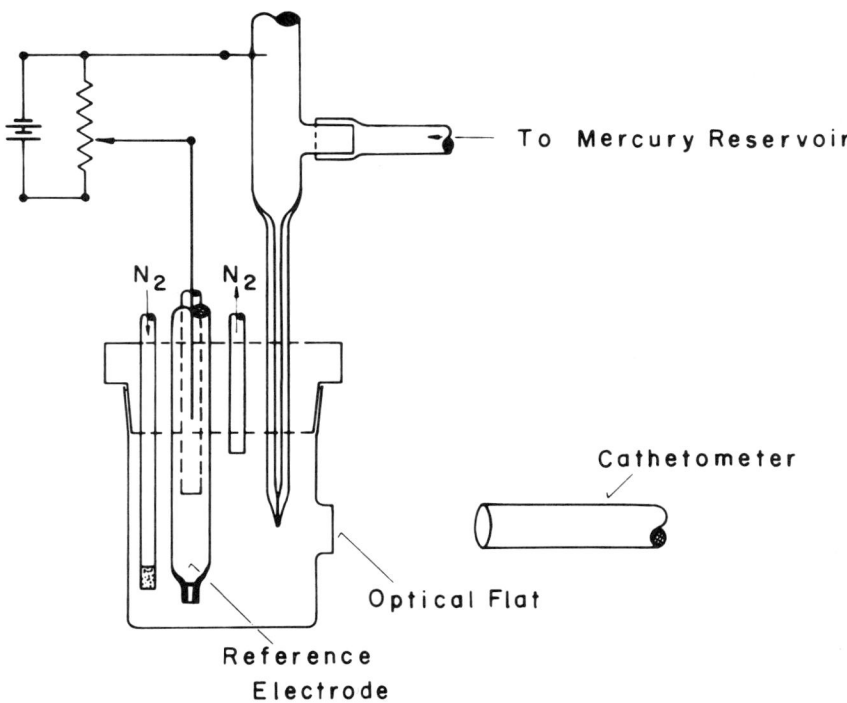

FIG. II 9   ELECTROCAPILLARY ELECTROMETER

proportional to the interfacial tension γ. A plot of ΔP vs the applied potential E can be easily converted to a plot of γ vs E after calibration in a solution of known interfacial tension. Tapering of the capillary is essential for the stability of the system. Thus, for a given value of γ there is only one set of values of ΔP, r_1 and r_2 which will satisfy equation [II.36]; that is, the mercury will be stable only at one particular position in the capillary.

A faster but less accurate way of determining γ is by the drop-time or drop-weight method. This is based on the assumption that a drop growing at the end of a capillary will be detached when the gravitational force on its mass just equals the force due to the interfacial tension.

Thus

$$V_d = \frac{2\pi r \gamma}{\Delta \rho g} \qquad [\text{II.37}]$$

or

$$\tau = \frac{2\pi r \gamma}{V \Delta \rho g} \qquad [\text{II.38}]$$

where V_d is the volume of a drop, V the volume delivered per unit time, τ is the drop time, $\Delta \rho$ is the difference in density between the dropping liquid and the medium (e.g., between mercury and an aqueous solution) and r is the radius of the capillary. Thus the volume of a drop, V_d, and the drop time, τ, are both proportional to the interfacial tension γ. Equations [II.37] and [II.38] are best applied in the manner of equation [II.36], by calibration in a system of known γ.

Close examination of the process of detachment of a drop from the end of a capillary (by fast photography) reveals a rather complex phenomenon, which cannot be described rigorously by equation [II.37]. It is found, however, that by application of a small correction factor which is a function of $(r/V_d^{1/3})$, accurate values of γ can be obtained.

A method which has gained importance in recent years (although it has been known for a very long time) is the maximum-bubble-pressure method. In this technique, the capillary is

4.2 MEASUREMENT OF γ

turned upwards and the pressure on the mercury is slowly increased. When a new drop begins to form, its radius of curvature is large and the pressure ΔP required to make it grow further is relatively low (cf. equation [II.36]). The radius of curvature reaches a minimum when the drop is hemispherical, and starts to grow again as the drop grows further. Thus, beyond this critical size the system is unstable and the drop bursts. The corresponding pressure is linearly related to the interfacial tension γ and its variation with applied potential can serve to obtain the electrocapillary curve, after suitable calibration in solutions of known γ/E behavior. The maximum-bubble-pressure method has been automated recently by Mohilner and combined with an "on-line" computer which can both run the experiment and analyze the results. Although the inherent accuracy of the method is not improved in this way, the final derived results (curves of Γ_i, C or q_M vs E) tend to be more accurate because measurements of γ are made at more closely spaced values of E and since random human errors are eliminated.

    Several other methods have been proposed for the determination of γ for the liquid-metal/aqueous-solution interphase. None of these equals the electrocapillary electrometer or the maximum-bubble-pressure method for accuracy or the drop-time method for convenience.

    The situation with respect to solid metals is much less satisfactory. Many methods have been attempted, all of which depend on determination of the surface free energy or some property related to it.

The accuracy and reproducibility of the γ values obtained for the solid-metal/aqueous-electrolyte interphase are at least two orders of magnitude lower than for mercury and the theory behind the experiment is often not well understood.

4.3 The Shape of the γ-E and C-E Curves in Different Systems

The shape of the electrocapillary curve for mercury in contact with an aqueous solution of a simple electrolyte (NaF), where neither the anion nor the cation is specifically adsorbed, is shown in Fig. II 10a. The corresponding C vs E curve is shown in Fig. II 10b. The absence of specific adsorption is indicated by the fact that the potential of zero charge is independent of the concentration of this electrolyte in solution. Fig. II 11a and II 11b show the effect of specific adsorption of the anion on the manner in which γ and C vary with potential. There is a marked decrease in γ at each potential on the positive (anodic) branch of the electrocapillary curve, and the potential of zero charge is shifted in the negative direction. The same is found, of course, when the cation is specifically adsorbed, but in this case, the decrease in γ is on the cathodic branch of the curve and E_z is shifted in the positive direction.

The effect of adsorption of neutral organic molecules is shown in Figs. II 12a and II 12b. Adsorption is restricted to a well defined potential region on both sides of the potential of zero charge. Outside this region, the γ vs E curve coincides with that obtained in the absence of the organic adsorbate. The limits of the adsorption region are marked by two rather sharp peaks in

4.3 SHAPE OF THE γ - E AND C - E CURVES

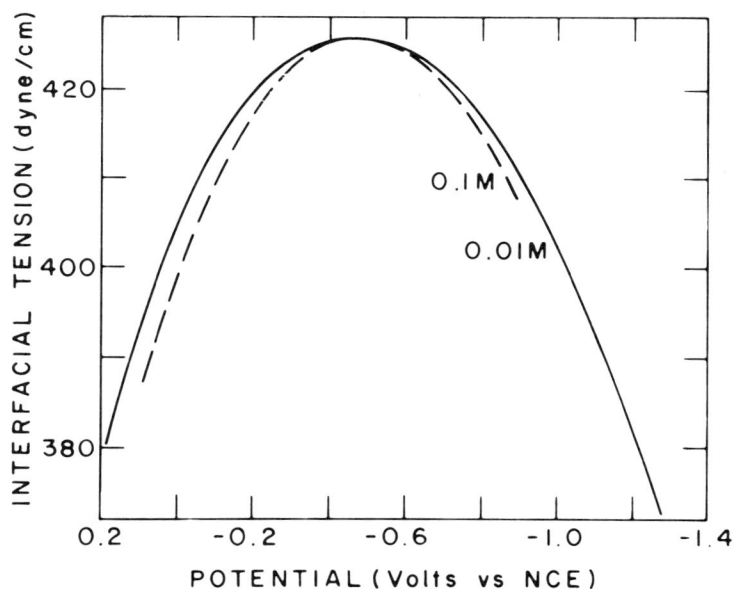

FIG. II 10a ELECTROCAPILLARY CURVES FOR NaF SOLUTIONS
(D.C. Grahame and B.A. Soderberg, Tech. Report No. 14, ONR (1954); J. Lawrence, R. Parsons and R. Payne, J. Electroanal. Chem. 16, 193 (1968)).

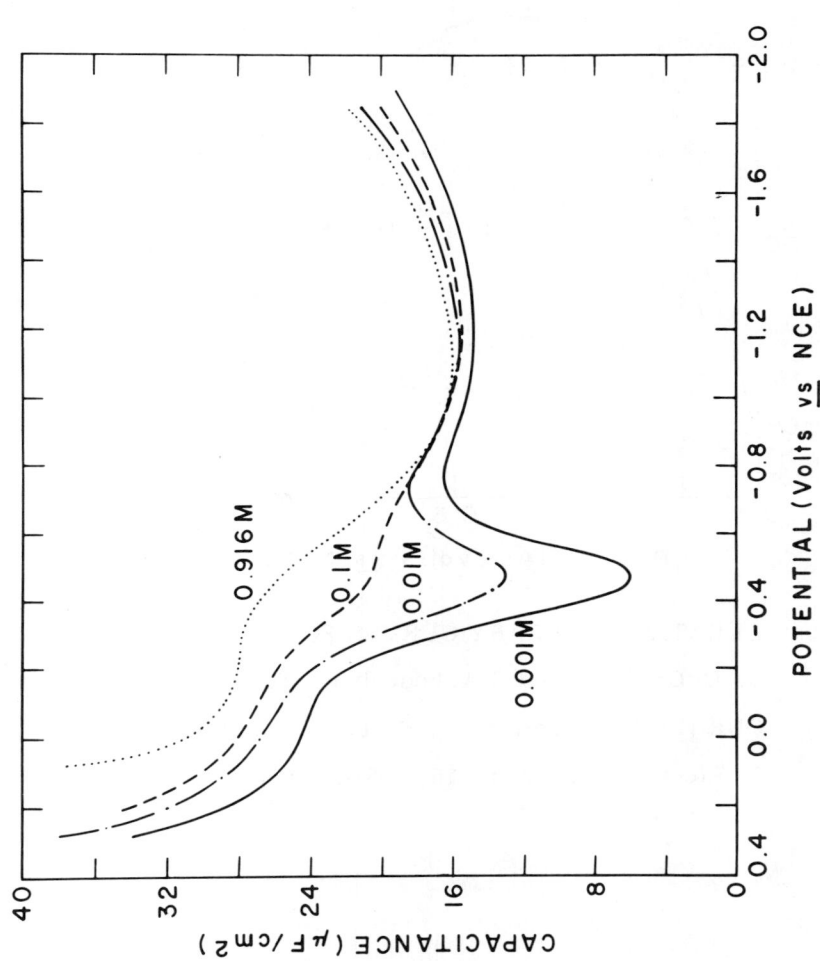

FIG. II 10b CAPACITY/POTENTIAL CURVES FOR THE SYSTEM SHOWN IN FIG. II 10a

4.3 SHAPE OF THE γ - E AND C - E CURVES

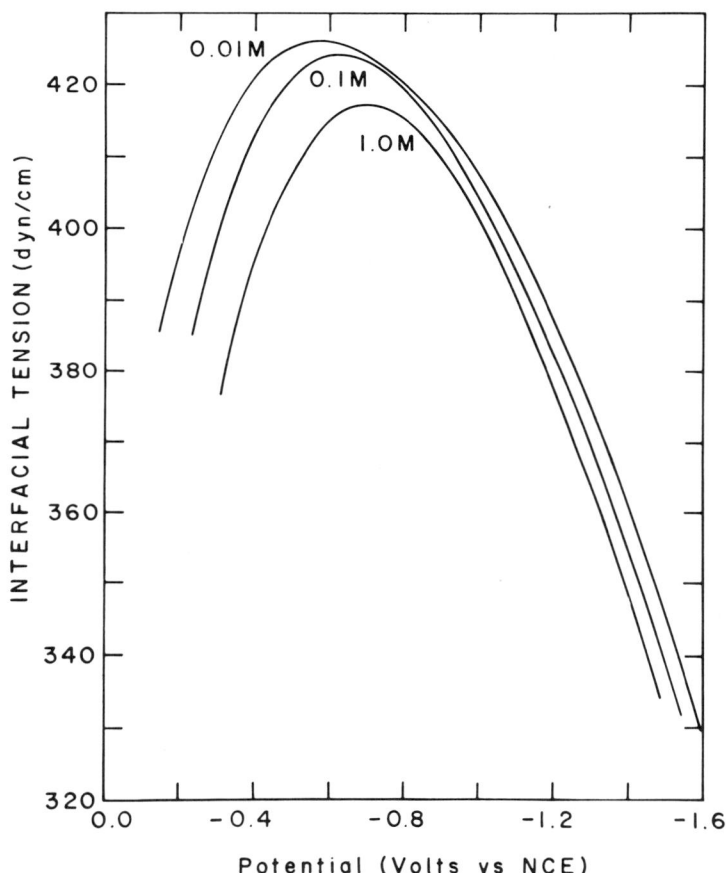

FIG. II 11a ELECTROCAPILLARY CURVES FOR KBr SOLUTIONS
(J. Lawrence, R. Parsons and R. Payne, J. Electroanal. Chem. 16, 193 (1968)).

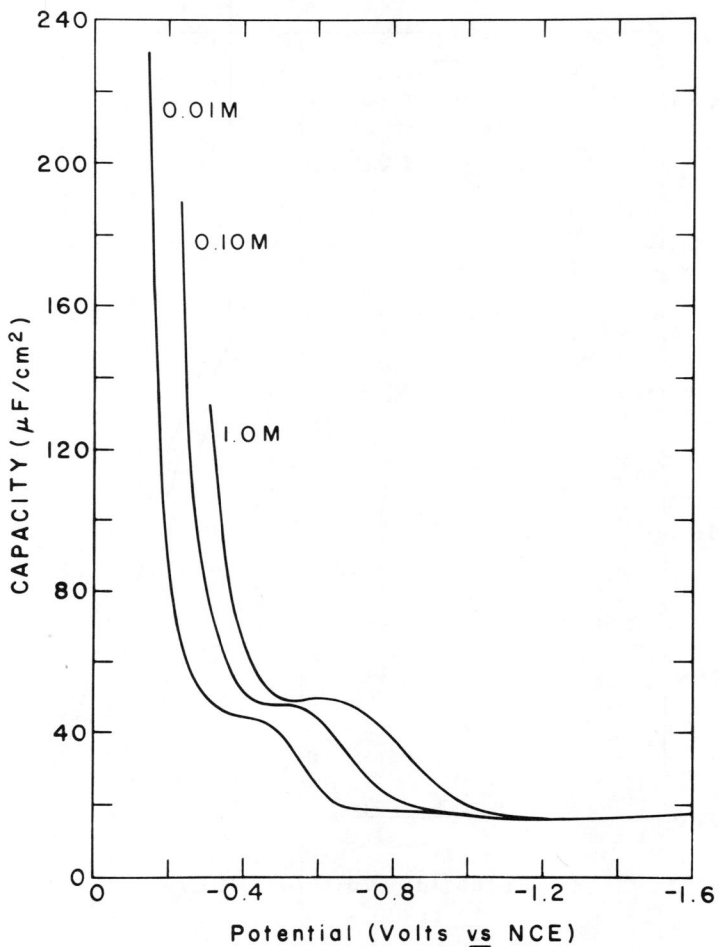

FIG. II 11b CAPACITY/POTENTIAL CURVES FOR THE SYSTEM SHOWN IN FIG. II 11a

4.3 SHAPE OF THE γ - E AND C - E CURVES   41

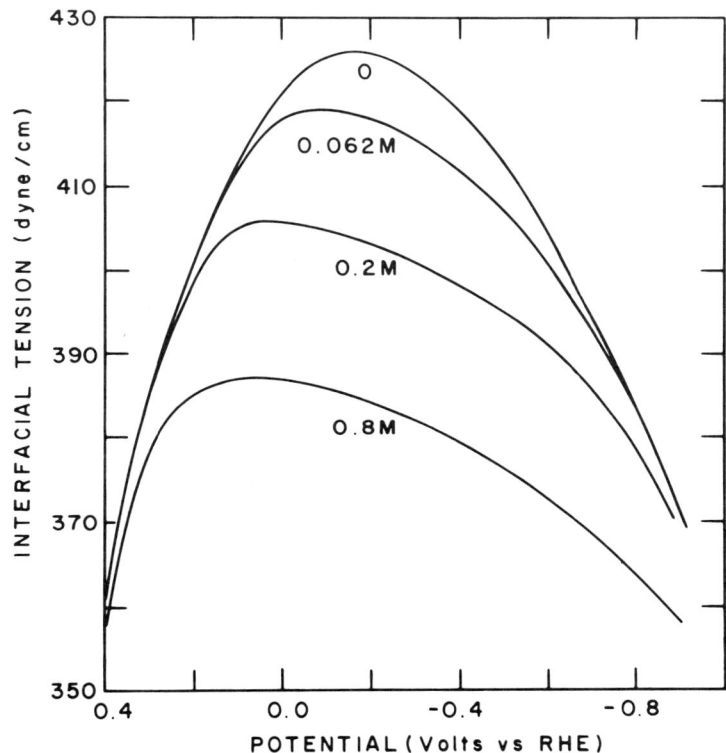

FIG. II 12a   ELECTROCAPILLARY CURVES FOR n-BUTANOL IN 0.1 M HCl SOLUTIONS (K. Muller, Ph. D. Dissertation, Philadelphia (1965)).

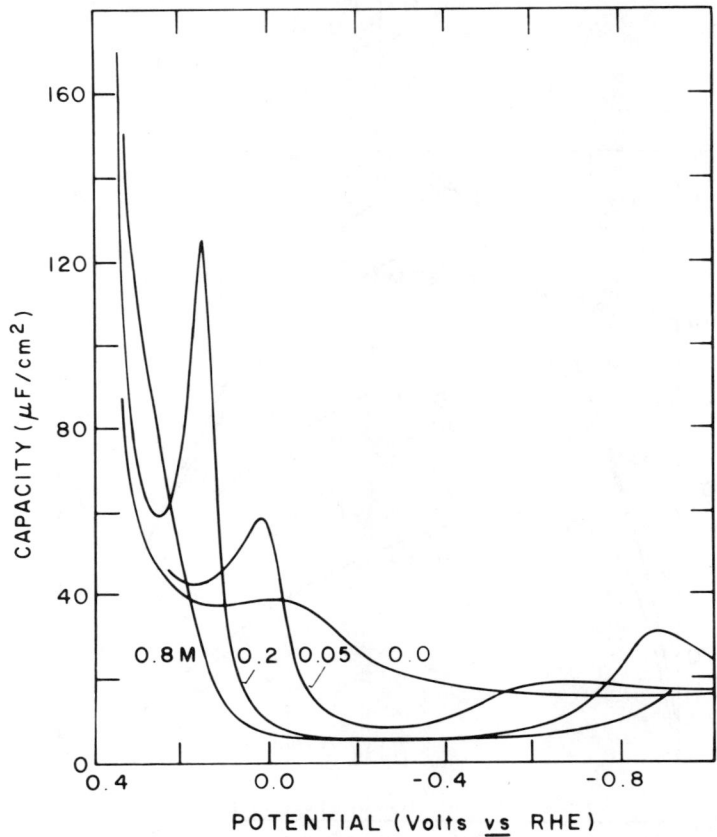

FIG. II 12b   CAPACITY/POTENTIAL CURVES FOR THE SYSTEM SHOWN IN FIG. II 12a

the capacity-vs-potential curve (Fig. II 12b). These peaks are referred to as the adsorption-desorption peaks. Between the peaks, in the region of adsorption, the capacity has a low, essentially constant, value. The potential region over which adsorption occurs increases with increasing bulk concentration of the organic species.

5.1 INTRODUCTION

5. Electrode Kinetics

5.1 Introduction

A simple relationship between the current density i and the overpotential η was proposed by Tafel in 1905, on the basis of his experimental investigations of the hydrogen-evolution reaction. This is known as the <u>Tafel equation</u>

$$\eta = a - b \log i \qquad [\text{II.39}]$$

The overpotential η is the difference between the potential E measured when a current density i is passed across the interphase and the reversible potential E_r corresponding to the system at equilibrium with no current passing.

$$\eta = E - E_r \qquad [\text{II.40}]$$

The parameter <u>b</u> is the Tafel slope

$$b = \pm \left(\frac{\partial \eta}{\partial \log i}\right)_{T,P,\mu} \quad \text{or} \quad b' = \pm \left(\frac{\partial \eta}{\partial \ln i}\right)_{T,P,\mu} \qquad [\text{II.41}]$$

The parameter <u>a</u> can be written as

$$a = b \log i_o = b' \ln i_o \qquad [\text{II.42}]$$

where i_o is the exchange-current density, to be discussed below. The Tafel equation should preferably be written in exponential form as

$$i = i_o \exp\left(\pm \frac{\eta}{b'}\right) = i_o 10^{\pm \eta/b} \qquad [\text{II.43}]$$

The positive and negative signs in equations [II.41] and [II.43] refer to anodic and cathodic reactions, respectively.* The numerical value of the Tafel slope depends on the mechanism of the electrode reaction and the experimental determination of this parameter is one of the important tools in the evaluation of reaction mechanisms.

The Tafel equation, which was originally proposed as a purely empirical relationship, has since been given a sound theoretical basis. It has been shown that the equation applies when one of the steps in a reaction sequence is rate-determining and the overpotential is high ($|\eta/b| \geqslant 1$).

5.2   Equations for a One-Step Reaction

The rate of a chemical reaction, according to the absolute rate theory, is given approximately by

$$v = \frac{kT}{h} \prod_i C_i \exp\left(-\frac{\Delta G^{o \neq}}{RT}\right) \qquad [\text{II. 44}]$$

where $\prod_i C_i$ is the product of the concentrations of the reactants, $\Delta G^{o \neq}$ is the standard free energy of activation for the reaction, and h is Planck's constant. To apply equation [II. 44] to an electrode reaction, that is, one which involves charge transfer, two modifications are made. First, since for each mole of reactant consumed or of product formed, nF coulombs of electricity will cross the interphase,

---

* The overpotential η has a negative sign for a cathodic reaction and a positive sign for an anodic reaction. The parameter b as defined in equation [II. 41] is always positive.

5.2 EQUATIONS FOR ONE-STEP REACTION

$$i = nFv \qquad [\text{II}.45]$$

Secondly, the standard free energy of activation $\Delta G^{o\neq}$ must be replaced by the standard electrochemical free energy of activation $\Delta \bar{G}^{o\neq}$. The relationship between these two quantities is similar to that between the chemical and the electrochemical potential (cf. equation [II.14]).

$$\Delta \bar{G}^{o\neq} = \Delta G^{o\neq} - \beta F \Delta \phi \qquad [\text{II}.46]$$

Thus, for a single electron transfer, equation [II.44] takes the form

$$i = F \frac{kT}{h} \prod_i C_i \exp(-\Delta G^{o\neq}/RT) \exp(\beta F \Delta \phi / RT) \qquad [\text{II}.47]$$

the parameter β in the last two equations is the <u>symmetry factor</u> which will be discussed in some detail in the next section. Equation [II.47] may be rewritten in the short form

$$i = F k_a^o \prod_i C_i \exp(\beta F \Delta \phi / RT) \qquad [\text{II}.48]$$

where

$$k_a^o = \frac{kT}{h} \exp(-\Delta G^{o\neq}/RT) \qquad [\text{II}.49]$$

The parameter k_a^o is the specific rate constant for the anodic reaction under zero-field conditions. The net anodic current density measured is the difference between the currents in the anodic and cathodic directions, hence

$$i = \vec{i} - \overleftarrow{i} \qquad [\text{II. 50}]$$

$$= Fk_a^o \prod_i C_i \exp(\beta F \Delta\phi/RT) - Fk_c^o \prod_j C_j \exp\left[-(1-\beta) F \Delta\phi/RT\right]$$

where $\vec{i}$ and $\overleftarrow{i}$ are the partial current densities for the anodic and cathodic reactions, respectively, $\prod_i C_i$ is the product of the concentrations of the reactants and $\prod_j C_j$ is the product of the concentrations of the products.* The quantity $\Delta\phi$ in equations [II. 46] to [II. 50] can be split into two terms

$$\Delta\phi = \Delta\phi_r + \eta \qquad [\text{II. 51}]$$

where $\Delta\phi_r$ is the absolute metal-solution potential difference when the system is at equilibrium.** Combining equations [II. 50] and [II. 51], one has

$$i = i_o \left\{ \exp(\beta F \eta/RT) - \exp\left[-(1-\beta) F \eta/RT\right] \right\} \qquad [\text{II. 52}]$$

---

* Note that this equation is written in such a way that the net anodic current density will be positive while the net cathodic current density will be negative.

** Equation [II. 51] is equivalent to equation [II. 40] where η is defined as $\eta = E - E_r$. The absolute values of $\Delta\phi$ and $\Delta\phi_r$ in equation [II. 51] cannot be measured, but the difference $\delta \Delta\phi = \Delta\phi - \Delta\phi_r = \eta$ is measurable. By combining a polarizable and a non-polarizable electrode, it was shown above (cf. section II. 3) that all the difference in the cell potential $\Delta E = E - E_r$ occurs at the polarizable interphase, hence $E - E_r = \Delta\phi - \Delta\phi_r = \eta$.

5.2 EQUATIONS FOR ONE-STEP REACTION

where i_o, the exchange-current density, is given by

$$i_o = Fk_a^o \prod_i C_i \exp(\beta F \Delta \phi_r / RT) = Fk_c^o \prod_j C_j \exp[-(1-\beta) F \Delta \phi_r / RT] \quad [\text{II.53}]$$

The exchange-current density is an electrochemical specific rate constant multiplied by a concentration term.

Two limiting cases of equation [II.52] are of special interest. At high positive values of the overpotential, when $(\beta F \eta / RT) \geq 1$ the second term on the right-hand side of equation [II.52] becomes negligible with respect to the first and one has

$$i = i_o \exp(\beta F \eta / RT) \quad [\text{II.54}]$$

This is identical with the Tafel equation in exponential form (equation [II.43]) if one writes $b = RT/\beta F$. The experimental Tafel equation is thus a special case of a more general rate equation, equation [II.52], applicable at relatively high values of the overpotential.

The other limiting case corresponds to systems very close to equilibrium. At low values of the overpotential, when $|\beta F \eta / RT| \leq 0.1$, equation [II.52] can be put into linear form by expanding the exponent and taking only the first two terms:

$$i = \frac{i_o F}{RT} \eta \quad [\text{II.55}]$$

At this point, it is appropriate to define more precisely the faradaic resistance R_F which was mentioned in section II.3 (cf. Figs. II.7; II.8) in connection with the discussion of some simple electrical equivalents (or equivalent circuits) of the metal/

solution interphase. Since, in general, R_F may be a function of potential, it is defined as the average

$$R_F = \frac{1}{i}\int \left(\frac{\partial \eta}{\partial i}\right)_\mu di \qquad [\text{II.}56]$$

In the Tafel region (i.e., in the region where the Tafel equation [II. 54] applies), the differential coefficient $(\partial \eta/\partial i)_\mu$ decreases exponentially with increasing overpotential.* Thus, for an anodic or cathodic process we have, by combining equations [II. 52] and [II. 56]

(cathodic) $\qquad R_F = -(b_c/i_c)\log i \qquad [\text{II.}57a]$

(anodic) $\qquad R_F = (b_a/i_a)\log i \qquad [\text{II.}57b]$

At potentials close to equilibrium, equation [II. 55] yields a faradaic resistance R_F which is a constant, independent of potential

$$R_F = \frac{RT/F}{i_o} \qquad [\text{II.}58]$$

This may be regarded as the "ohmic region" where a linear relationship between applied potential and current density is maintained to a good approximation.

---

* For a cathodic reaction, the quantities i and η are negative by definition, thus the faradaic resistance R_F always has a positive value.

5.3 The Symmetry Factor β

The origin of the symmetry factor, β, has been the subject of lengthy discussions in electrochemistry. Consider as an example the discharge step in the hydrogen-evolution reaction.

$$H_3O^+ + e_M + M = MH + H_2O \qquad [\text{II.59}]$$

where M is a site on the metal surface and e_M is an electron in the metal. The standard electrochemical free energy of the reaction, $\Delta \bar{G}^o$, is the difference in standard electrochemical potentials between the final and initial states

$$\Delta \bar{G}^o = (\bar{\mu}^o_{MH} + \bar{\mu}^o_{H_2O}) - (\bar{\mu}^o_M + \bar{\mu}^o_{H_3O^+} + \bar{\mu}^o_{e_M}) \qquad [\text{II.60}]$$

For the neutral species $\bar{\mu}^o = \mu^o$. Using equation [II.13] for the charged species, one obtains for $\Delta \bar{G}^o$,

$$\Delta \bar{G}^o = \Delta \mu^o + F(\phi_M - \phi_s) = \Delta \mu^o + F\Delta \phi \qquad [\text{II.61}]$$

where $\Delta \mu^o$ is a constant, independent of potential. Thus, the free energy of the initial state depends on the inner potential difference $\Delta \phi$ across the interphase. In the present example, a positive potential difference will stabilize the initial state, while a negative potential difference will make it more labile. Now, the activated complex represents a state which is intermediate between the initial and final states. The standard electrochemical free energy of activation $\Delta \bar{G}^{o\neq}$ is given by

$$\Delta \bar{G}^{o\neq} = \bar{\mu}^o_{\neq} - (\bar{\mu}^o_M + \bar{\mu}^o_{H_3O^+} + \bar{\mu}^o_{e_M}) \qquad [\text{II.62}]$$

where $\bar{\mu}_{\neq}^{o}$ is the standard electrochemical potential of the activated complex. Equation [II.62] can be written in the form

$$\Delta \bar{G}^{o\neq} = \Delta \mu_{\neq}^{o} + F\Delta\phi_{\neq} \qquad [II.63]$$

or

$$\Delta \bar{G}^{o\neq} = \Delta \mu_{\neq}^{o} + \beta F\Delta\phi \qquad [II.64]$$

where $\Delta\phi_{\neq} = \phi_M - \phi_{\neq}$ is the potential difference between the metal and the site of the activated complex. The symmetry factor β is the ratio between the changes in $\Delta\phi_{\neq} = \phi_M - \phi_{\neq}$ is the potential difference between the metal and the site of the activated complex. The symmetry factor β is the ratio between the changes in $\Delta \bar{G}^{o\neq}$ and $\Delta \bar{G}^{o}$ caused by a given change in the metal/solution potential difference, $\delta(\Delta\phi)$.

$$\beta = \frac{\delta(\Delta \bar{G}^{o\neq})}{\delta(\Delta \bar{G}^{o})} \qquad [II.65]$$

In a sense, β may also be viewed as an efficiency factor, since it represents the fraction of the total free energy $\delta(\Delta \bar{G}^{o})$ added to the system which is used to increase the reaction rate.

The arguments presented above may be further clarified by considering the potential energy diagram in Fig. II.13. It is seen that the increase in free energy of the initial state is accompanied by a smaller increase in the free energy of the activated complex. The difference between the changes in these quantities represents the decrease in free energy of activation. It can be seen from simple geometrical considerations that β is

5.3 SYMMETRY FACTOR, β   51

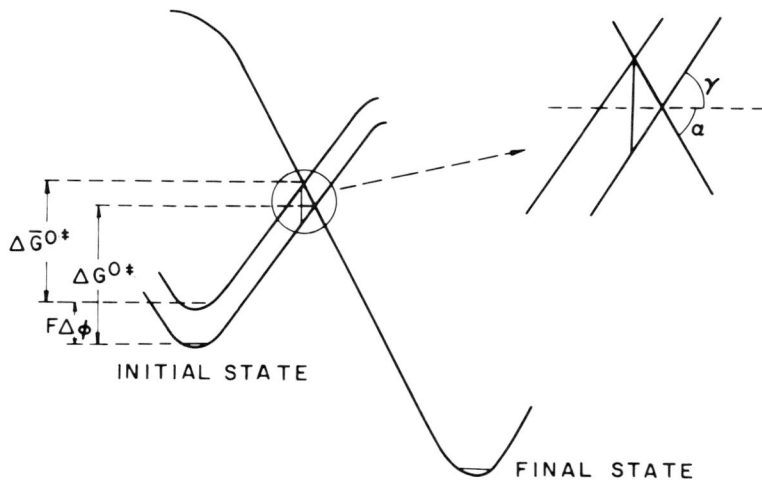

FIG. II 13 POTENTIAL ENERGY DIAGRAM FOR A CHARGE TRANSFER PROCESS

related to the slopes of the two potential energy curves at their intersection:

$$\beta = \frac{\tan \gamma}{\tan \alpha + \tan \gamma} \qquad [\text{II. 66}]$$

where $\tan \alpha$ and $\tan \gamma$ are the slopes of the two curves as shown in Fig. II.13.

Some further insight into the significance of β can be obtained by a detailed examination of the process of charge transfer. Consider the charged species involved; the initial state consists of an H_3O^+ ion in the outer Helmholtz plane and an electron in the metal. The final state consists of a discharged H atom adsorbed on the surface and a water molecule. Thus, at some point in the process, an electron will be removed from the metal to neutralize the approaching ion. This process can occur only when the initial and final states for the electron have exactly the same energy, namely, at the top of the barrier shown in Fig. II 13, where the two potential energy curves intersect.* The two structures "just before" and "just after" the electron transfer have equal probability in the overall resonance structure of the activated complex, which is the species represented by the top of the energy barrier. Thus, the activated complex may be said to be half-charged and its energy will change by half as much as that of the initial state, as a result of an applied electric field.

The last interpretation gives rise to a value of $\beta = 0.50$, while the first two interpretations (equations [II. 65] and [II. 66]) do not suggest any <u>a priori</u> value for β. It is commonly accepted, on both experimental and theoretical grounds, that the symmetry factor β should be close to one half, and this value will be assumed in the subsequent discussion of kinetic parameters.**

5.4   Equations for Multistep Processes

So far, the discussion has been restricted to the case of an electrode reaction taking place in a single step with the transfer of one electron. Most electrode reactions, however, take place in

---

* Actually, the potential energy curves do not intersect. Instead, resonance splitting occurs and the curve for the initial state merges smoothly into the curve for the final state.

** It is possible to envisage situations in which β will be far from the value of 0.5 (and would even approach zero or unity). In such cases, calculations show that β should be strongly dependent on potential. If a linear Tafel region is observed experimentally, it is most probable that β is constant in the same potential region and a value of $\beta \cong 0.5$ may be used to analyze the kinetics of the reaction.

5.4 EQUATIONS FOR MULTISTEP PROCESSES

several consecutive steps, and rate equations for the more general case have been derived. In considering such reaction sequences, it is generally assumed that charges cross the interphase one at a time, that is, that only one electron can be transferred in each elementary step. However, charge transfer may occur in any of the consecutive steps of the overall reaction. A new quantity, the stoichiometric number ν, is defined. It is the number of times a given step must take place for the overall reaction to occur once. The general equation will be

$$i = i_o \left\{ \exp[\beta(n - \vec{n} - \overleftarrow{n}) + \vec{n}] (F\eta/\nu RT) - \exp - [(1-\beta)(n - \vec{n} - \overleftarrow{n}) + \overleftarrow{n}](F\eta/\nu RT) \right\} \quad [\text{II. 67}]$$

where n is the total number of electrons transferred and $\vec{n}$ and $\overleftarrow{n}$ are the numbers of electrons transferred before and after the rate-determining step, respectively, all for one act of the overall reaction.

Equation [II. 67] is applicable provided two conditions are met. First, the coverage of the electrode surface by adsorbed intermediates formed in the reaction sequence must be low (cf. section II. 7 below) and second, the heat of adsorption must be independent of the coverage. As in the case of the single-step reaction (equation [II. 52]), two extreme cases are considered. At high overpotentials (the so-called "Tafel region"), equation [II. 67] reduce to

$$i = i_o \exp \left\{ [\beta(n - \vec{n} - \overleftarrow{n}) + \vec{n}](F\eta/\nu RT) \right\} \quad [\text{II. 68}]$$

which is a more general form of the Tafel equation, with a Tafel slope of

$$b = \frac{2.3(\nu RT/F)}{\beta(n - \vec{n} - \overleftarrow{n}) + \vec{n}} \qquad [\text{II.69}]$$

At low overpotential (the "linear" or "ohmic" region)

$$i = \frac{i_o nF\eta}{\nu RT} \qquad [\text{II.70}]$$

Occasionally, equation [II.68] is written in the form

$$i = i_o \exp\left(\frac{\alpha F \eta}{RT}\right) \qquad [\text{II.71}]$$

In this equation, α is an empirical parameter called the <u>transfer coefficient</u>. It is related to the experimentally observed slope in the linear Tafel region by the equation

$$\alpha = \pm \frac{2.3RT}{F}\left(\frac{\partial \log i}{\partial \eta}\right)_\mu = \frac{2.3RT}{F} \cdot \frac{1}{b} \qquad [\text{II.72}]$$

Comparison of equations [II.68] and [II.71] shows that α is related to the symmetry factor β and the other kinetic parameters by the equation

$$\alpha = \frac{\beta(n - \vec{n} - \overleftarrow{n}) + \vec{n}}{\nu} \qquad [\text{II.73}]$$

It is obvious, then, that the value of the transfer coefficient α in equation [II.71] cannot generally be assumed to be one-half, although the assumption that $\beta = 0.5$ is a reasonable one, as

5.4 EQUATIONS FOR MULTISTEP PROCESSES

discussed above* (cf. section II.5.3). It will be noted that the Tafel slope (equation [II.69]) depends on parameters which are determined by the reaction mechanism and serves as a useful criterion in mechanism determination.

The use of equation [II.68] may be clarified by application to a specific case. Consider, for example, the oxygen-evolution reaction proceeding by the reaction sequence shown below.

$$OH^- + M \longrightarrow MOH + e_M \qquad [II.74]$$

$$MOH + OH^- \longrightarrow MO + H_2O + e_M \qquad [II.75]$$

$$2MO \longrightarrow 2M + O_2 \qquad [II.76]$$

The following table lists the values of the relevant parameters and the resulting Tafel slope, according to equation [II.68], for each of the steps assumed, in turn, to be rate-determining. Four electrons are transferred in the overall reaction (n = 4) and $\beta = 0.5$ is assumed.

---

* The discussion in the last few sentences may appear superfluous to the reader, since the difference between α and β in equation [II.73] is obvious. The point is nevertheless made because so often in past and even present literature in electrochemistry, a confusion between these quantities is found, and the transfer coefficient in equations like [II.71] is erroneously assumed to equal one-half.

Rate deter-mining step	$\vec{n}$	$\overleftarrow{n}$	ν	b
[II.74]	0	2	2	2.3(2RT/F)
[II.75]	2	0	2	2.3(2RT/3F)
[II.76]	4	0	1	2.3(RT/4F)

5.5 Double-Layer Effects: The Frumkin Correction

In the discussion of the current/potential relationship presented up to this point, the concentration term $\prod_i C_i$ (cf. equations [II.50] and [II.53]) was not dealt with specifically and was, in fact, "hidden" in the exchange current density i_o. In this section we shall discuss the effect of the diffuse-double-layer structure on the concentration of ionic species at the electrode surface. In section II.7 we shall deal with the dependence on potential of the concentration of adsorbed intermediate and of free sites on the electrode surface. The resulting Tafel slopes will also be discussed.

Consider the first discharge step in the oxygen-evolution reaction equation [II.74] discussed above. The initial state in this reaction is represented by an OH^- ion at a potential ϕ_2 in the OHP, rather than in the bulk of the solution. The concentration of ions in the outer Helmholtz plane is related to their concentration in the bulk of the solution by the Boltzmann equation

$$C(s) = C^o \exp(-ze_o \phi_2 / kT) \qquad [II.77]$$

where $C(s)$ is the concentration of OH^- ions at the OHP, C^o is

5.5 DOUBLE-LAYER EFFECTS: FRUMKIN CORRECTION

their bulk concentration and $ze_o\phi_2$ is the electrostatic work required to take an ion from the bulk to the OHP. The rate equation (equation [II. 48]) will now be

$$i = k_a F(1-\theta)C^o \exp\left(\frac{F\phi_2}{RT}\right) \exp\left[\frac{\beta F(\phi_M - \phi_2)}{RT}\right] \quad \text{[II. 78]}$$

In this equation $(1-\theta)$ is a quantity proportional to the concentration of free sites on the surface (cf. section II. 7), $z = 1$ and (e_o/k) has been substituted by (F/R). Note that the exponential term in potential now includes $\phi_M - \phi_2$ rather than $\phi_M - \phi_s = \Delta\phi$ in equation [II. 48] since the charge-transfer process is aided only by this part of the total metal/solution potential difference.

Equation [II. 78] may be used to correct for the diffuse-double-layer effect in the absence of specific adsorption in systems where the dependence of ϕ_2 on ϕ_M can be calculated.

The implications of equation [II. 78] are particularly interesting in the case of electrochemical reduction of anions. Equation [II. 78] can be rewritten in the form

$$i = k_c F(1-\theta)C^o \exp\left[\frac{(\beta - z_i)\phi_2 F}{RT}\right] \exp\left(\frac{-\beta F \phi_M}{RT}\right) \quad \text{[II. 79]}$$

where z_i, the valency of the anion is a negative integer. Now, in the absence of specific adsorption, ϕ_M and ϕ_2 always change in the same direction i.e. $d\phi_2/d\phi_M > 0$. Thus, as the potential is made more negative, the first exponential term in equation [II. 79] decreases while the second increases. In dilute solution, at potentials near the potential of zero charge (where $d\phi_2/d\phi_M$ has its maximum value), this effect has been found to be large enough to cause a decrease in cathodic current with increasing cathodic potential.

The Frumkin correction is often neglected in electrode-kinetic studies. This is justified when high electrolyte concentrations are used and the variation of ϕ_2 with ϕ_M is negligible. In reaction-order studies the ionic strength is usually held constant by the use of an excess of inert electrolyte, which maintains all double-layer effects constant as far as possible.

5.6 Types of Overpotential

The rate equations presented above were derived with the tacit assumption that the overall heterogeneous process considered is controlled by one of the activation-controlled steps taking place across the electrode/solution interphase. To be accurate, the overpotential developed in such a process should be denoted η_a, the activation overpotential.

Another source of overpotential in electrode reactions is the finite rate of transport of reactants to the interphase or of products from the interphase. This overpotential will be denoted η_d, the diffusion overpotential, even though in many cases, mass transport by convection plays a major role.

When mass transport to the interphase cannot keep up with the rate of electrochemical reaction taking place, the concentration at the electrode surface $C(s)$,* of the substance undergoing electrochemical change, will be lower than that in the bulk. Under steady-state conditions the two concentrations are related

---

* More precisely, $C(s)$ is the concentration at the OHP.

5.6 TYPES OF OVERPOTENTIAL

through the equation

$$\frac{C(s)}{C^o} = 1 - \frac{i}{i_L} \qquad [\text{II.80}]$$

where i_L is the limiting current density, determined by the maximum rate of mass transport to the electrode surface in a given experimental setup. Assuming, for the moment, that the activation overpotential is negligible, the electrode will react reversibly to the concentration at its surface to give a potential

$$E = E^o + \frac{RT}{nF} \ln C(s) \qquad [\text{II.81}]$$

When there is no current passing, the concentration on the surface equals that in the bulk and hence

$$E_r = E^o + \frac{RT}{nF} \ln C^o \qquad [\text{II.82}]$$

From the last three equations the diffusion overpotential is obtained as

$$\eta_d = E - E_r = \frac{RT}{nF} \ln \frac{C(s)}{C^o} = \frac{RT}{nF} \ln \left(1 - \frac{i}{i_L}\right) \qquad [\text{II.83}]$$

When both activation and diffusion overpotentials are operative, the total measured overpotential η will be the sum of these quantities.

An additional overpotential may be observed, which arises from the solution resistance between the working electrode and the edge of the capillary which leads to the reference electrode. This is the resistance overpotential η_R, which is proportional to the current density in a given solution and depends on the cell configuration.

The measured overpotential is thus given by

$$\eta' = \eta_a + \eta_d + \eta_R \qquad [\text{II. 84}]$$

Experimental methods are available for distinguishing between these contributions to η, or for making η_d and η_R negligible in comparison with η_a. Unless otherwise stated, it will be assumed that the latter procedure has been successfully followed and the measured overpotential is essentially equal to the activation overpotential.

6. Evaluation of Kinetic Parameters

6.1 The Overall Reaction

The first step in understanding a reaction mechanism is the determination of the overall reaction. This may appear to be a trivial matter in simple cases such as the hydrogen- or oxygen-evolution reactions, but presents a major problem in more complex cases, like the oxidation or reduction of organic molecules at electrodes. A definite answer can be obtained only by analysis of the products formed, but electrochemical methods are available which allow a determination of n, the number of electrons transferred per act of the overall reaction and this may often help in identifying the product.

The simplest method for the determination of n would seem to be the measurement of the polarographic diffusion current i_d, given by the Ilkovic equation

6.1 OVERALL REACTION

$$i = 708 nm^{2/3} t^{1/6} D^{1/2} C^o \qquad [\text{II}.85]$$

where m is the rate of flow of mercury in mg/sec, t is the drop time, D the diffusion coefficient and C^o the concentration in millimoles per liter. The numerical constant is calculated for 25°C and i_d is given in microamperes.

Polarography is commonly used as an analytical tool to determine C^o in solution. It can also be used to evaluate n in solutions of known concentration, provided the diffusion coefficient D is known with sufficient accuracy. For best results, equation [II.85] is not used directly for the calculation of n. Instead, the system is calibrated with a compound of known n.*

Another method of determining n is that of constant-potential electrolysis. The potential is set at a value which corresponds to a high overpotential and the current is followed as a function of time, while the solution is efficiently stirred. The current then decreases exponentially with time according to the equation

$$i(t) = i^o \exp(-i^o A/nFC^o V)t \qquad [\text{II}.86]$$

where V is the volume of the solution, A is the area of the electrode and i^o is the initial current density. The parameter n

---

* Similarly, when polarography is used as an analytical tool, a calibration curve of i_d vs C^o is first obtained for the compound studied and the unknown concentration in the solution which is analyzed is obtained from i_d with the aid of the calibration curve.

can be obtained from the slope of a plot of log i vs t. The product $i^o A$ is simply the total initial current, thus it is not necessary to measure A. The method is applicable both to fast and slow reactions. The experiment must be continued until at least 10% of the reactant initially present in solution has been consumed, that is, until the current has decreased to 90% or less of its initial value.

A similar method, based on following the variation of current with time at constant potential is the potential-step method. As above, the potential is stepped to a value where the reaction is controlled by mass transport only (that is, to the region of limiting current) but in this case the solution is not stirred and an excess of supporting electrolyte is added to ensure that mass transport occurs only by diffusion. Solution of the diffusion equation (Fick's second law) with the boundary conditions corresponding to the experimental conditions selected here, yields*

$$i_d = nFD^{1/2}C^o(\pi t)^{-1/2} \qquad [\text{II. 87}]$$

Thus n can be evaluated from a plot of i vs $t^{-1/2}$ provided that D is known. The method is applicable to fast (reversible) or slow (irreversible) reactions provided that the potential is properly chosen, (in the region of diffusion limitation) in each system. The fundamental difference between the last two methods is that in the former, the current decays as a result of the decrease in con-

---

* Assuming conditions of semi-infinite linear diffusion.

6.1 OVERALL REACTION

centration in the whole solution (which is well stirred) while in the latter only the concentration very near the electrode surface is affected, while the bulk concentration C^o remains essentially unchanged during the transient.

Another method by which n can be determined is chronopotentiometry, or constant-current potentiometry. If the electrode reaction is fast and controlled only by diffusion to the surface, (care being taken to eliminate mass transport by convection), the potential changes rapidly at first (just after the current is turned on), then very slowly. A second sharp change in potential occurs when the concentration of reactant at the electrode surface has reached zero.

The electrode potential varies with time according to the equation

$$E = E_{1/2} + \frac{2.3RT}{nF} \log \frac{\tau^{1/2} - t^{1/2}}{t^{1/2}} \qquad [\text{II.88}]$$

where $E_{1/2}$ is the polarographic half-wave potential and τ is the transition time, that is, the time taken until the sharp change in potential occurs (cf. Fig. II.14). The number of electrons transferred in the overall reaction can be evaluated from the slope of a plot of E vs log [$(\tau^{1/2} - t^{1/2})/t^{1/2}$]. This is convenient because it is not necessary to know the diffusion coefficient or the concentration in solution. On the other hand, the method is limited to fast electrochemical reactions with no kinetic complications. The current is adjusted so that τ will not be more than, say, ten seconds, in order to justify the assumption of mass transport by diffusion only.

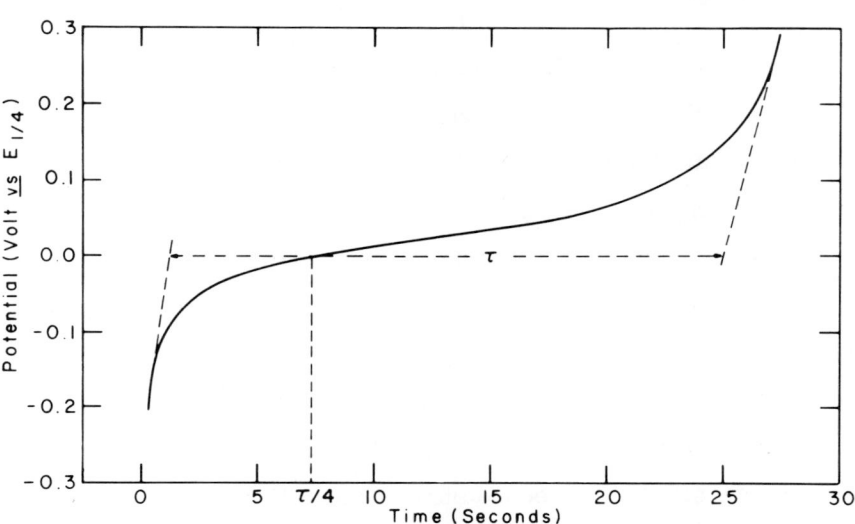

FIG. II 14 SCHEMATIC REPRESENTATION OF CHRONOPOTENTIOGRAM FOR REVERSIBLE PROCESS INVOLVING ONE ELECTRON

The transition time is related to the applied current through the equation

$$\tau^{1/2} = \frac{\pi^{1/2} nFD^{1/2} C^o}{2i} \qquad [II.89]$$

where i is the applied current density. This equation holds even if the reaction is slow (provided there are no kinetic complications) and from it n can be evaluated, if D and C^o are known.

A modification of this technique is thin-layer chronopotentiometry. The essence of this method consists in arranging two parallel-plate electrodes so that the distance between them is small compared to the thickness of the Nernst diffusion layer in unstirred solutions. The transition time τ corresponds here to the time required for total electrolysis and hence, instead of equation [II.89], one has

6.1 OVERALL REACTION

$$\tau = \frac{nFLC^o}{i} \qquad [\text{II. 90}]$$

where L is the distance between the two electrodes. Well defined transition times are obtained in thin-layer chronopotentiometry experiments and n can be calculated in cases where the value of D is not known.

Finally, we should mention the linear-potential-sweep method. This is a variation of constant-potential coulometry, in that the potential is still the externally controlled variable, but it is made to change with time at a constant rate.

Under diffusion-controlled conditions, a peak in the current/potential curve is observed. If the reaction is very fast (negligible activation overpotential), the peak current density will be given by

$$i_p = 2.72 \times 10^5 n^{3/2} D^{1/2} C^o v^{1/2} \qquad [\text{II. 91}]$$

where v is the rate of change of potential with time. Thus n can be evaluated from a plot of i_p vs $v^{1/2}$ if D and C^o are known.

If the reaction is partially controlled by activation, the peak current density is

$$i_p = 3.01 \times 10^5 n \left(\frac{2.3RT}{bF}\right)^{1/2} D^{1/2} C^o v^{1/2} \qquad [\text{II. 92}]$$

In this case, the potential E_p at which the peak current is observed also varies with the sweep rate, and it can be shown that

$$\frac{dE_p}{d \log v} = b/2 \quad ^* \qquad [\text{II.93}]$$

With the value of b from equation [II.93], the number of electrons n can be obtained by utilizing equation [II.92] if D and C^o are independently known.** When the reaction proceeds along several parallel paths to produce a number of different products, a non-integral value of n may be obtained. Analysis of reaction products then becomes necessary, and the <u>faradaic efficiency</u>, which is the fraction of the current used to produce each of the end products, is calculated. It is an essential feature of electrode reactions that the faradaic efficiency is generally a function of the inner potential difference $\Delta \phi$ across the interphase. It may thus be possible to produce one specific product at high efficiency or to suppress the formation of an undesirable side product by proper adjustment of the potential.***

---

* This and the previous equation are given in a form different from that commonly used (cf. Experiment 9). The transformation can easily be made by substituting $b = (2.3RT/\alpha n_a F)$. (cf. section II.10.3).

** Here again it must be assumed that there are no preceding chemical reactions in the homogeneous phase or other kinetic complications.

*** It is well to remember that adjustment of the potential is not the only means available for controlling the course of an electrode reaction. Choice of electrode material, of electrolyte and of solvents as well as the use of small amounts of certain additives in the solution, can serve the same purpose.

6.2 TAFEL SLOPE b 67

Once the overall reaction is known, the task of determining the reaction mechanism is reduced to distinguishing between a small number of reaction paths which are energetically plausible, and determining the rate-determining step.[*] The various kinetic parameters which can be determined experimentally and used for the elucidation of the reaction mechanism will now be discussed.

6.2 The Tafel Slope b

The Tafel slope is obtained from the linear portion of a plot of η <u>vs</u> log i. It is expressed in decimal scale as

$$b = \frac{2.3(\nu RT/F)}{\beta(n - \vec{n} - \overleftarrow{n}) + \vec{n}} \qquad [\text{II.69}]$$

Every combination of reaction path and rate-determining step gives rise to a definite value of the Tafel slope. However, several different mechanisms often produce the same numerical value of b. Thus the Tafel slope alone, just like any other kinetic para-

---

[*] The approach taken here towards the question of mechanism determination is the intuitive approach most commonly used in research. If the general formal equations are written, a large number of possible reaction paths result. Moreover, the total number of independently measurable quantities is always less than the total number of equations and the reaction path cannot be uniquely determined. Fortunately, most reaction paths can be rejected on the basis of chemical intuition (e.g. some paths will require the formation of extremely unstable intermediates). Thus one is left to choose among a small number of reaction paths and this choice can usually be made on the basis of experimentally available parameters.

meter, cannot be used to determine uniquely the mechanism of a reaction. Some examples of the value of the Tafel slope were given above (cf. section II.5.4). A different method for the calculation of b will be discussed in section II.7.2.

6.3 The Exchange-Current Density i_o

The exchange-current density is the product of the electrochemical specific rate constant and a concentration term (cf. equation [II.53]). As such, it does not help directly in mechanism determination, although it is useful in comparing the catalytic activities of different electrode materials (electrocatalysts) for a given reaction. The exchange-current density is obtained experimentally by extrapolation of the linear portion of the $\eta/\log i$ plot to the reversible potential, where $\eta = 0$. It is also possible, in principle, to measure i_o directly by measuring the isotopic exchange rate. The latter method is, however, rather tedious and is rarely applied.

6.4 The Stoichiometric Number ν

The stoichiometric number for the rate-determining step is the number of times this step occurs per act of the overall reaction. It can be calculated with the aid of equation [II.70] from measurement of the $i - \eta$ relationship in the low-overpotential linear region.

$$i = \left(\frac{i_o nF}{\nu RT}\right)\eta \qquad [\text{II.70}]$$

6.5 REACTION ORDERS   69

Such measurements require very accurate determinations of i_o which cannot always be achieved. Furthermore, equation [II.70] can only be used for reactions having a high exchange-current density, for which the reversible potential is easily reached and maintained, making measurements at low overpotentials of a few millivolts possible.

6.5 Reaction Orders in Electrode Kinetics

Most electrode-kinetic measurements are performed under quasi-zero-order conditions. Often the solvent itself serves as reactant (e.g. hydrogen and oxygen evolution in aqueous solutions) and in other cases the concentration of reactant is kept constant by maintaining a steady supply (e.g. by bubbling a gaseous reactant through the cell at constant partial pressure) or by working at low total current and/or time, so that the amount of reactant consumed is negligible. Measurements under quasi-zero-order conditions are convenient because the current density is proportional to the reaction rate, which can therefore be measured without effecting any concentration changes in the solution.

Reaction-order studies in electrode reactions involve several complications which are not found in the case of other heterogeneous kinetic studies. First, the effect of the double-layer structure on the reaction order must be considered. Diffuse-double-layer effects (cf. equation [II.78]) can be taken into account by application of theoretical equations to the calculation of the potential ϕ_2 in the OHP and with it evaluation of the relationship between surface and bulk concentrations. Whenever possible, this effect is eliminated by the use of a high concentration of supporting electrolyte. In this way, the structure of the compact double layer can also be maintained constant and all complications due to changes in adsorption in both the inner and outer Helmholtz planes are avoided.

In reaction-order studies, it must be remembered that the reversible potential depends on the concentration of reactant. Thus, two experimental parameters relating to reaction orders may be defined, one at constant overpotential

$$\rho_1(k) = \left(\frac{\partial \log i}{\partial \log C_k}\right)_\eta \qquad [\text{II. 94}]$$

and one at constant potential E with respect to a fixed reference electrode, (and hence at constant metal-solution potential difference).

$$\rho_2(k) = \left(\frac{\partial \log i}{\partial \log C_k}\right)_{\Delta\phi} \qquad [\text{II. 95}]$$

In both cases, the temperature, pressure and the concentrations of all other species in solution are also maintained constant.*
Sometimes the parameter $(\partial \log i_o/\partial \log C_k)$ is determined. This, however, is not different from $\rho_1(k)$ defined by equation [II. 94] since i_o is simply the current measured at a fixed overpotential of $\eta = 0$. Another derivative $(\partial \eta/\partial \log C_k)_i$ is sometimes measured.

---

* For simplicity of printing, these subscripts are omitted here and in further equations.

6.5 REACTION ORDERS

However, this is related to $\rho_1(k)$ since by the laws of partial derivatives

$$\left(\frac{\partial \eta}{\partial \log C_k}\right)_i = -b\rho_1(k) \qquad [\text{II.96}]$$

An example will serve to illustrate the use of the parameters $\rho_1(k)$ and $\rho_2(k)$. Consider the discharge of azide ions (N_3^-) to form molecular nitrogen,

$$N_3^- + M \longrightarrow MN_3 + e_M \qquad [\text{II.97}]$$

$$MN_3 \longrightarrow MN + N_2 \qquad [\text{II.98}]$$

$$2MN \longrightarrow 2M + N_2 \qquad [\text{II.99}]$$

and assume that the first step is rate-determining. Equation [II.50] applied to the high-overpotential region (the linear Tafel region) will take the form

$$\begin{aligned}
i &= Fk_a^o(1-\theta)C_{N_3^-} \exp\left(\frac{\beta F \Delta \phi}{RT}\right) \\
&= Fk_a^o(1-\theta)C_{N_3^-} \exp\left(\frac{\beta F \Delta \phi_r}{RT}\right) \exp\left(\frac{\beta F \eta}{RT}\right)
\end{aligned} \qquad [\text{II.100}]$$

Hence $\rho_2(N_3^-) = 1$ and $\rho_1(N_3^-) = 1 + \dfrac{\beta F}{RT} \cdot \dfrac{\partial \Delta \phi_r}{\partial \ln C_{N_3^-}}$

Under reversible conditions, the absolute metal/solution potential difference $\Delta \phi_r$ may be expressed as

$$\Delta \phi_r = \Delta \phi_r^o - \frac{RT}{F} \ln C_{N_3^-} \qquad [\text{II.22}]$$

and therefore, $\rho_1(N_3^-) = 1 - \beta$

In the example worked out above, double-layer effects have been neglected. The results, therefore, apply only if the structure of the double layer is maintained essentially constant by the use of an excess of supporting electrolyte. If the reacting species is the only electrolyte (e.g. an aqueous solution of NaN_3 in the above example) and its concentration is varied over several orders of magnitude in order to obtain reliable reaction-order data, the structure of the double layer must be considered. This was done by Frumkin in 1933 for the case of the hydrogen-evolution reaction. Equation [II.78] was used to calculate the reaction-order parameters $\rho_1(H_3O^+)$ and $\rho_2(H_3O^+)$ and diffuse-double-layer theory was applied to calculate ϕ_2 in this equation as a function of concentration in solution.

It should be noted that reaction orders found for electrode reactions also depend on the fractional surface coverage θ and on the type of adsorption isotherm applicable to the system. This feature is common to all heterogeneous reactions and will be discussed in section II.7.3 below.

6.6 The Effect of pH

The effect of pH is a special case of reaction-order effect, present when H_3O^+ or OH^- ions are involved in the reaction. Thus, for the hydrogen-evolution reaction, with the assumption that the first charge-transfer step is rate-determining, an equation such as [II.100] can be written, from which one obtains

6.7 ISOTOPE EFFECTS

$$\left(\frac{\partial \log i}{\partial pH}\right)_{\Delta\phi} = -1 \quad ; \quad \left(\frac{\partial \log i}{\partial pH}\right)_{\eta} = -(1-\beta) \qquad [\text{II}.101]$$

if the activity coefficient of H_3O^+ ions is assumed, as a first approximation, to be independent of concentration. Care must be taken when studies in the pH range 4-10 are conducted in the absence of a buffer system, since the concentration of H_3O^+ or OH^- ions near the electrode surface may depart substantially from that in the bulk as a result of the reaction taking place.

6.7   Isotope Effects

It has been known for a long time that electrolysis of an aqueous solution of, say, H_2SO_4, leads to enrichment of the remaining liquid in the heavier isotopes of hydrogen, while the gas coming off has a higher percentage of the light isotope. The separation factor for deuterium, S_D, is defined as

$$S_D = \frac{(C_H/C_D)_{gas}}{(C_H/C_D)_{solution}} \qquad [\text{II}.102]$$

Separation arises from the different masses of hydrogen and deuterium, which give rise to slightly different zero-point energies and bond strengths in the $(H-H_2O)^+$ and $(D-H_2O)^+$ ions, and different probabilities for the tunnelling of the positive species across the interphase.

Separation factors have been measured for the hydrogen-evolution reaction under a great variety of experimental conditions. The values found for S_D (or S_T for tritium) fall into three or four well defined groups, each of which can be identified with a distinct reaction mechanism. Values of S_D and S_T calculated theoretically for the various mechanisms are in agreement with the experimental results. Thus, isotopic separation factors can serve as powerful tools in the elucidation of reaction mechanisms. So far, this has been applied mostly to the isotopes of hydrogen for which, because of the high ratio of the masses, the rates of reaction for the two isotopes are quite different and relatively easily measurable. This, however, is simply a matter of experimental convenience and application to the study of other reaction mechanisms is possible.

6.8 Apparent Energy of Activation

The apparent energy of activation for an electrode reaction is given by

$$E_a^{\neq} = 2.3R \left(\frac{\partial \log i}{\partial (1/T)} \right)_{\eta, C} \qquad [\text{II}.103]$$

The variation of the reversible potential E_r with temperature[*] is measurable and, in many cases, can be found in standard tables, so that the condition of constant overpotential in equation [II.103] can be realized. The corresponding definition in terms of constant $\Delta\phi$ is somewhat more complicated, because it requires a correction for the variation with temperature of $\Delta\phi$ at the reference electrode. Since single potential differences across interphases cannot be measured, this quantity can only be calculated

---

[*] With respect to a reference electrode maintained at constant temperature.

approximately, and an uncertainty in the value of $E_a^{\neq}$ at constant $\Delta\phi$ results.

The apparent energy of activation is a function of overpotential and decreases with increasing overpotential according to the equation

$$E_{a(\eta)}^{\neq} = E_{a(\eta=0)}^{\neq} - 2.3RT(\eta/b) \qquad [\text{II.}104]$$

if measured in the linear Tafel region. The numerical value of $E_a^{\neq}$ does not give much information concerning the mechanism of a single reaction, although it may help to rule out certain reaction paths by consideration of the energies of the intermediates involved. A comparison of values of $E_a^{\neq}$ for two similar reactions may indicate whether the same step controls the rates of both reactions.

7. Adsorption of Intermediates

7.1 The Quasi-Equilibrium Assumption

The adsorption of intermediates formed in consecutive steps in an electrode reaction plays an important role in the determination of the observed kinetic parameters.

Let us consider, as an example, the anodic evolution of bromine which proceeds through the steps

$$Br^- + M \underset{k_{-1}}{\overset{k_1}{\rightleftarrows}} MBr + e_M \qquad [\text{II.}105]$$

$$2MBr \xrightarrow{k_2} 2M + Br_2 \qquad [\text{II.}106]$$

and let us assume that the second step is rate-determining. The constants k_1, k_{-1} and k_2 are electrochemical specific rate constants and MBr represents the intermediate, bromine atom, adsorbed on the surface.

Steady state is reached in this reaction sequence when both consecutive steps occur at the same rate

$$i = \vec{i}_1 - \overleftarrow{i}_1 = \vec{i}_2 - \overleftarrow{i}_2 \qquad [\text{II}.107]$$

The assumption that the second step is rate-determining is equivalent to the statement that the first step proceeds in both directions at a rate which is very high compared to the steady-state reaction rate.

Under these circumstances, the first step may be said to be in quasi-equilibrium and treated mathematically to a good approximation, as though it were actually at equilibrium. The situation is such that the net current density i is small compared with $i_{o,1}$, the exchange-current density for the first step; yet it is large with respect to $i_{o,2}$, the exchange-current density for the second step. As a result, equilibrium in the first step is hardly disturbed, while the second step is far from equilibrium and is in the linear Tafel region, where its reverse rate may be neglected.

The concept of quasi-equilibrium is a disturbing one, as it seems self-contradictory to assume that a reaction is at equilibrium and yet proceeds at a net finite rate in one direction   To justify and clarify this assumption further we shall resort to the equations of irreversible thermodynamics.

The net rate of a reaction close to equilibrium is given

7.1 QUASI-EQUILIBRIUM ASSUMPTION

by

$$v_i = v_i^e \left(\frac{A_i}{RT} \right) \quad [\text{II.108}]$$

where v_i^e is the exchange rate of the reaction, that is, the rate at which it proceeds in both directions at equilibrium, when no net reaction is observed. The affinity A_i of the reaction is the distance of the system from equilibrium on the free-energy scale, or the free energy which can be gained by returning the system to equilibrium. When several consecutive reactions are involved and the system is at steady state, all steps occur at the same rate and one has

$$v_1^e A_1 = v_2^e A_2 = \ldots\ldots = v_i^e A_i = \text{const} \quad [\text{II.109}]$$

If the exchange rate for the ith step is much lower than that for any other, this step will be the rate-determining one and the corresponding affinity A_i will be much higher than that of any other step.

$$A_1, A_2, \ldots\ldots \ll A_i \quad [\text{II.110}]$$

Thus, although all steps in the sequence proceed at the same rate, the affinities of all the steps, except the rate-determining step, are near zero and these steps may be considered to be effectively at equilibrium. Since the rate-determining step occurs ν times per act of the overall reaction, $A_i = A/\nu$ and equation [II.108] can be written in the form

$$v = v_{rds}^e \frac{A}{\nu RT} \qquad [\text{II.111}]$$

where v_{rds}^e is the exchange rate for the rate-determining step.

Applied to an electrode reaction, equation [II.111] reads

$$i = \frac{i_o nF\eta}{\nu RT} \qquad [\text{II.70}]$$

since the affinity in this case is equal to the electrical energy per mole which is applied to remove the system from its equilibrium potential. It is, incidentally, gratifying to find that this equation, arrived at from considerations of irreversible thermodynamics, is identical to that obtained by taking the linear approximation of the exponential dependence of current on potential expressed by equation [II.67].

7.2 Calculation of Tafel Slopes with the Use of the Quasi-Equilibrium Assumption

Consider now the rate equations for the discharge step and its reverse in bromine evolution (equation [II.105]).

$$\vec{i}_1 = k_1 C_{Br^-}(1 - \theta)\exp\left(\frac{\beta F \Delta\phi}{RT}\right) \qquad [\text{II.112}]$$

$$\overleftarrow{i}_1 = k_{-1} \theta \exp\left[-\frac{(1-\beta)F\Delta\phi}{RT}\right] \qquad [\text{II.113}]$$

The fractional coverage θ is proportional to the surface concentration of adsorbed intermediates and the fraction $(1 - \theta)$ is proportional to the concentration of free sites on the surface, which act as reactants in the adsorption step. Since quasi-equilibrium is assumed, the forward and reverse currents of

7.2 CALCULATION OF TAFEL SLOPES

equations [II.112] and [II.113] can be equated. This gives rise to an adsorption isotherm relating the fractional surface coverage θ to the potential difference $\Delta\phi$ across the interphase

$$\frac{\theta}{1-\theta} = KC_{Br^-} \exp\left(\frac{F\Delta\phi}{RT}\right) \qquad [II.114]$$

where $K = k_1/k_{-1}$. Equation [II.114] is the Langmuir adsorption isotherm modified to apply to the case of adsorption involving charge transfer.

The rate of bromine evolution was assumed to be governed by the second step in the reaction sequence — the recombination of two adsorbed atoms on the surface (equation [II.106])

$$i \cong i_2 = k_2 \theta^2 \qquad [II.115]$$

Inserting the value of θ from equation [II.114] yields a rather complex expression

$$i = \frac{k_2 K^2 C_{Br^-}^2 \exp(2F\Delta\phi/RT)}{[1 + KC_{Br^-} \exp(F\Delta\phi/RT)]^2} \qquad [II.116]$$

At the limit of very low coverage where $1 - \theta \cong 1$ and $KC_{Br^-} \exp(F\Delta\phi/RT) \ll 1$, the coverage is a simple exponential function of $\Delta\phi$ and the rate equation takes the more convenient form

$$i = k_2 K^2 C_{Br^-}^2 \exp\left(\frac{2F\Delta\phi}{RT}\right) \qquad [II.117]$$

with a corresponding Tafel slope of $b = 2.3(RT/2F)$. It can be easily ascertained that this is in agreement with the value of b obtained by application of equation [II. 69] to the reaction sequence discussed here.

As the coverage increases beyond, say, $\theta = 0.1$, the plot of $\Delta\phi$ vs log i will depart from linearity and will eventually reach a limiting value as θ approaches unity. The shape of the $\Delta\phi$ vs log i plot is shown in Fig. II 15 together with the variation of θ with $\Delta\phi$.

Consider a somewhat more complex case of the oxygen-evolution reaction discussed above, viz.

$$OH^- + M \rightleftharpoons MOH + e_M \qquad [\text{II. 74}]$$

$$MOH + OH^- \rightleftharpoons MO + H_2O + e_M \qquad [\text{II. 75}]$$

$$2MO \longrightarrow 2M + O_2 \qquad [\text{II. 76}]$$

and assume that the last recombination step is rate-determining. As a result of quasi-equilibrium in the first two steps and assuming low coverage by both adsorbed intermediates, we have

$$\theta_{MOH} = K_1 C_{OH^-} \exp\left(\frac{F\Delta\phi}{RT}\right) \qquad [\text{II. 118}]$$

$$\theta_{MO} = K_2 C_{OH^-} \theta_{MOH} \exp\left(\frac{F\Delta\phi}{RT}\right) \qquad [\text{II. 119}]$$

The net measured current will be given by

$$i = \vec{i}_3 = k_3 \theta_{MO}^2 = k_3 (K_1 K_2)^2 C_{OH^-}^4 \exp\left(\frac{4F\Delta\phi}{RT}\right) \qquad [\text{II. 120}]$$

The corresponding Tafel slope is $b = 2.3\, RT/4F$. The reaction order predicted for this mechanism is also of interest. The

7.2 CALCULATION OF TAFEL SLOPES

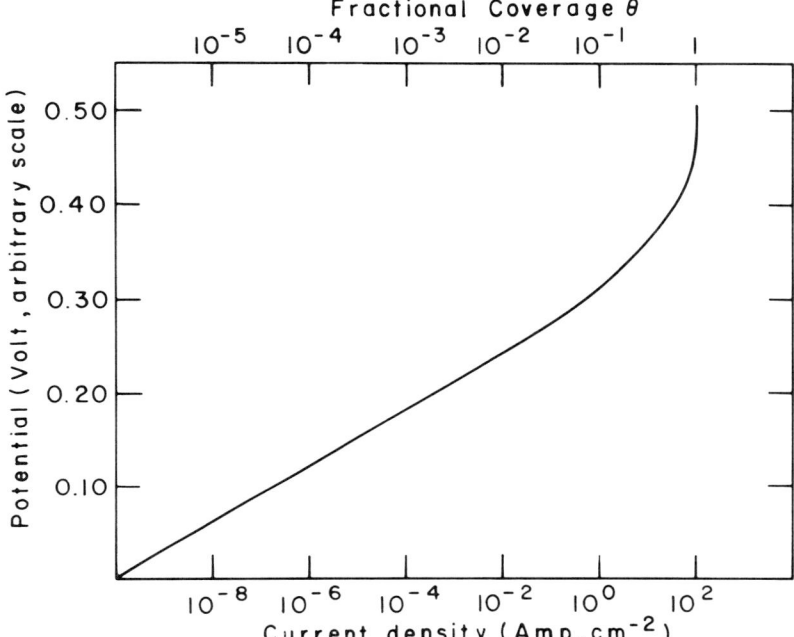

FIG. II 15 PLOTS OF CURRENT DENSITY AND FRACTIONAL COVERAGE VS POTENTIAL ASSUMING A RATE-DETERMINING RECOMBINATION STEP

parameter $\rho_2(OH^-)$ is obtained directly from equation [II.120] as

$$\rho_2(OH^-) = \left(\frac{\partial \ln i}{\partial \ln C_{OH^-}}\right)_{\Delta\phi} = 4 \qquad [\text{II.121}]$$

while the reaction order at constant overpotential can be shown (cf. section II.6.5) to be

$$\rho_1(OH^-) = \left(\frac{\partial \ln i}{\partial \ln C_{OH^-}}\right)_{\eta} = 0 \qquad [\text{II.122}]$$

In both cases the presence of an excess of supporting electrolyte is assumed. The importance of defining precisely the conditions obtaining in reaction-order studies is obvious.

The use of the quasi-equilibrium assumption for the prediction of kinetic parameters may be summarized as follows. The rate equation for the rate-determining step is first written. Unless this happens to be the first charge-transfer step in the sequence, it includes terms for the concentration of one or more adsorbed intermediates formed in previous steps. These concentration terms are evaluated under the assumption that quasi-equilibrium exists in all steps prior to the rate-determining step. Upon introducing these terms into the rate equation, one obtains the dependence of current on potential and on the concentration of ionic species in solution.

7.3 The Effect of the Adsorption Isotherm on the Kinetic Parameters

The Langmuir adsorption isotherm used up to this point in the derivation of kinetic parameters represents an idealized limiting case. It is based on the assumptions that the surface is homogeneous and that lateral-interaction effects are negligible. This leads to an adsorption constant (the equilibrium constant K in equation [II.114] above) which is independent of the fractional coverage θ.

In many systems of practical interest, a better approximation is obtained if it is assumed that the equilibrium constant K decreases exponentially with increasing coverage

$$K = K^o \exp\left(-\frac{r\theta}{RT}\right) \qquad [\text{II}.123]$$

An equivalent, often preferred way of stating the same assumption, is to say that the apparent standard free energy of adsorption

7.3 EFFECT OF ADSORPTION ISOTHERM

varies linearly with coverage.

$$\Delta G^o_\theta = \Delta G^o_{\theta=0} + r\theta \qquad [\text{II}.124]$$

This assumption gives rise to the Frumkin isotherm

$$\frac{\theta}{1-\theta} \exp\left(\frac{r\theta}{RT}\right) = K^o C \exp\left(\frac{F\Delta\phi}{RT}\right) \qquad [\text{II}.125]$$

from which the Langmuir isotherm is obtained as a special case for $r = 0$. Fig. II 16 shown the variation of θ with $\Delta\phi$ for three values of the parameter r.

At intermediate values of the coverage $(0.2 < \theta < 0.8)$, the exponential term in equation [II.125] predominates and upon taking logarithms of both sides and neglecting the term $\ln\left(\frac{\theta}{1-\theta}\right)$ one obtains

$$\theta = \frac{RT}{r} \ln K^o C + \frac{F\Delta\phi}{r} \qquad [\text{II}.126]$$

This is known as the Temkin isotherm.* It is characterized by a linear dependence of θ on the potential $\Delta\phi$ and a logarithmic dependence on the concentration C.

Now, in the case of the bromine-evolution reaction, the adsorption constant is $K = k_1/k_{-1}$. Since we have assumed an exponential dependence of K on θ (equation [II.123]), we may reasonably extend this assumption of exponential dependence of θ to both k_1 and k_{-1}. It is clear that the rate of the adsorption step

---

* The Temkin isotherm was originally written in the form $\theta = (RT/r)\ln K^o C$. The term $(F\Delta\phi/r)$ arises when adsorption also involves charge transfer (cf. equation [II.114]).

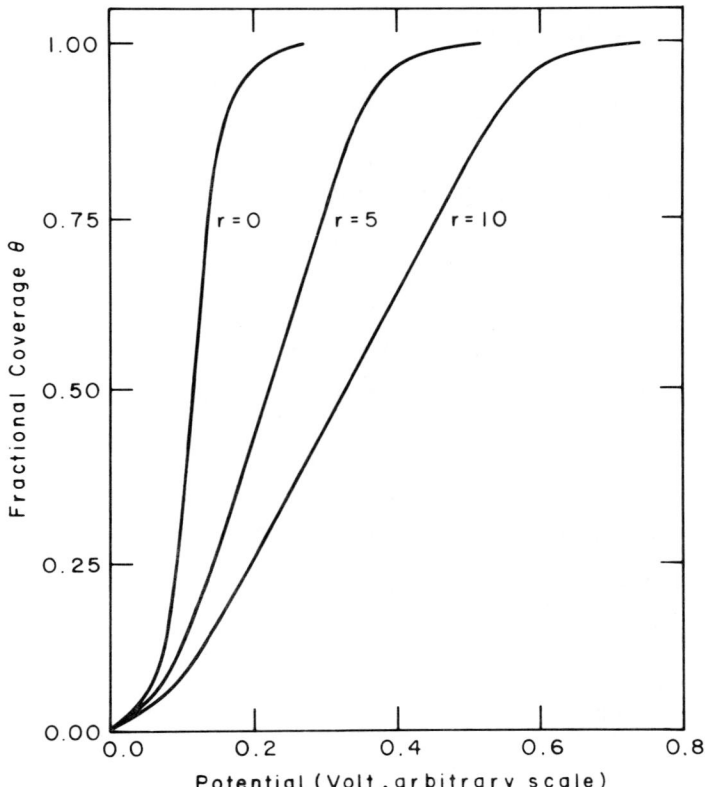

FIG. II 16 FRACTIONAL SURFACE COVERAGE BY ADSORBED INTERMEDIATES AS A FUNCTION OF POTENTIAL (r in kcal/mole)

decreases while the rate of the desorption step increases with θ. Thus we obtain

$$k_1 = k_1^o \exp\left(-\frac{\beta r \theta}{RT}\right) \qquad \text{[II.127]}$$

and

7.3 EFFECT OF ADSORPTION ISOTHERM

$$k_{-1} = k^o_{-1} \exp\left[\frac{(1-\beta)r\theta}{RT}\right] \qquad [\text{II}.128]$$

and for the combination step (cf. equation [II.115]).

$$k_2 = k^o_2 \exp\left[\frac{2(1-\beta)r\theta}{RT}\right]^* \qquad [\text{II}.129]$$

If we introduce these values of k_1 and k_{-1} into equations [II.112] and [II.113], respectively, and neglect all pre-exponential terms in θ, we obtain equation [II.126] (the Temkin isotherm). Substituting this linear relationship between θ and $\Delta\phi$ into the rate equation [II.115] and using the expression for k_2 given in equation [II.129] we find** (neglecting pre-exponential terms in θ),

$$i = \vec{i}_2 = k^o_2 K^o C_{Br^-} \exp\left(\frac{F\Delta\phi}{RT}\right) \qquad [\text{II}.130]$$

instead of equation [II.117] for the Langmuir case derived previously. It is seen by comparison of these equations that both the reaction order and the Tafel slope depend on the nature of the adsorption isotherm involved.

For the second example discussed in the preceding section (equations [II.74]-[II.76]), the situation is more complex since two adsorbed intermediates are formed and an equilibrium step such as [II.75] does not involve an exponential term in θ as it leaves the

---

* The factor 2 in the exponent arises because two species are desorbed for each occurrence of this step.

** The exponent of the concentration term in equation [II.130] is $2(1-\beta)$. This was taken to be unity since we assume $\beta = 0.50$ in kinetic analysis. In most cases, application of the Temkin isotherm leads to fractional reaction-order values.

number of occupied sites unaltered. If one assumes that
$\theta_{MO} \gg \theta_{MOH}$ and that θ_{MO} is in the intermediate region, one
obtains the rate equation

$$i = \vec{i}_3 = k_3^o K_1^o C \exp\left(\frac{\Delta\phi F}{RT}\right) \qquad [\text{II.131}]$$

instead of equation [II.120] derived above for Langmuir conditions
at low total coverage.

It is important to note that the exact form of dependence
on θ of the apparent standard free energy of adsorption does not
affect the observed kinetic behavior. If, instead of equation
[II.124], one writes

$$\Delta G_\theta^o = \Delta G_{\theta=0}^o + r\theta^n \qquad [\text{II.132}]$$

where n is some small number, equations [II.125] through
[II.129] will be modified by adding the power n to all θ terms in
the exponents. Thus, the final rate equations [II.130], [II.131],
based on the Frumkin (or Temkin) isotherm, will remain unaltered
even if ΔG_θ^o does not vary in an <u>exactly</u> linear manner with θ.
Any small deviation from linearity in the final rate equation will
be cancelled.

8.   The Adsorption Pseudocapacity

8.1   Physical Significance of the Adsorption Pseudocapacity

The formation of adsorbed intermediates in a reaction such
as [II.74] or [II.105] is associated with charge transfer across the
interphase. When a potential is suddenly applied, a transient
faradaic current must flow in order to establish the equilibrium

8.1 PHYSICAL SIGNIFICANCE

value of θ. The total charge during the transient is proportional to θ or to the change in θ caused by the change in potential.

$$q_F = k'\theta \qquad [\text{II.133}]$$

where k' is the charge needed to form a complete monolayer of adsorbed intermediates. Its numerical value is of the order of 200 $\mu C/cm^2$ for a monovalent species occupying a single site on the surface. Since θ is generally a function of potential (cf. equations [II.125], [II.126]), so is the charge, and a new type of differential capacity may be defined

$$C_\phi = \left(\frac{\partial q_F}{\partial \Delta\phi}\right)_C = k'\left(\frac{\partial \theta}{\partial \Delta\phi}\right)_C \qquad [\text{II.134}]$$

This is termed the <u>adsorption pseudocapacity</u>. It is noted that an adsorption pseudocapacity <u>cannot</u> exist at an ideal polarized electrode. On the contrary, it typically occurs when a step in a reaction sequence is at quasi-equilibrium and is followed by a slow, rate-determining step. The prefix "pseudo" is used because C_ϕ refers to a leaky capacitor; indeed so leaky, that it cannot exist unless the interphase leaks, that is, unless charge is transferred across it.

The capacitive nature of the change of θ with $\Delta\phi$ can be further illustrated in the following manner. Suppose that, to take a specific case, the bromine-evolution reaction proceeds at a steady rate at a potential $\Delta\phi$ and that a corresponding equilibrium value of θ in the discharge step (equation [II.105]) is maintained. Now, let a sinusoidal voltage of small amplitude $\delta\Delta\phi$ be superimposed on $\Delta\phi$ and assume further that k_1 and k_{-1} in equation [II.105] have very high values, so that θ always has its equilibrium value corresponding to the momentary value of the potential

$$\Delta\phi = \Delta\phi_{t=0} + (\delta\Delta\phi)\sin(2\pi ft) \qquad [\text{II.135}]$$

where f is the frequency of the sine wave. The rate of change of coverage with time is proportional to the rate of change of the potential

$$\frac{d\theta}{dt} = \frac{d(\Delta\phi)}{dt} = (\delta\Delta\phi)2\pi f \cos(2\pi ft) \qquad [\text{II.136}]$$

but from equation [II.133] we find

$$i = \frac{dq_F}{dt} = k'\frac{d\theta}{dt} \qquad [\text{II.137}]$$

hence

$$i = i_m \cos(2\pi ft) = i_m \sin(2\pi f + \frac{\pi}{2})t \qquad [\text{II.138}]$$

where i_m is the amplitude of the current wave. The phase shift of $\pi/2$ between current and potential is the characteristic a.c. response of a pure capacitor.

If a very high-frequency a.c. signal is used or if the rate constants k_1 and k_{-1} in equation [II.105] are not large enough, the change in θ cannot keep up with the change in $\Delta\phi$. The phase shift between the current and potential becomes smaller. This is described in the equivalent-circuit representation of the reaction (cf. section II.8.2) as a resistor (the faradaic resistance in equation [II.105]) in series with the capacitor.

8.2 The Equivalent Circuit and Frequency Response

The equivalent circuit for a two-step reaction where an adsorbed intermediate is formed, is shown in Fig. II 17 in which R_{F_1} and R_{F_2} are the faradaic resistances for the discharge and recombination steps, respectively. The frequency response of the equivalent circuit is of interest. At very high frequencies, the circuit behaves as a pure capacitor having the double-layer capacity C_{dl}. At very low frequencies it behaves as a pure resistor having a resistance $R_{F_1} + R_{F_2}$. At intermediate frequencies, one finds regions in which the circuit behaves as a capacitor C_ϕ in series with a resistor R_{F_1} or as two capacitors, C_ϕ and C_{dl}, in parallel. The impedance Z of the equivalent circuit shown in Fig. II 17 for the case of $R_{F_2} \gg R_{F_1}$ (i.e. when the recombination step is rate-determining) is given by

$$Z = \frac{R_{F_2}(1 + \omega R_{F_1} C_\phi)}{R_{F_2} \omega(C_\phi + C_{dl} + \omega R_{F_1} C_\phi C_{dl}) + 1} \qquad [\text{II.139}]$$

where $\omega = 2\pi f$ is the angular velocity of the applied signal. A plot of log Z vs log ω is shown in Fig. II 18. The following numerical values have been chosen for the circuit parameters: $R_{F_1} = 10$ ohm.cm^2, $R_{F_2} = 10^4$ ohm.cm^2, $C_{dl} = 20$ μF/cm^2, $C_\phi = 2 \times 10^3$ μF/cm^2. The horizontal portions of the curve represent pure resistive behavior while a slope of -1 represents pure capacitive behavior.

It should be remembered that Fig. II 17 represents a non-linear electrical circuit, that is, one in which the values of all components may vary with potential. This is particularly true for

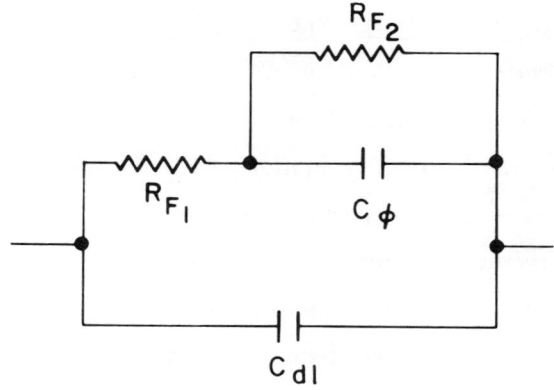

FIG. II 17   EQUIVALENT CIRCUIT FOR SYSTEM EXHIBITING AN ADSORPTION PSEUDOCAPACITANCE

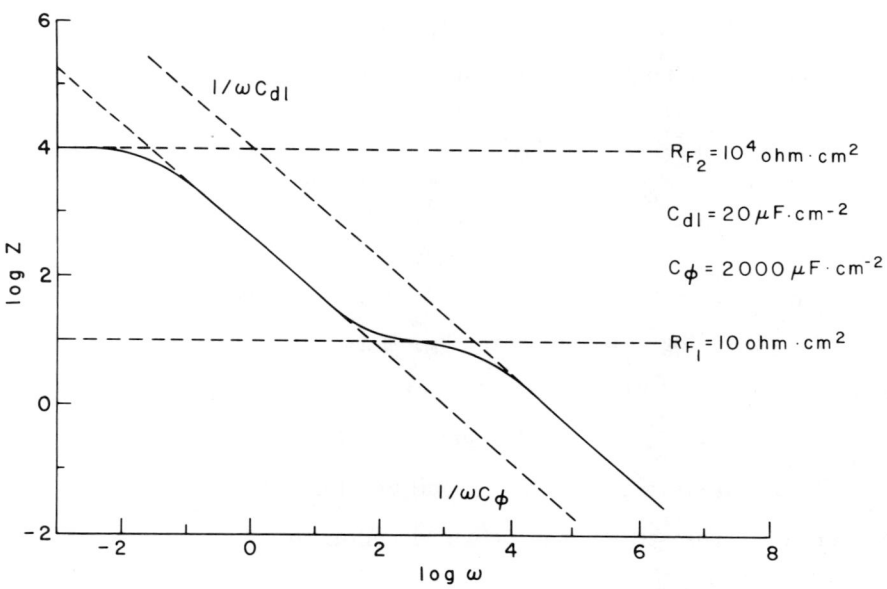

FIG. II 18   FREQUENCY RESPONSE OF CIRCUIT SHOWN IN FIG. II 17

8.3 EQUATIONS

R_{F_2} which is an exponential function of $\Delta\phi$ (cf. equation [II.57]) and C_ϕ which assumes appreciable values only over a limited potential range, as shown in the next section.

8.3 Equations for Adsorption Pseudocapacity

The adsorption pseudocapacity is proportional to the rate of change of coverage with potential. Under Langmuir conditions, equation [II.114] may be written in the form

$$\theta = \frac{KC\exp(F\Delta\phi/RT)}{1 + KC\exp(F\Delta\phi/RT)} \qquad [\text{II}.140]$$

from which C_ϕ is found to be

$$C_\phi = \frac{k'F}{RT}\left[\frac{KC\exp(F\Delta\phi/RT)}{[1 + KC\exp(F\Delta\phi/RT)]^2}\right] \qquad [\text{II}.141]$$

Alternatively, C_ϕ can be expressed as a function of coverage

$$C_\phi = \frac{k'F}{RT}[\theta(1-\theta)] \qquad [\text{II}.142]$$

When the Frumkin isotherm (equation [II.125]) applies, θ cannot be written as an explicit function of $\Delta\phi$ and so C_ϕ can be expressed only as a function of θ

$$\frac{1}{C_\phi} = \frac{1}{k'}\frac{d(\Delta\phi)}{d\theta} = \frac{RT}{k'F}[\theta(1-\theta)]^{-1} + \frac{r}{k'F} \qquad [\text{II}.143]$$

hence

$$C_\phi = \frac{k'F}{RT}\left[\frac{\theta(1-\theta)}{1 + (r/RT)\theta(1-\theta)}\right] \qquad [\text{II}.144]$$

The dependence of C_ϕ on $\Delta\phi$ can then be obtained numerically by calculating $\Delta\phi$ as a function of θ from equation [II.125]. This is shown in Fig. II 19 for three values of r. The curves for C_ϕ have a maximum at $\theta = 0.5$ and are symmetrical with respect to both potential and coverage. The maximum is given by

$$C_\phi(\max) = \frac{k'F}{4RT}\left[1 + \frac{r}{4RT}\right]^{-1} \quad [\text{II.145}]$$

This may reach values of the order of 2000 $\mu F/cm^2$ for $r = 0$ (Langmuir conditions) but falls very rapidly with increasing r. For $k' = 200\ \mu C/cm^2$, $C_\phi = 8\ \mu F/cm^2$ for $\theta = 0.001$ or $\theta = 0.999$, independent of r for all reasonable values of this parameter. If we compare it with typical values of C_{dl} (20–40 $\mu F/cm^2$), we see that the adsorption pseudocapacitance will play an appreciable and even predominant role over this wide range of coverage. The corresponding range of potential depends strongly on r as seen in Figs. II 16 and II 19.

8.4 Open-Circuit Decay

Measurement of the rate of decay of overpotential immediately following polarization can serve as an important tool in the study of electrode kinetics, and particularly in the detection of the existence of adsorbed intermediates.

Consider first the simple equivalent circuit shown in Fig. II 7c. During steady-state polarization, a constant current density i flows through the resistor R_F and the overpotential η

8.4 OPEN-CIRCUIT DECAY

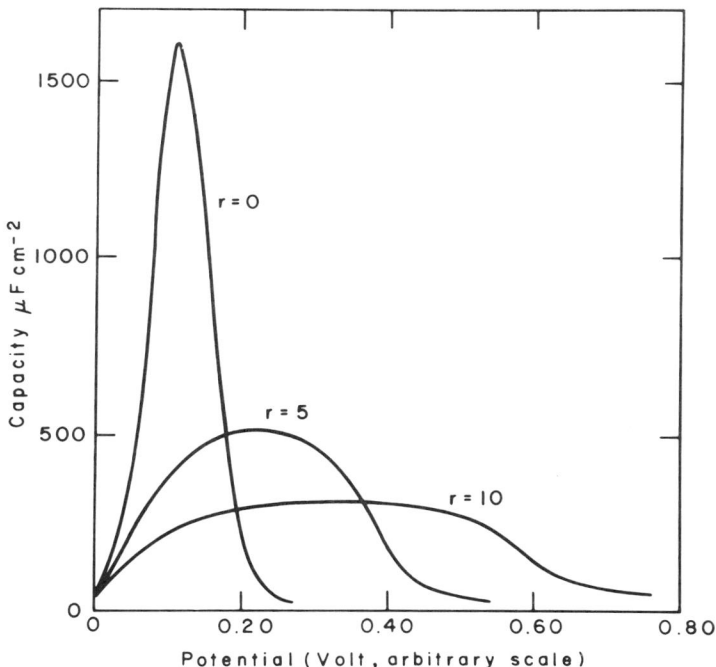

FIG. II 19 THE ADSORPTION PSEUDOCAPACITY AS A FUNCTION OF POTENTIAL (r in kcal/mole)

is developed across the terminals of the double-layer capacitor.* When the external source of polarization is disconnected, the double-layer capacitor is discharged through the faradaic resistance R_F and the overpotential decreases with time. The double-layer capacity can be obtained experimentally from the decay curve as a function of potential, since at any time during

---

* The additional potential $\Delta\phi_r$ which is determined by the components of the system but is independent of the current is not important for the purpose of this discussion.

decay, the equation

$$i = -C d\eta/dt \qquad [\text{II.146}]$$

holds, where i is the current density measured during steady-state polarization at the same overpotential. If we assume that the capacity is constant in the range of overpotential measured and substitute the Tafel expression for the current density,

$$i = i_o \exp \eta/b'_a \qquad [\text{II.147}]$$

we have

$$\int \exp(-\eta/b'_a) d\eta = -\frac{i_o}{C} \int dt \qquad [\text{II.148}]$$

hence the overpotential η during decay is given as a function of time by the equation

$$\eta = a - b_a \log(t + \tau) \qquad [\text{II.149}]$$

in which

$$a = -b_a \log(i_o/b_a C) \qquad [\text{II.150}]$$

and τ is a constant of integration, given by

$$\tau = b_a C/i(t=0) \qquad [\text{II.151}]$$

$i(t=0)$ being the measured steady-state polarization current density at the overpotential $\eta(t=0)$ from which open-circuit decay begins. Thus, the slope of the open-circuit-decay curve is equal to the Tafel slope provided that the capacity is constant in the range of potential over which measurements are made.

8.4 OPEN-CIRCUIT DECAY

A more interesting situation arises if the capacity varies with potential in the manner shown in Fig. II 19, that is, when there is a substantial concentration of adsorbed intermediates and an adsorption pseudocapacity is added in parallel to the double-layer capacity, as shown in Fig. II 17. The expression for η as a function of t becomes very complex but the qualitative behavior can be predicted in a simple way. As the capacity increases, $d\eta/dt$ decreases and <u>vice versa</u>. Thus, if open-circuit decay begins from an overpotential where $\theta \cong 1$, it will decrease rapidly at first, more slowly at intermediate values of θ and rapidly again for values of θ approaching zero. The range of potentials over which the value of $d\eta/dt$ is small depends on the type of isotherm governing the system and on the value of the interaction parameter r in the Frumkin isotherm, as seen in Fig. II 19. For large values of r, the pseudocapacity is nearly constant over a wide potential region and equation [II.149] may be applied.

It is important to note that the numerical value of the pseudocapacity is large compared to the double-layer capacity (at $\theta = 0.01$, $C_\phi \cong 80~\mu F/cm^2$), so that the shape of the open-circuit-decay curve is sensitive to small concentrations of adsorbed intermediates and can serve to detect and, under certain conditions, determine the surface concentration of such species. The shape of the η - t curves for three initial values of θ is shown schematically in Fig. II 20.

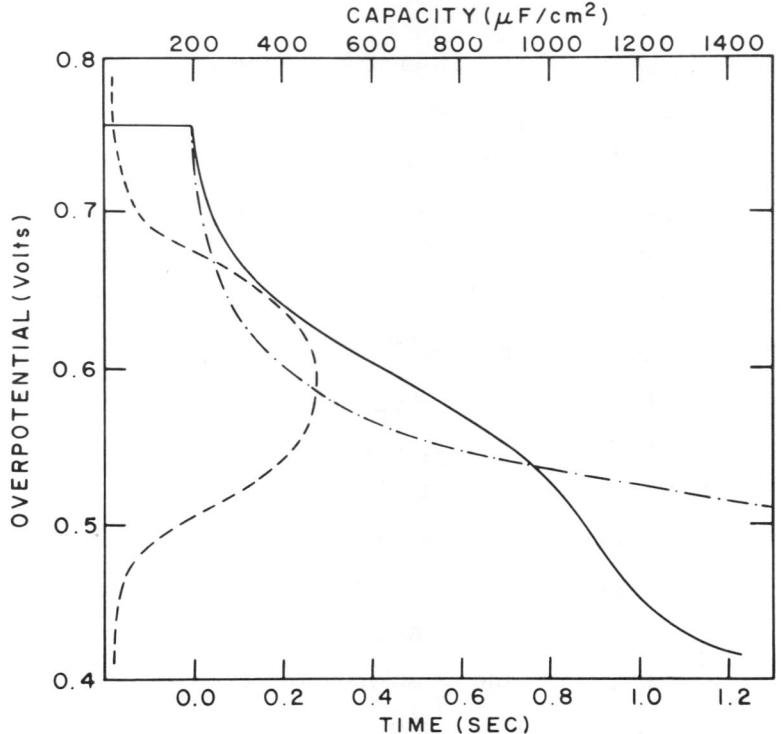

FIG. II 20   SCHEMATIC OPEN-CIRCUIT-DECAY PLOTS FOR DIFFERENT SYSTEMS
–·–·–   NEGLIGIBLE CONCENTRATION OF ADSORBED INTERMEDIATE
———   HIGH CONCENTRATION OF ADSORBED INTERMEDIATE FOLLOWING FRUMKIN ISOTHERM WITH r = 5 kcal/mole
– – – –   CORRESPONDING PSEUDOCAPACITY/POTENTIAL PLOT

9.   Adsorption of Neutral Species

9.1   General

Four major types of adsorption are of importance in determining the rate and mechanism of electrode reactions. The adsorption of ions in the inner and outer Helmholtz planes determines the structure of the ionic double layer and the potential

9.1 GENERAL

drop across various segments of the interphase. It may also affect the free energy of adsorption of reactants, products or intermediates. The effect is greatest when the reactant itself is an ion. Extensive studies of the adsorption of various ions on mercury have been reported, but work on solid electrodes has been more limited because of the difficulties involved, in spite of the great technological importance of such data.

The adsorption of intermediates and its effect on the kinetic parameters of electrode reactions have been dealt with in section II. 7. The influence of these intermediates on the a. c. response of the interphase is discussed in section II. 8. 2.

The solvent plays a major role in the adsorption of any species in the metal/solution interphase. Since the electrode is solvated, electrosorption* is always a replacement reaction. The observed values of the thermodynamic properties (ΔG, ΔH and ΔS of electrosorption) are the differences between the corresponding quantities for the adsorbate and for a number of solvent molecules m replaced by each adsorbate molecule. The potential dependence of solvent adsorption is often the cause of the observed potential dependence of the electrosorption of neutral species, as seen in the next section.

The solvent also acts as a dielectric in the interphase. This affects the magnitude of the electric field and modifies the lateral electrostatic interactions between ions or dipoles.

---

* The term "electrosorption" is used for the potential-dependent adsorption of a species from solution on the surface of a solvated electrode.

However, in this respect, it must be remembered that the effective dielectric constant in the interphase tends to be much lower than in the bulk of the solvent, as discussed in section II.1.5 above.

The adsorption of neutral species is of importance in many branches of electrochemistry. In fuel cells the organic fuel is first electrosorbed and then undergoes electrochemical oxidation. In electroorganic syntheses both reactants and products (and, for that matter, any stable intermediates formed in the reaction sequence) may be electrosorbed on the electrode surface and thus affect the course of the reaction. Small amounts of adsorbed impurities can decrease the reaction rates by several orders of magnitude. It is interesting to note that most of these "electrochemical poisons" are also potent physiological poisons, because of their ability to retard vital enzymatic reactions taking place in the organism. Electrochemical poisons can sometimes be used to advantage, to decrease the rate of an undesired side reaction. This is the basis for the use of corrosion inhibitors as a means of lowering the corrosion rate in chemical plants and installations.

In the following section the adsorption of neutral species in relation to solvent adsorption will be discussed.

9.2 Adsorption Isotherms for the Electrosorption of Neutral Species

Electrosorption differs from adsorption on nonconducting substrates in that an additional degree of freedom, the potential $\Delta\phi$, exists. Two types of isotherms must, therefore, be con-

9.2 ADSORPTION ISOTHERMS

sidered; one describing the change of coverage with concentration in the bulk at constant potential, and the other describing the change of θ with $\Delta\phi$ at constant bulk concentration.

The electrosorption of a neutral species RH on a hydrated electrode may be represented by the equilibrium

$$RH_{soln} + mH_2O_{ads} \rightleftharpoons RH_{ads} + mH_2O_{soln} \qquad [II.152]$$

Equating the chemical potentials of reactants and products one obtains the isotherm

$$\frac{\theta}{(1-\theta)^m} \frac{[\theta + m(1-\theta)]^{m-1}}{m^m} = KC_{RH}/C_{H_2O} \qquad [II.153]$$

The constant K is determined by the standard free energy of electrosorption ΔG^o_{ads} which, in turn, is given by

$$\Delta G^o_{ads} = \left(\mu^o_{RH_{ads}} - \mu^o_{RH_{soln}} \right) - m \left(\mu^o_{H_2O_{ads}} - \mu^o_{H_2O_{soln}} \right) \qquad [II.154]$$

Equation [II.153] is derived with the tacit assumption that K is independent of θ; in other words, it is the equivalent of the Langmuir isotherm, applied to the case of electrosorption. The corresponding Frumkin isotherm can be obtained simply by substituting $K^o \exp(-r\theta/RT)$ for K. It must be remembered, though, that r will now be the rate of change of the apparent standard free energy of electrosorption with coverage, that is, the difference between the corresponding quantities for RH and for m water molecules.

The potential dependence of the adsorption of neutral molecules is the combined effect of the change with potential of

$\overset{o}{\mu}_{RH_{ads}}$ and of $m\overset{o}{\mu}_{H_2O_{ads}}$ in equation [II.154]. In many cases $\overset{o}{\mu}_{RH_{ads}}$ may be regarded as constant and the variation of $\overset{o}{\mu}_{H_2O_{ads}}$ with $\Delta\phi$ (due to the interaction between the dipole moment of water molecules $\vec{\mu}$ and the field $\vec{F}$) is sufficient to account for the variation of θ_{RH} with $\Delta\phi$. The discussion which follows is based on this assumption. It is justified because the dipole moment of water molecules is relatively large and it is in the region of the interphase where the field is the strongest. Organic molecules, on the other hand, tend to be oriented on the surface with their dipolar functional groups away from the electrode beyond the OHP, where the electric field is smaller by a factor of five or ten.

Adsorbed water molecules can take up one of two positions with their dipole vectors perpendicular to the electrode surface; the dipoles will point either in the direction of the field or in the opposite direction. The electrostatic energies of interaction with the field $\vec{F}$ for the two positions are equal in magnitude but opposite in sign. At the potential of zero charge where the field is zero, there should be no preference for either position. We can define a quantity B by

$$B = \frac{N\uparrow - N\downarrow}{N\uparrow + N\downarrow} \qquad [II.155]$$

where $N\uparrow$ and $N\downarrow$ are the numbers of water molecules per cm^2 in the "up" and "down" positions, respectively. As q_M departs from zero in either direction, the water dipoles tend to align themselves with the field. However, two neighboring molecules pointing in the same direction repel each other while if they point in opposite directions, the force is one of attraction. As the field is in-

9.2 ADSORPTION ISOTHERMS

creased, the tendency for the dipoles to turn in the direction of the field and their tendency to form "up" and "down" pairs counteract each other and B in equation [II.155] varies gradually with the electric field. This variation can be expressed approximately by the equation

$$B = \tanh\left(\frac{\vec{\mu}\vec{F} - BUg}{kT}\right) \quad [II.156]$$

where g is the number of nearest neighbors of a water molecule on the surface and U is the energy of interaction between two nearest neighbors. The average free energy of adsorption of a water molecule depends on B and so the dependence of θ on the field can be expressed by writing the equilibrium constant K in equation [II.153] as

$$K = K^o \exp\left[-mB\left(\frac{\vec{\mu}\vec{F} - BUg}{kT}\right)\right] \quad [II.157]$$

The above theory predicts a bell-shaped curve of θ <u>vs</u> q_M with a maximum at the potential of zero charge where $N\uparrow = N\downarrow$ or $B = 0$. This is indeed found experimentally (cf. Fig. II 21) except that the potential of maximum adsorption does not coincide with the potential of zero charge. The shift is believed to be due to the preferred orientation of water molecules with the oxygen atoms towards the surface at zero field. Thus, a small negative field must be applied to reach the point where $N\uparrow = N\downarrow$, corresponding to maximum adsorption.

The above theory was proposed by Bockris, Devanathan and Muller in 1963. An earlier phenomenological theory was

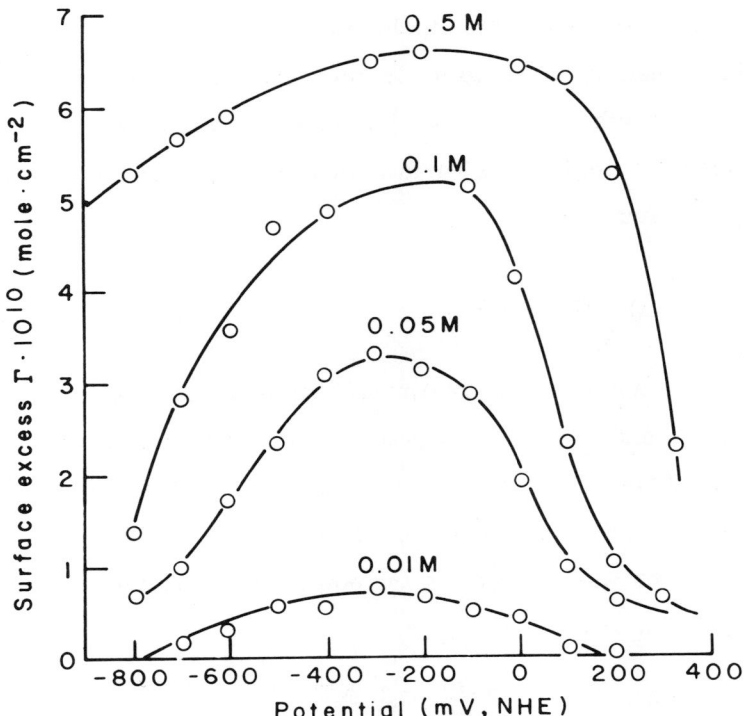

FIG. II 21 SURFACE EXCESS OF BUTANOL AS A FUNCTION OF POTENTIAL (E. Blomgren, J. O'M. Bockris and C. Jesch, J. Phys. Chem. 65, 2004 (1961)).

proposed by Frumkin in 1926 and has been further developed and modified by Frumkin and Damaskin. According to this theory the variation of θ with electrode potential is due to the change in the energy of the double-layer capacitor which results when water is replaced by organic molecules having a lower dielectric constant. Instead of equation [II. 157] this theory predicts the relationship

$$K = K^o \exp\left[-\frac{1}{2}(C_o - C_1)\overline{E}^2 + \overline{E}\ E_N C_1 \right] \qquad [\text{II. 158}]$$

where C_0 and C_1 are the values of the double-layer capacity at $\theta = 0$ and $\theta = 1$, respectively, $\overline{E}$ is the potential on the rational scale ($\overline{E} = E - E_z$) and E_N is the shift in the potential of zero charge as θ goes from zero to unity.

Equations [II.157] and [II.158] are based on different models and their validity should be tested by comparison with experiment. The former has the advantage that it is based on a molecular model and thus gives insight to the detailed structure of the interphase. The latter makes use of experimentally determinable quantities (C_0, C_1, E_N) but does not enlighten us as to the relationship between the numerical values of these quantities and the molecular structure of the system.

9.3 The Combined Adsorption Isotherm

A case of interest arises when the adsorbed intermediate is relatively large and takes the place of several solvent molecules on the surface, for example

$$CH_3 \underset{O}{\overset{||}{C}}H + H^+ + e_M \rightleftharpoons CH_3 \underset{OH}{\overset{M}{C}}H \qquad [\text{II.159}]$$

$$2\, CH_3 \underset{OH}{\overset{M}{C}}H \xrightarrow{\text{r.d.s.}} CH_3 \underset{OH}{\overset{H}{C}} - \underset{OH}{\overset{H}{C}} CH_3 \qquad [\text{II.160}]$$

The dependence of coverage on potential will be governed by charge transfer on the one hand and by competition with the solvent molecules on the other. The resulting isotherm is derived by combining equations [II.157] and [II.114]. A simple expression applicable at high field strength has been derived by Gileadi and coworkers. If $\vec{\mu}\vec{F}/kT \gg 1$, we can write equation [II.157] in the approximate form

$$K = K'_o \exp(-m\vec{\mu}\vec{F}/kT) \qquad [II.161]$$

since under these conditions, most water molecules will be oriented with their dipoles in the direction of the field and hence $B \rightarrow 1$.

The field $\vec{F}$ is written in terms of the potential difference across the interphase divided by a thickness parameter d, taken to be equal to the diameter of a water molecule.

$$\vec{F} = (\Delta\phi - \Delta\phi_z)/d = \Delta\phi/d + \text{const} \qquad [II.162]$$

where $\Delta\phi_z$ is the potential difference across the interphase at $q_M = 0$.* Combining the last two equations, we have

$$K = K_o \exp\left(-\frac{m\vec{\mu}\Delta\phi}{kTd}\right) = K_o \exp\left(-\frac{m\vec{\mu}\Delta\phi}{kTd} \cdot \frac{F}{e_o N}\right) = K_o \exp\left(-\frac{m\vec{\mu}}{e_o d} \frac{\Delta\phi F}{RT}\right) \qquad [II.163]$$

To obtain the combined isotherm, it is necessary to substitute into equation [II.114] the value of K derived above. Thus,

---

* In the detailed theory one writes, instead of $\Delta\phi_z$ in equation [II.162], the potential difference $\Delta\phi_{max}$ at the potential of maximum adsorption. Since both are constants this finer detail of the theory can be safely ignored here.

10.1 STATEMENT OF PROBLEM

$$\frac{\theta}{1-\theta} = K_o C \exp(\Delta\phi F/RT)(1 - m\vec{\mu}/e_o d) \qquad [\text{II.164}]$$

If, instead of the Langmuir isotherm, the Frumkin isotherm (cf. equation [II.125]) were used, only the left-hand side of equation [II.164] would be affected. Indeed the combined isotherm may be written in more general form as

$$f(\theta) = K(\Delta\phi)C \exp(\Delta\phi F/RT) \qquad [\text{II.165}]$$

in which $f(\theta)$ and $K(\Delta\phi)$ are functions of θ and $\Delta\phi$, respectively and should be written in the form applicable to the system being studied.

It should be obvious that introduction of a new isotherm also affects the values of the Tafel slopes calculated for each mechanism. Moreover, the numerical values of the slopes will now depend on the size of the adsorbed intermediate, that is, on m, the number of water molecules displaced from the surface by each molecule of intermediate adsorbed. Such behavior has indeed been observed experimentally; however, a detailed discussion of this effect is outside the scope of this section.

10. Analysis of Reactions Occurring Under Mixed Activation and Diffusion Control

10.1 Statement of the Problem

We have seen in section [II.5.6] above that when an electrochemical reaction takes place at a finite rate, $C(s)$, the surface concentration of the reactant, departs from its bulk concentration C^o. The ratio between these quantities is given at steady state by

$$\frac{C(s)}{C^o} = 1 - \frac{i}{i_L} \qquad [\text{II}.80]$$

in which i_L is the limiting current density, determined by the maximum rate of mass transport to the electrode under the experimental conditions chosen. If an excess of supporting electrolyte is added and the solution is not agitated in any way, that is, if conditions are maintained such that mass transport occurs only by diffusion, equation [II.80] still holds, except that i_L is replaced by i_d, the diffusion-limited current density. As long as $i/i_d \ll 1$, the concentration at the surface does not depart significantly from the concentration in the bulk and the reaction may be said to be activation-controlled. As i approaches i_d, the surface concentration $C(s)$ approaches zero and the reaction may be said to be controlled entirely by diffusion. At intermediate values of the current density (say, when $0.05 < (i/i_d) < 0.95$) the reaction is under mixed activation and diffusion control and the ratio of concentrations $C(s)/C^o$ departs significantly from unity. It is possible to analyze the experimental results observed in this region by solving the diffusion equation for given boundary conditions. An outline of the method for the case of a potential-step function applied to the interphase will be given below, following the derivation of Delahay.

10.2 The Variation of Current with Time at Constant Potential

Consider a simple reaction of the type

10.2 VARIATION OF CURRENT

$$\text{Ox} + \text{ne}_M \xrightarrow{k_c} \text{Red} \qquad [\text{II.166}]$$

taking place at an overpotential high enough that only the rate of the forward reaction need be considered. Two differential equations, which describe the variation of C_{Ox} and C_{Red} as functions of time and distance from the electrode surface, must be solved simultaneously:

$$\frac{\partial C_{Ox}(x,t)}{\partial t} = D_{Ox} \frac{\partial^2 C_{Ox}(x,t)}{\partial x^2} \qquad [\text{II.167}]$$

$$\frac{\partial C_{Red}(x,t)}{\partial t} = D_{Red} \frac{\partial^2 C_{Red}(x,t)}{\partial x^2} \qquad [\text{II.168}]$$

The boundary conditions, assuming that the species Red is not present initially in solution, are:

$$C_{Ox}(x,0) = C^o \quad ; \quad C_{Red}(x,0) = 0 \qquad [\text{II.169}]$$

$$C_{Ox}(\infty,t) = C^o \quad ; \quad C_{Red}(\infty,t) = 0 \qquad [\text{II.170}]$$

In addition, the sum of the fluxes of Ox and Red must equal zero at $x = 0$.

$$D_{Ox}\left[\frac{\partial C_{Ox}(x,t)}{\partial x}\right]_{x=0} + D_{Red}\left[\frac{\partial C_{Red}(x,t)}{\partial x}\right]_{x=0} = 0 \qquad [\text{II.171}]$$

The relation to the activation-controlled kinetic parameter k_c is introduced through the sixth boundary condition

$$D_{Ox}\left[\frac{\partial C_{Ox}(x,t)}{\partial x}\right]_{x=0} = k_c C_{Ox}(0,t) \qquad [\text{II.172}]$$

which states that the flux of reactant (Ox) at the surface of the electrode (x = 0) is equal to the specific rate constant multiplied by the concentration of the reactant at the surface. (The current density is equal to the flux of reactant at x = 0 multiplied by nF).

The solution of the diffusion equations [II.167] and [II.168] under the above boundary conditions, yields an expression for the current as a function of time

$$i = nFC^o D_{Ox}^{1/2} t^{-1/2} \lambda \exp \lambda^2 \, \text{erfc}\, \lambda \qquad [\text{II.173}]$$

in which

$$\lambda = k_c t^{1/2}/D_{Ox}^{1/2} \qquad [\text{II.174}]$$

Combining this with equation [II.87] for the diffusion-limited current density

$$i_d = nFD_{Ox}^{1/2} C^o (\pi t)^{-1/2} \qquad [\text{II.87}]$$

gives

$$i/i_d = \pi^{1/2} \lambda \exp \lambda^2 \, \text{erfc}\, \lambda \qquad [\text{II.175}]$$

This equation allows a determination of λ, and hence k_c as a function of potential from measurement of the current density/potential relationship provided i_d can be measured in the same experimental setup.

10.3 SIGNIFICANCE OF THE αn_a TERM

The weakness of this treatment lies in the use of equation [II.172] as one of the boundary conditions for solving the differential equations. This is valid only if the reaction studied is first-order with respect to the reactant (Ox) and if there are no kinetic or catalytic complications. The equation can be suitably modified for, say, a second-order reaction, but in any case the mechanism must be at least partly known in order that k_c may be evaluated as a function of potential; while the purpose of such evaluation is to determine the mechanism. Nevertheless this type of treatment is useful in cases where the reaction is known to follow the simple scheme represented by equation [II.166] and it has been used (in slightly modified form) mainly in polarography.

10.3 The Significance of the αn_a Term in Irreversible Polarography, Chronopotentiometry and Linear Potential Sweep

We note that the time-dependent current density $i(t)$ is given by (cf. equation [II.172])

$$i(t) = nFk_c C_{Ox}(0,t) \qquad [II.176]$$

The activation-controlled current density at the same potential is given by

$$i_{ac} = nFk_c C^o \qquad [II.177]$$

Hence a plot of $\log k_c$ (or, for that matter, $\log \lambda$) against potential is nothing but the usual Tafel plot, and we can write (cf. equation [II.43])

$$k = k_s \exp\left(\pm \frac{E - E^o}{b'}\right) \qquad [\text{II}.178]$$

where k_s is the value of the specific rate constant at the standard reversible potential E^o. In the treatment of reactions under mixed activation and diffusion control, it is common to write the potential dependence of k in the form

$$k = k_s \exp[\pm \alpha n_a F(E - E^o)/RT] \qquad [\text{II}.179]$$

in which n_a is said to be the number of electrons taking part in one act of the rate-determining step, and α is a transfer coefficient. While equation [II.179] is formally correct, it led to a great deal of confusion when α was mistakenly assigned a value of one-half and n_a was calculated from the measured value of αn_a. Comparison with equations [II.68] and [II.71] shows that α, as defined in equation [II.179]* does not have a simple physical significance and it cannot be given any a-priori numerical value. The product αn_a is related to the Tafel slope, as we note by comparing equations [II.178] and [II.179].

$$\alpha n_a = 2.3RT/bF \qquad [\text{II}.180]$$

It is recommended, in order to avoid confusion, that in all equations relating to reactions under mixed diffusion and activation control, this last relationship should be referred to in order to replace αn_a by the corresponding Tafel slope b. In the table

---

* This is clearly not the same α which appears in equation [II.71].

11.1 CORROSION RATES AND MIXED POTENTIALS

below, the important equations are given in terms of αn_a and in terms of b, for easy reference.

Method	Experimental Parameter	in terms of b	in terms of αn_a
Chronopotentiometry	$dE/d \log [1 - (t/\tau)^{1/2}]$	$-b_a ; +b_c$	$\mp 2.3 \, RT/\alpha n_a F$
	$dE_{1/4}/d \log i$	$+b_a ; -b_c$	$\pm 2.3 \, RT/\alpha n_a F$
Linear Potential Sweep	$dE/d \log v$	$b_a/2 ; -b_c/2$	$\pm 2.3 \, RT/2\alpha n_a F$
	$di_p/d(v^{1/2})$	$3.01 \times 10^5 \, n \left(\dfrac{2.3 \, RT}{bF}\right)^{1/2} D^{1/2} C°$	$3.01 \times 10^5 n (\alpha n_a)^{1/2} D^{1/2} C°$
Irreversible Polarography	$dE/d \log [(i_d - i)/i]$	$0.916 \, b_a ; -.916 \, b_c$	$\pm 2.3 \times 0.916 \, RT/\alpha n_a F$
	$dE_{1/2}/d \log t$	$-b_a/2 ; +b_c/2$	$\mp 2.3 \, RT/2\alpha n_a F$

11. Fundamentals of Corrosion

11.1 Corrosion Rates and Mixed Potentials

All metals except gold are thermodynamically unstable with respect to their oxides. They are mined in the form of various compounds, but rarely in the native form. Corrosion represents the tendency of pure metals and alloys to return to thermodynamically more stable compounds. The fundamental process of corrosion is the anodic dissolution of the metal to form its ions

$$M \longrightarrow M^{+z} + ze_M \qquad \text{[II.181]}$$

The ions may be carried away into the bulk of the solution in contact with the metal, or they may form insoluble salts which precipitate on its surface. A layer of oxide or hydroxide may be formed on the surface and this is sometimes further oxidized to a higher oxide. Such layers can be porous, allowing further corrosion (as with iron), or they may be compact and nonporous, forming a protective layer which essentially eliminates further corrosion (as with aluminum). In this section we shall disregard these secondary processes and focus our attention on the corrosion reaction itself.

When an isolated piece of metal is in contact with a corrosive medium (e.g., the hull of a boat in sea-water, a pipeline in moist soil, etc.) the corrosion reaction represented by equation [II.181] cannot proceed unless another process takes place at the same rate to carry away the electrons accumulated in the metal.*

In most cases, spontaneous corrosion is accompanied by hydrogen evolution or oxygen reduction, although other cathodic reactions may serve the same purpose of consuming the electrons released in the anodic dissolution of the metal.

When steady state has been reached, the metal attains a constant potential, E_{corr}, which is termed the <u>mixed</u> or <u>corrosion potential</u>. At this potential there is an anodic overpotential

---

* Simple calculation shows that if as little as 10^{-8} gm/cm^2 of iron were dissolved and the electrons released were not used up in a cathodic process, a surface charge of about 20 μC/cm^2 would be created which, in turn, would change the potential $\Delta\phi$ across the interphase by about one volt, enough to stop the corrosion process altogether.

11.1 CORROSION RATES AND MIXED POTENTIALS

$$\eta_a = E_{corr} - E_{r,a} \qquad [\text{II.}182]$$

driving the metal-dissolution process, and a cathodic overpotential

$$\eta_c = E_{corr} - E_{r,c} \qquad [\text{II.}183]$$

driving the cathodic process. The value of these overpotentials and hence the corrosion potential E_{corr} are determined by the condition that the anodic and cathodic processes must occur at exactly the same rate.

The current/potential curves for an anodic and cathodic process are shown schematically in Fig. II 22. At the point of intersection, the two processes occur at the same rate, i_{corr}, which is the steady-state corrosion rate. The potential corresponding to this point is the corrosion potential, E_{corr}.

Assuming that both processes are in the linear Tafel region at the corrosion potential, we may write

$$i_{corr} = i_{o,a} \exp\left(\frac{E_{corr} - E_{r,a}}{b'_a}\right) = i_{o,c} \exp\left(-\frac{E_{corr} - E_{r,c}}{b'_c}\right) \qquad [\text{II.}184]$$

where the subscripts a and c refer to the anodic and cathodic process, respectively. It should be noted that the corrosion potential is determined by the reversible potentials of the two processes taking place as well as by their exchange-current densities and Tafel slopes, in other words, by both thermodynamic and kinetic factors.

If the corroding metal is connected to a source of current or potential and polarized anodically or cathodically with respect

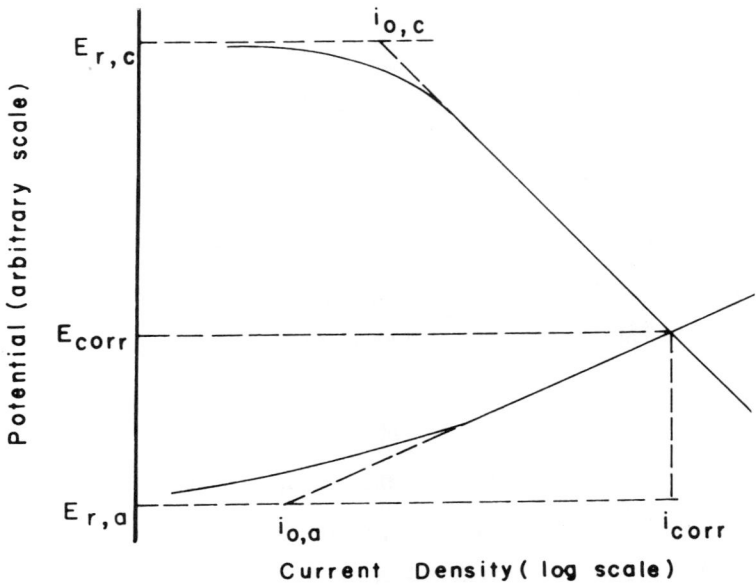

FIG. II 22   MIXED POTENTIAL AND CORROSION CURRENT IN SPONTANEOUS CORROSION

to its corrosion potential, the current which flows is

$$i = i_{o,a} \exp\left(\frac{E - E_{r,a}}{b'_a}\right) - i_{o,c} \exp\left(-\frac{E - E_{r,c}}{b'_c}\right) \qquad [\text{II.}185]$$

Combined with equation [II.184] this yields

$$i = i_{corr} \left[\exp\left(\frac{E - E_{corr}}{b'_a}\right) - \exp\left(-\frac{E - E_{corr}}{b'_c}\right)\right] \qquad [\text{II.}186]$$

Equation [II.186] is of great interest here because it is identical in form to the general rate equation for a single-electrode reaction (cf. equation [II.52]), except that i_o and E_r

are replaced by i_{corr} and E_{corr}, respectively, and the anodic and cathodic Tafel slopes refer to different electrode reactions. It permits the electrochemical determination of the corrosion rate i_{corr} by extrapolation of the $E/\log i$ plot from high values of $E - E_{corr}$ on either the anodic or cathodic side to the corrosion potential, or by measurements of i as a function of E in the vicinity of E_{corr}.

The anodic and cathodic reactions occur on different parts of the metal, which may be termed anodic and cathodic areas. These are determined by the structure and composition of the metal, mechanical strain, (highly strained areas tend to be more anodic), partial contact with the corrosive medium (through faulty coating), variation in composition of the medium, etc. It should be kept in mind that under steady-state corrosion, the total anodic and cathodic currents must be equal, while the current densities may differ substantially depending on the relative sizes of the anodic and cathodic areas. One well known and troublesome result of this is pit-corrosion which occurs when the anodic areas are very small and hence sustain a high current density. This causes deep pits in the metal under conditions where the total average rate of corrosion is relatively small.

11.2   Passivity

Passivity is the name given to the phenomenon shown by certain metals which are able to withstand corrosion in media where they are thermodynamically unstable. Chemical passivity was discovered over a century ago, when it was observed that iron will dissolve freely in dilute nitric acid while in concentrated solution a protective film is formed and dissolution stops immediately. Many other metals and alloys (e.g., Ni, Cr, Ta, Al) in contact with oxidizing media form passive films spontaneously. Certain ions, notably chloride, tend to cause breakdown of passivity and it may become very difficult or even impossible to form a passive film in their presence.

Electrochemical passivation is similar in many respects to chemical passivation. As the metal is polarized anodically, its rate of dissolution first increases until a certain critical potential is reached. Beyond this potential, the rate of dissolution drops sharply (often by several orders of magnitude). A typical current/potential curve observed for an electrode undergoing passivation is shown in Fig. II 23. The current in the transpassive region may be due to another process taking place (e.g., oxygen evolution) or to conversion of the passive film into a higher-valence state which is soluble and allows free dissolution of the metal to take place again.

The nature of the passivating film on different metals is still the subject of controversy. A layer 50-150 Å thick is found to be sufficient, but it has been argued that the surface is already rendered passive when the film is only a monolayer.

Passivation never takes place instantaneously. When a constant current is applied, the time required to form or remove the passive film can be used to measure its thickness. In most cases the passive film is a nonstoichiometric oxide which has a

11.3 POTENTIAL/pH DIAGRAMS

high electronic conductivity. The electric field in the film is therefore very low and ionic currents which would cause further dissolution of the metal (or film thickening) are negligible. An exception to this are the valve metals (Al, Ta) on which a non-conducting film can form when very high potentials are applied.*
In this case, high-field ionic conductance occurs and the metal cannot be said to be truly passivated, since a relatively thick film may be formed.

11.3 Potential/pH Diagrams

A useful aid in the description of metal stability, corrosion and passivation is the potential-pH diagram developed by Pourbaix.

If all the chemical and electrochemical reactions which can take place in a system and the corresponding standard electro-chemical potentials are known, it is possible to draw lines representing constant concentration ratios of reactants and products, as a function of potential and pH.

Consider the simple example of pure water which can undergo one chemical and two electrochemical reactions

$$H_2O \rightleftharpoons H^+ + OH^- \qquad [\text{II.187}]$$

$$H_2 \rightleftharpoons 2H^+ + 2e_M \qquad [\text{II.188}]$$

$$2H_2O \rightleftharpoons O_2 + 4H^+ + 4e_M \qquad [\text{II.189}]$$

---

* These are the films used for the production of electrolytic capacitors.

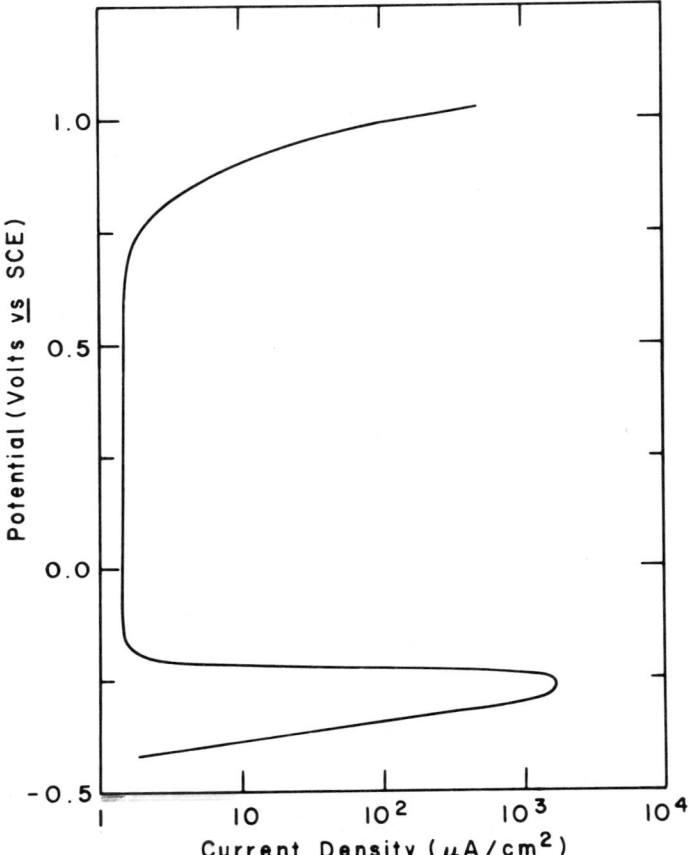

FIG. II 23 TYPICAL PASSIVATION CURVE FOR STAINLESS STEEL IN H_2SO_4

The corresponding equations relating to the concentrations of the species involved to potential and pH are

$$E_r = -\frac{2.3RT}{F}pH - \frac{2.3RT}{2F}\log p_{H_2} \qquad [II.190]$$

and

11.3 POTENTIAL/pH DIAGRAMS

$$E_r = 1.228 - \frac{2.3RT}{F}pH + \frac{2.3RT}{4F}\log p_{O_2} \qquad [\text{II}.191]$$

for the hydrogen- and oxygen-evolution reactions, respectively, where the potentials are on the normal-hydrogen scale. Fig. II 24 is a graphical representation of equations [II.190] and [II.191]. The lower set of lines represents the reversible potentials for hydrogen evolution as a function of pH at different partial pressures of hydrogen and the upper lines are their equivalent for oxygen evolution. The lowest and the highest lines are marked on all other potential/pH diagrams in aqueous solutions. The area between these lines represents the region of thermodynamic stability of water saturated with hydrogen and oxygen each at a partial pressure of one atmosphere.

The dotted lines in Fig. II 24 divide the domains of stability and predominance of various species in solution. On the line itself, the concentrations of the two species involved are equal. The vertical dotted line is related to the dissociation of water (equation [II.187]). The horizontal line in the acid region represents the simple equation

$$H^- \rightleftharpoons H^+ + 2e_M \qquad [\text{II}.192]$$

for which

$$E_r = -1.125 + \frac{2.3RT}{2F}\log\frac{a_{H^+}}{a_{H^-}} \qquad [\text{II}.193]$$

The sloping line in the alkaline region corresponds to the equilibrium

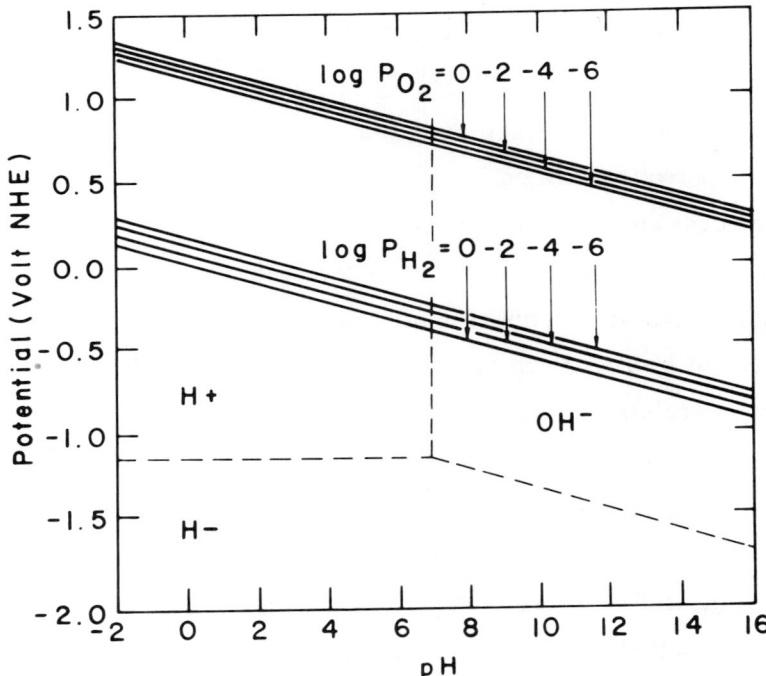

FIG. II 24   POTENTIAL-pH DIAGRAM FOR WATER[17]

$$H^- + H_2O \rightleftharpoons OH^- + 2H^+ + 2e_M \qquad [II.194]$$

for which

$$E_r = -0.711 - \frac{2.3RT}{F}pH + \frac{2.3RT}{2F}\log\frac{a_{OH^-}}{a_{H^-}} \qquad [II.195]$$

A complete description of this system would also include species such as H_2O_2, HO_2^- and perhaps O_3, with their relevant chemical and electrochemical equilibria.

A relatively simple potential/pH diagram for a metal is

11.3 POTENTIAL/pH DIAGRAMS

shown in Fig. II 25. The solid phases involved here are metallic titanium and its three stable oxides, TiO, Ti_2O_3, TiO_2, which are in equilibrium with 10^{-6} molar solutions of the divalent and trivalent ions.* The equations for the nine equilibria involved in this diagram should be obvious and will not be listed here. The region of stability of water is confined by the two dotted lines $\underline{a}$ and $\underline{b}$. A third dotted line at the standard potential for the Ti^{++}/Ti^{+++} couple divides the regions of predominance of these two ions. As in the case of Fig. II 24, the complete diagram would contain lines corresponding to additional species such as TiO^{++}, TiO_2^{++}, etc. These have been omitted for clarity.

The regions in which the metal, its solid oxides and its soluble ions are stable, are termed regions of immunity, passivity and corrosion, respectively. It must be realized, however, that potential/pH diagrams are drawn on the basis of purely thermodynamic data and their usefulness is correspondingly restricted. Thus, it can be stated flatly that no corrosion

---

* In simplified potential/pH diagrams a concentration of 10^{-6} $\underline{M}$ of the soluble species is arbitrarily chosen for drawing the line between the region of stability of a solid phase and that of the soluble species in equilibrium with it. Taking a different concentration would only cause a slight parallel displacement of the lines, but would not alter the shape of the diagram significantly.

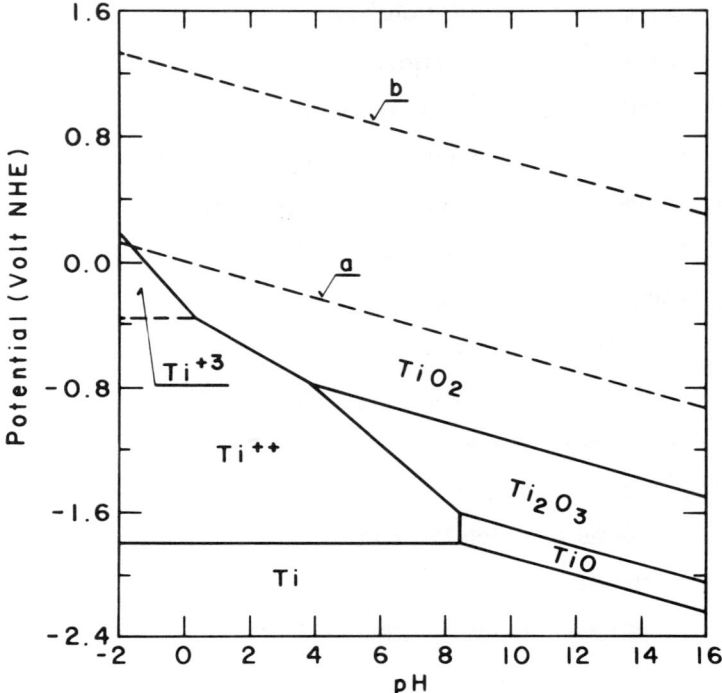

FIG. II 25   POTENTIAL-pH DIAGRAM FOR TITANIUM [17]

can occur in the region of immunity.* In the corrosion region, on the other hand, whether measurable corrosion is or is not observed depends on kinetic parameters. Likewise, the term "passivity", in the sense used in the potential/pH diagrams, simply means that a solid phase is stable in that region.

---

* This statement may sometimes appear to be disproved experimentally. In such cases one is urged to re-examine the systems and find whether the addition of new components (e.g., a complexing agent) may not require the preparation of a revised potential/pH diagram.

11.4 CORROSION PREVENTION

Whether this will stop the corrosion of the metal or allow continued growth of the oxide layer depends on the nature of the layer formed. In the case of titanium, a compact protective layer is formed which prevents further corrosion. Thus, although this metal in its pure form is highly reactive and should rapidly decompose water, in practice it does not corrode except over a restricted potential and pH region and is listed by Pourbaix between platinum and palladium in the list of "practical" nobility of metals.

11.4 Corrosion Prevention

The methods of corrosion prevention are many and varied. Only three of these, which are electrochemical in nature, will be considered here. The electrochemical methods of corrosion prevention described below are widely used in practice.

It should be remembered that economic as well as scientific considerations are involved in any corrosion problem. To quote an extreme example, one could reduce corrosion very substantially in a municipal water-supply system if stainless-steel pipelines were used; obviously the cost of such an installation would be prohibitive. Initial cost has to be weighed against maintenance expenses and the best solution is not necessarily the one giving rise to the lowest corrosion rate. In some specialized electronic equipment where reliability is essential (e.g., in space and military applications) electrical contacts are gold-coated to prevent corrosion. In this case corrosion must be prevented at all costs. The objective of the corrosion engineer and scientist is to use the least expensive materials and corrosion-prevention techniques which may be applicable in each situation.

(i) Cathodic protection

The electrochemical basis for cathodic protection can be clearly seen by considering Fig. II 26 (cf. Fig. II 22). The open-circuit potential of the metal is E_{corr} and the corrosion rate is i_{corr}. An external d.c. power source and an auxiliary anode are used to pass a net cathodic current i, which reduces the corrosion rate to i'_{corr} as shown in Fig. II 26. By application of a suitable cathodic current, it is thus possible to bring the corrosion rate down to the desired level. However, with large structures (e.g. pipelines) the electric power consumed and the installations involved may become expensive.

An alternative method based on the same principle is used in the protection of the hulls of ships and boats from corrosion. Here an active metal, generally zinc, is used as a "sacrificial anode" in contact with the corrodable metal, and no external power source is required. The two metals in contact form a cell, the terminals of which have been shorted. The zinc serves as the anode and is slowly dissolved while the other metal is cathodically protected by the current passed, as shown in Fig. II 26.

(ii) Anodic protection

Anodic protection can be used successfully only in certain well defined cases which occur commonly in the chemical industry and in the storage of certain corrosive chemicals. The structure to be protected (e.g. a metallic container for the storage of sulfuric acid) is made the anode, and its potential is set in the

11.4 CORROSION PREVENTION

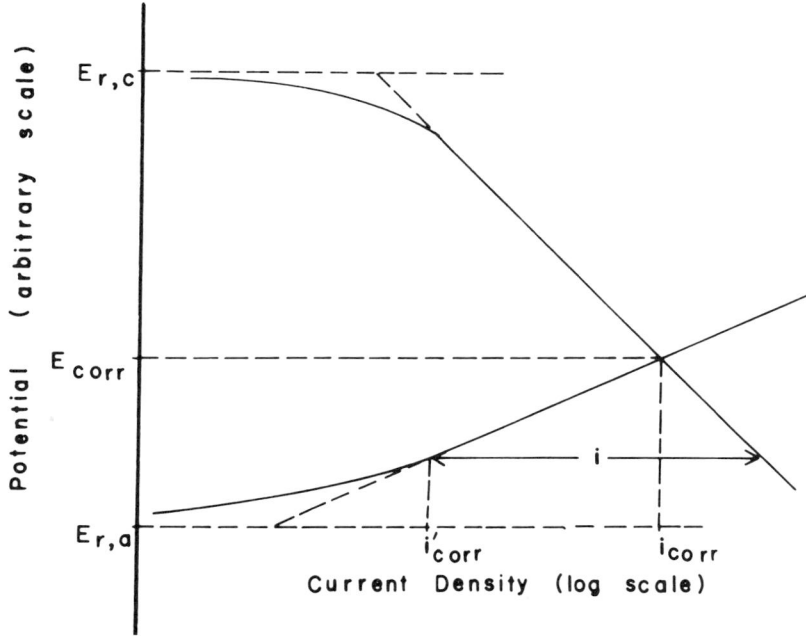

FIG. II 26 GRAPHICAL REPRESENTATION OF CATHODIC PROTECTION

passive region (Fig. II 23). The nature of the auxiliary electrode is of little importance in this case since it is the cathode and is cathodically protected.

The current density in the passive region is extremely small, and so the electric power needed for anodic protection is two or three orders of magnitude smaller than that required for cathodic protection. For this reason, it is preferred to cathodic protection whenever there is a choice. Unfortunately, passivation does not readily occur on iron and on ferrous alloys in the presence of chloride ions and this severely limits the application of this method.

(iii) Corrosion inhibitors

From Figs. II 22 and II 26 it is clear that any electrochemical poison which retards either the cathodic or the anodic reactions involved in the spontaneous corrosion process, can retard the process itself.

In deaerated aqueous solutions, the corrosion rate is controlled mainly by the cathodic hydrogen-evolution reaction. Many compounds which are known to poison this reaction (e.g. As_2O_3, pyridine, piperidine, some nitrogen-containing alkaloids) are found to be effective corrosion inhibitors. The action of such inhibitors is twofold. They are adsorbed on the surface of the metal (up to a monolayer or less) and physically block the sites on the surface at which hydrogen is ordinarily evolved. The structure of the double layer is altered by the adsorption of the inhibitor (the dielectric constant is usually decreased and the surface dipole or "chi-potential" in equation [II.16] is changed). Thus, the specific electrochemical rate constant for both the anodic and cathodic reactions may be changed even on sites which are not directly blocked by the adsorbed inhibitor.

Another class of corrosion inhibitors includes oxidizing agents such as Pb_3O_4 or $Cr_2O_7^=$. These form redox couples which bring the potential on the surface of the metal to the passive region (Fig. II 23) and thus stop corrosion. Oxidizing or passivating corrosion inhibitors are very effective but often have to be used in substantial concentrations in order to maintain their potency. Their main application is in relatively low-volume closed systems, such as chemical plants, boilers, heat exchangers, etc.

12. Batteries and Fuel Cells

12.1 Introduction

Batteries and fuel cells are devices in which chemical free energy is converted into electrical energy. In batteries the chemical free energy is stored in the electrode material and released when a suitable load is connected across the terminals. Primary batteries can be used only once, while secondary batteries can be recharged and used again many times. The latter thus act as energy-storage devices. Fuel cells differ from batteries in that the chemical free energy is generally stored outside the cell and the electrodes act as catalysts for the electrochemical oxidation and reduction reactions taking place. This is the case in systems like the hydrogen/oxygen, hydrazine/air or methanol/air fuel cells. Other systems such as the high-temperature $Li/LiCl/Cl_2$ or the low-temperature $Zn/ZnBr_2(aq)/Br_2$ cells behave to some extent as secondary batteries for energy storage. Here the metal electrode is used up when energy is withdrawn from the cell and the electrolyte formed during operation is electrolyzed in a central plant. The halogen and the metal may be fed continuously into the cell as in a typical fuel cell. The main purpose of such devices, however, is to store energy and make it available in relatively small mobile units.*

---

* A high price is commonly paid for electrical energy stored in small packages. We now pay \$20-\$50 per kwh of electrical energy used in flashlights or portable radios, the sum depending on the type of battery employed. At this rate, it would cost about two dollars to warm a cup of tea!

12.2 Batteries

Some of the batteries manufactured and used today, like the Leclanché cell and the lead acid battery, have been in use for over a hundred years. As a result of the exacting requirements in military and space applications, a renewed effort in the development of high-performance battery systems has been made in recent years.

The cathode in most batteries consists of a higher oxide of a metal (e.g. MnO_2, $NiO(OH)$, PbO_2) in intimate contact with an inert-metal grid which serves as a current collector. The bottleneck during operation of the battery (and, incidentally, the most fascinating aspect from the point of view of the physical chemist) is the reaction taking place in the solid state. The active material is usually prepared in the form of fine particles pressed on the current collector, sometimes mixed with a conductor (e.g. metal powder) to decrease resistance. The electrode is thus made up of particles, each of which contains $10^5 - 10^{12}$ molecules of the oxide which must be accessible for quick reduction and oxidation during the discharge-charge cycle.* When the surface of, for example, a NiOOH particle is reduced to $Ni(OH)_2$ the diffusion of Ni^{3+} ions from the interior is too slow to maintain a reasonable discharge rate for the battery.

The heavy nickel ions need not move, however, in order to allow reduction of the molecules deep in the particle. This can

---

* Simple calculation shows that in a particle of 1μ radius, roughly 1% of the molecules will be in the ten outermost layers, while 10% will be buried under a thousand or more atomic layers.

be achieved by a proton-transfer reaction in the solid state as shown schematically below.

O=Ni(OH)-----HO-Ni(HO) → Ni(OH)-OH-----O=Ni(HO)

A hydrogen atom bound covalently to one oxygen atom and by a hydrogen bond to the other, has to change position, so that it forms a hydrogen bond with the first oxygen atom and a covalent bond with the second. A "chain" process is obviously possible as shown schematically below.

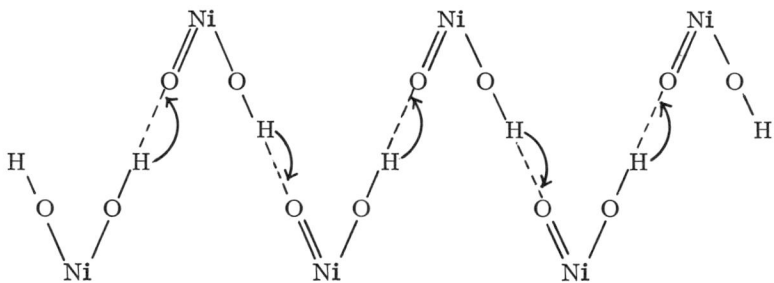

This mechanism of proton conduction probably accounts for the high rates of charge and discharge possible in Ni - Cd and Ni - Fe batteries. A similar mechanism may be suggested (but has not been proved) for the MnO_2/MnOOH reaction in the Leclanché and alkaline Mn - Zn batteries.

Even with these relatively fast pathways for charge transfer in the solid state, a battery cannot be discharged too fast without substantial loss in capacity, i.e. in the total charge available. In Fig II 27, the dependence of potential on charge during constant-current discharge of a NiOOH electrode at different rates is shown schematically.* If a rapidly discharged battery is allowed to stand on open circuit for some time, an additional amount of charge becomes available, as shown by the dotted lines in Fig. II 27. Behavior such as shown in this figure will be qualitatively observed for any battery. The decrease in available charge with increasing discharge rate depends primarily on the structure of the electrode (grain size), on the rate of the solid-state reaction and on previous history (numerous charge/discharge cycles can change the grain structure of the electrode).

An important feature of a battery electrode is the rate of change of potential in the plateau region. A very flat plateau with little dependence on remaining charge is advantageous in many applications where a constant voltage is needed. Such batteries, however, tend to run out of charge suddenly and there is often no way of telling how much charge is left in the battery at any time.

Many other properties determine the usefulness of primary and secondary batteries and often special systems have been designed to meet specific requirements. For example, when high efficiency with prolonged shelf life is desired in marine weather operations, a sea-water activated cell is used. This is stored in the dry state and activated just before use by immersion in sea water. Metallic lithium in a nonaqueous solvent is used when high energy density per unit weight is a crucial feature. The

---

* The discharge rate of a battery is often defined in terms of the number of hours that would be required to discharge it completely, assuming that all the charge is available. Thus, a battery may be discharged in 99.5 hr. at the 100 hr. rate, in 9 hr. at the 10 hr. rate and in 35 min. at the 1 hr. rate.

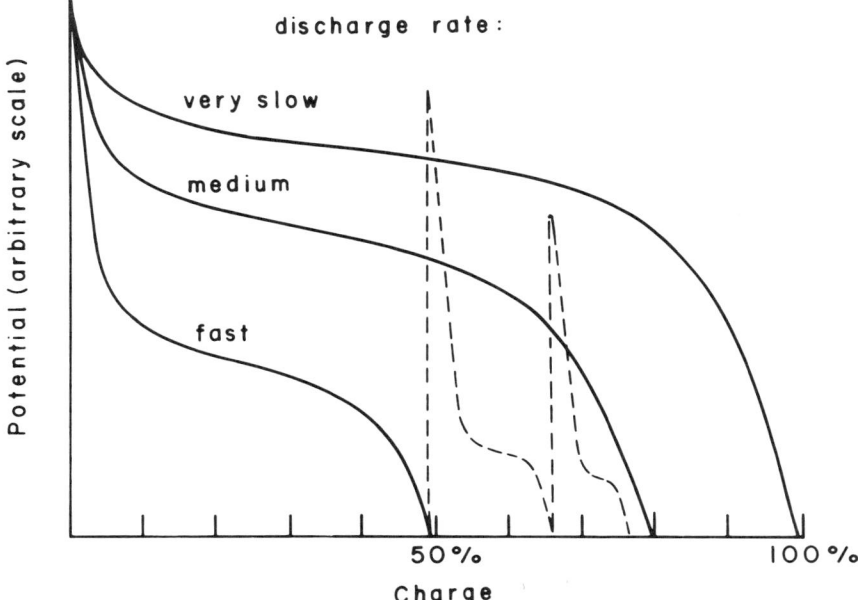

FIG. II 27   DISCHARGE CURVE FOR BATTERY

Zn/KOH/HgO battery (known as the mercury cell) is used when low volume, constant potential and long shelf life are important. The type of separators used in secondary batteries affects the performance quite markedly and the useful lifetime of the battery (number of charging cycles) is often limited by the deterioration of the separator. These and other aspects of battery design and construction will not be discussed here.

12.3   Fuel Cells

Although the possibility of obtaining electrical energy by reducing oxygen and oxidizing hydrogen at two electrodes in a cell was mentioned as early as the middle of the last century, the development of fuel cells into practical devices dates back only to the early 1950s. The first system developed and operated successfully (in space missions) was the hydrogen/oxygen fuel cell in an alkaline electrolyte at medium (150-200°C) temperatures. The main advantage of this system is the energy available per unit weight. For prolonged operation (e.g. a few weeks) the weight of the cell becomes negligible compared with that of the fuel and its container. In comparison, the weight of fuel (active material) is always a small fraction of the total weight of a battery. Fig. II 28 shows the number of watt-hours per pound as a function of operating time for one-kilowatt power units. For short operation, batteries are preferable, but as the time (i.e. the overall energy) required is increased the figure for batteries is only slightly improved while the energy density of the fuel cell increases rapidly and approaches asymptotically the energy density of the fuel itself.

Fuel cells may be classified according to the fuel used, (hydrogen, hydrazine, hydrocarbons, etc.), the temperature (low, medium, high) or the type of electrolyte (aqueous, molten-salt, solid-electrolyte). Detailed discussions of the different types of fuel cells are to be found in specific monographs and will not be discussed here. The outstanding problem in fuel-cell technology is the preparation of efficient and inexpensive electrocatalysts for the common, less active fuels (mainly hydrocarbons and their mixtures). The most successful method of overcoming intrinsically low rates of reactions taking place at the electrode surface is that of increasing its surface area without substantially increasing its size. This is done by the use of porous electrodes, where the liquid partially penetrates the intricate pore structure

12.3 FUEL CELLS

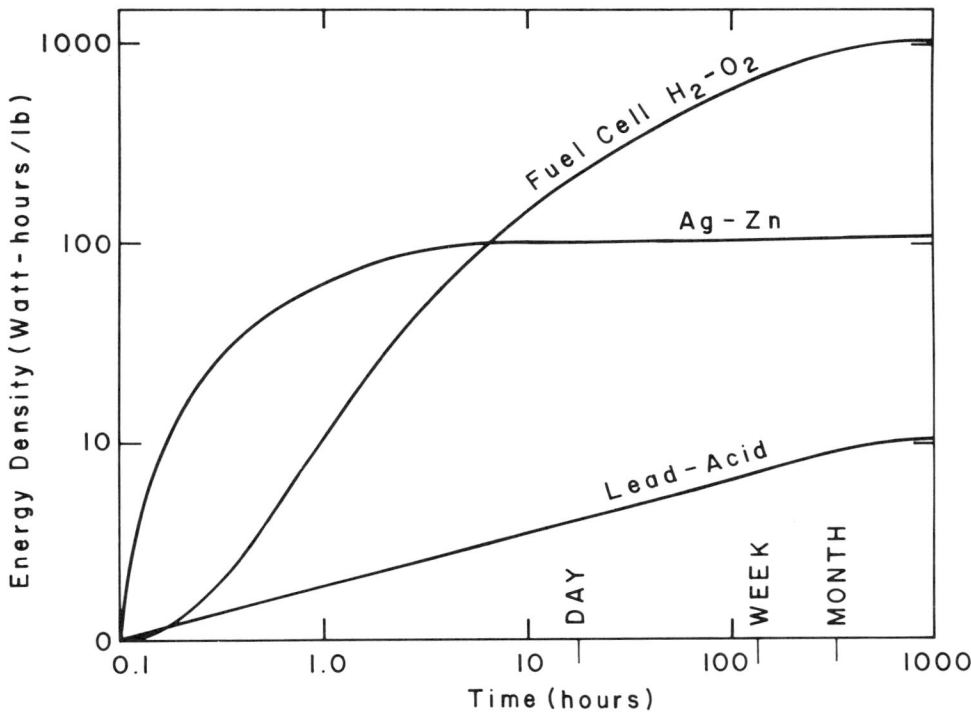

FIG. II 28   DEPENDENCE OF ENERGY DENSITY ON OPERATION TIME

and makes contact with the solid over large areas. The ratio between the real area of contact between the solid and the liquid (i.e. the real area of the interphase) and the apparent or cross-sectional area is termed the roughness factor. Its magnitude for practical porous electrodes is in the range of 10^2 to 10^4.

Three factors can control and limit the rate of reaction at a fuel-cell electrode: the energy of activation for the overall reaction, the rate of transport of fuel or oxidant to the electrode and the ohmic drop in solution. When the reactant is a gas, the reaction mainly takes place near the three-phase boundary between solid, liquid and gas. Further down in the bulk, the rate of diffusion of gaseous reactant to the surface is too slow to maintain a sufficiently high current density. Further up (near the edge of the liquid layer) the ohmic resistance would tend to limit the maximum current attainable. The situation is shown schematically in Fig. II 29.

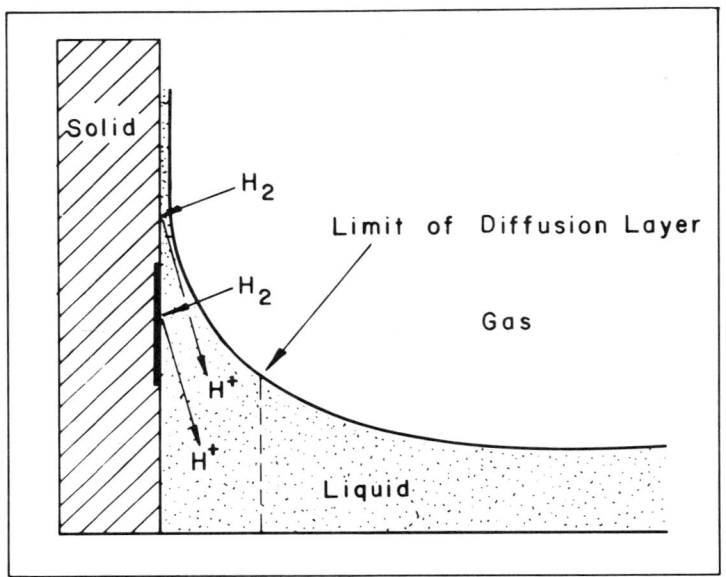

FIG. II 29   ELECTROCHEMICAL REACTION TAKING PLACE AT THE THREE-PHASE BOUNDARY (MOST ACTIVE AREA OF THE ELECTRODE IS MARKED BY HEAVY LINE)

The detailed structure of a given porous electrode may be very complex and impossible to treat rigorously. Two simplified models have been assumed and treated mathematically. In the simple pore model, Fig. II 30, a liquid meniscus is formed at the gas/liquid interphase and the reaction takes place at the solid/liquid inter-phase just behind the meniscus as shown by the arrows in Fig. II 29. The position of the meniscus is stabilized by the conical shape of the pore or by the use of double-pore electrodes.

12.3 FUEL CELLS

The thin-film model is based on the assumption that most of the inner part of the small pores is covered with a thin layer of liquid and reaction takes place at the surface of the electrode below this thin layer, as shown in Fig. II 31.

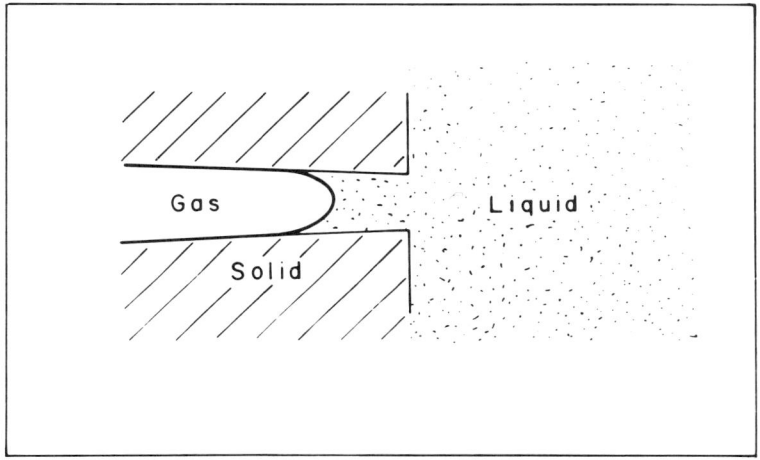

FIG. II 30   SIMPLE PORE MODEL OF POROUS ELECTRODE

The mathematical treatment of both models is rather complex and will not be given here. For a given set of kinetic and physical parameters, the plots of overpotential η against apparent current density for a planar electrode and for the two model porous electrodes are given in Fig. II 32. The intrinsic exchange-current density of $i_o = 10^{-6}$ Amp/cm^2 can be increased to an effective value of over 10^{-3} Amp/cm^2 for the case of the

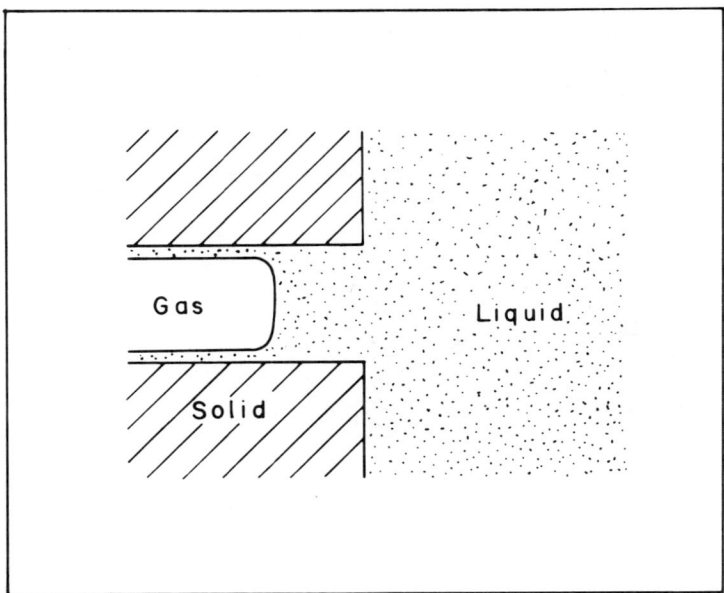

FIG. II 31 THIN LAYER MODEL OF POROUS ELECTRODE

thin-film model and the maximum diffusion-limited current can be increased by an even greater factor.

The theory of porous electrodes also yields the current-distribution profile on the electrode surface. This is of great practical value when an inert, inexpensive matrix (e.g. graphite) is used as the electrode and the expensive catalyst (e.g. a noble metal) need be applied only to the areas where reaction takes place.

12.3 FUEL CELLS

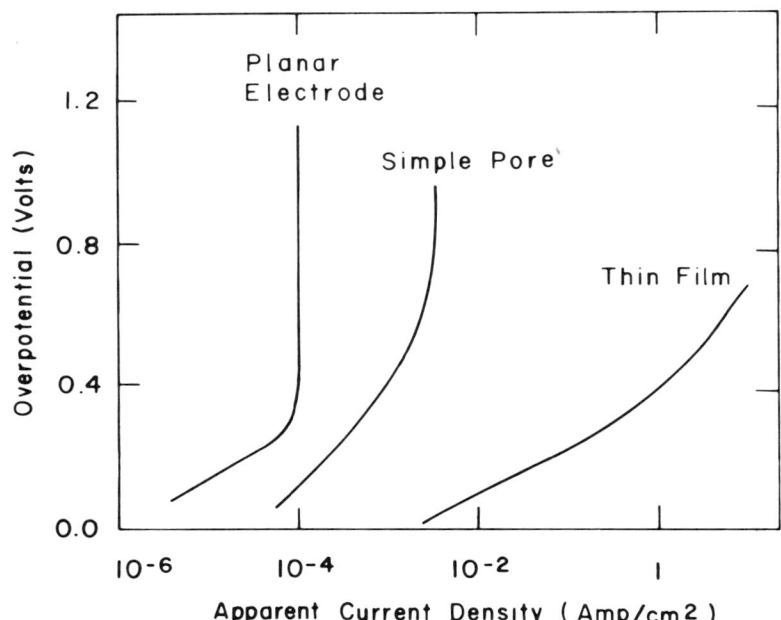

FIG. II 32 CALCULATED η/i RELATIONSHIP FOR TWO MODELS OF POROUS ELECTRODE COMPARED TO PLANAR ELECTRODE HAVING THE SAME KINETIC PARAMETERS. (S. Srinivasan and E. Gileadi, Handbook of Fuel Cell Technology, C. Berger, Editor, Prentice Hall 1968)

The commercial success of fuel cells (for example, as the source of energy in electric cars) hinges on the development of highly effective, stable and inexpensive electrocatalysts. At present, fuel cells have been used only for very specialized purposes (e.g. space flight) in which minimum weight has been of utmost importance.

13. Fundamentals of Electroplating

13.1 The Cathodic Discharge of Metal Ions

The discharge of a cesium ion on a mercury cathode to form an amalgam is perhaps the simplest metal-deposition process which can be envisaged.

$$Cs^+ + e_{Hg} \longrightarrow Cs(Hg) \qquad [II.196]$$

It differs in two important aspects from a typical electroplating process, for example, the discharge of cupric ions on an iron cathode.

$$Cu^{++} + 2e_{Fe} \longrightarrow Cu \qquad [II.197]$$

First, the energy of hydration of the relatively large Cs^+ ion is low and the ion may be regarded as being almost without a primary hydration shell in aqueous solution. In contrast, the cupric ion is heavily hydrated and should be written in the form $[Cu(H_2O)_n]^{++}$. From its initial state, the hydrated ion goes through several states of partial hydration until it loses all attached water molecules and becomes incorporated in the crystal lattice. Secondly, deposition on the liquid metal surface is a relatively simple process. The surface is uniform and structureless and it changes very slowly as metal is deposited, as a result of formation of the amalgam and increase in the concentration of the deposited metal.

When the substrate upon which deposition occurs is a

13.1 CATHODIC DISCHARGE OF METAL IONS

solid, several complications arise. First, the surface is not uniform and in all cases of practical interest it is polycrystalline. The rate of deposition may differ by orders of magnitude on the different crystal faces, thus causing preferential, nonuniform deposition in certain areas and directions.

If the crystal structure and dimensions of the deposited metal are different from those of the substrate, a region of high free energy, (i.e. low thermodynamic stability) may arise at the interface between the two metals. Epitaxial growth occurs in the first few layers of deposit while further away, the usual (most stable) crystal structure of the deposited metal is observed. The strain between the two crystal structures is relieved by lattice imperfections, impurity ions or by dislocations.

The structure of the surface has been made the subject of a considerable amount of study. It can be shown on purely thermodynamic grounds, that deposition of an ion on an ideally flat and perfect crystal surface would require a rather high negative potential, while in practice, electrodeposition can be effected at potentials very close to the reversible potential. The situation is similar to that encountered in the case of formation of very small droplets or crystallites from supersaturated vapors or solutions, respectively. It is believed that crystals (formed by chemical precipitation, cooling or by electrodeposition) grow primarily by incorporating new ions or atoms into kinks and steps on the surface and not by building new layers on top of flat surfaces. This posed an interesting problem, since, if growth occurred preferentially in steps and kinks, these should be rapidly used up, leaving an ideally flat surface which would require a much higher overpotential (or supersaturation) for crystal growth to continue. The answer to this problem came with the realization of the existence of screw dislocations which have the property of self-propagation. Thus, as more and more ions are incorporated at the step formed by this type of dislocation, the line of the step rotates and the surface can advance into the solution in a screw-like fashion.

Finally, electrodeposition on a foreign substrate is maintained only for a very short period. (At a typical plating current density of $10mA/cm^2$, a copper deposit ten atomic layers in thickness will be formed in about 0.5 sec.). Afterwards, the substrate is coated with the metal being deposited and further deposition is conducted on it.

13.2   Electrocrystallization and Electroplating

The process of electroplating is intimately linked with electrocrystallization since the various characteristics of the plated surface depend on the crystalline structure of the deposited metal, and particularly on the size and shape of the small crystals formed during electrodeposition.

When electrodeposition is in progress, two basic phenomena compete. Existing crystals tend to grow further (either outward into the solution or sideways on the surface) and new crystals are spontaneously formed. The latter process is called nucleation.

Two energetically opposing factors determine the relative rates of crystal growth and nucleation. On the one hand, the free energy of small nuclei on the surface is greater than the free

13.2 ELECTROCRYSTALLIZATION

energy of the larger existing crystallites on the surface.* On the other hand, an ion may have to lose only one water molecule from its primary hydration shell in order to be adsorbed on a flat part of the surface (and thus form the center for a new nucleus) while it may require the loss of two or more water molecules in order to be incorporated in a step or kink and thus increase the size of an existing crystal. It has been shown, both on theoretical grounds and experimentally, that the nucleation rate increases with increasing overpotential. Thus, the grain size of electrodeposits is inversely proportional to the overpotential. For baths operated at similar current densities, those having a low exchange-current density will produce deposits of smaller grain size. A similar effect can be achieved by using certain additives which adsorb on the surface and inhibit crystal growth without significantly affecting the nucleation rate. The conditions which favor high nucleation rate also favor outward rather than lateral growth of existing crystallites, since the former process is energetically equivalent to the formation of new nuclei. As the crystals grow outwards they sharpen and the electric field around them increases. As a result, electrodeposition on sharp protruding areas of the surface occurs at higher overpotentials and these areas tend to grow faster, eventually forming dendritic-type deposits. This effect can be largely eliminated if deposition is conducted from a bath containing the relevant ion in the form of a negatively charged

---

* Just as the free energy of a minute droplet of a liquid is greater than the free energy of the bulk liquid (as a result of the excess surface free energy).

complex (e.g. $[Cu(CN)_4]^{2-}$ or $[Ag(CN)_2]^{-}$), as will be discussed below.

13.3 The Properties of a Plating Bath
(i) Faradaic Efficiency

The faradaic or current efficiency during plating is defined as the fraction of the total charge passed which is used to form the deposit. In aqueous plating baths, the competing reaction is the hydrogen-evolution reaction but in general, any other parallel cathodic reaction can be the cause of a decrease in the faradaic efficiency.

The faradaic efficiency affects the plating process in several ways. The most obvious of these is the plating time, which increases with decreasing efficiency. Next, the way in which the faradaic efficiency depends on current density influences the uniformity of plating, as will be discussed in the next section. If the hydrogen-evolution reaction takes place simultaneously with metal deposition, there results hydrogen embrittlement by penetration of atomic hydrogen into the specimen being plated. This problem arises with ferrous alloys, and it becomes serious in the case of high-strength steels, which are particularly susceptible to hydrogen embrittlement. In nonaqueous baths, side reactions which lower the faradaic efficiency may also adversely affect the lifetime of the bath.

An examination of the potential/pH diagrams (cf. Fig. II 25) makes it clear that while the reversible potential of metal

13.3 PROPERTIES OF PLATING BATH

deposition is independent of pH^*, the value of E_r for hydrogen evolution becomes more cathodic as the pH is increased. Thus, the hydrogen-evolution reaction can be depressed (and faradaic efficiency increased) by making the bath more alkaline. The pH of the bath must be carefully controlled, however, since above a certain point hydroxides of the metal may form and will tend to be incorporated in the metal deposited. Furthermore, hydrogen evolution causes an increase in the local pH at the cathode over its average value in the bulk. This effect is pronounced in poorly buffered solutions and at low faradaic efficiencies and may cause a deterioration in the quality of plating obtained.

(ii) Throwing Power

If a well polished planar surface is to be electroplated, the counter electrode (viz. the anode) can easily be positioned in such a way as to cause an even current-density distribution on the surface of the cathode, and a correspondingly even electrodeposited layer. This is hardly ever the case in practice. Surfaces to be electroplated have recessed areas and protrusions and it is usually impractical to fabricate an anode of corresponding shape in order to obtain a uniform current distribution. The throwing power of a bath is a measure of its ability to form a deposit of uniform thickness irrespective of the shape of the surface being plated. The throwing power of a given bath can be measured quantitatively in a "Haring and Blum" cell shown schematically in Fig. II 33, in which the distances of the two cathodes from the

---

* This is not strictly true for amphoteric metals which can form hydroxy complexes even at relatively low pH values.

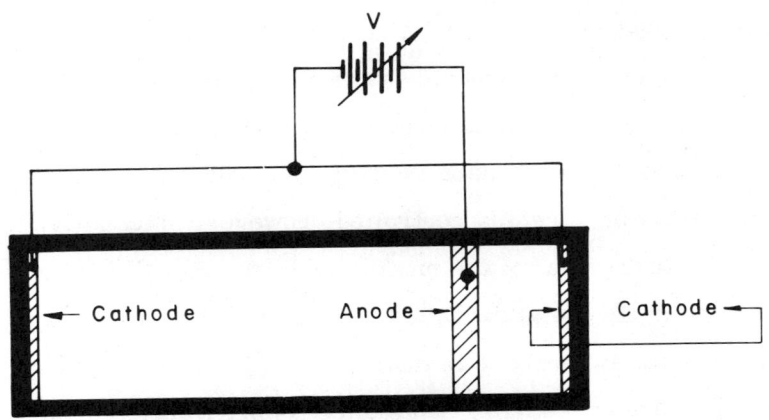

FIG. II 33   HARING AND BLUM CELL

common anode are in the ratio of 5:1. Since the same potential is applied to both half cells, the current density will be inversely proportional to the total resistance in each case. This resistance is the sum of the solution resistance R_{sol} and the faradaic resistance R_F in each half cell. Two extreme cases may now be considered. If the exchange-current density for the cathodic reaction is very high, one has $R_{sol} \gg R_F$ and the current will be controlled mainly by the solution resistance. The deposit on the nearer cathode will be five times as thick as that on the second cathode and in such a case the throwing power is very poor. If the exchange-current density is very low, one has a situation in which $R_{sol} \ll R_F$ and deposits of equal thickness will be formed on both cathodes, corresponding to ideal throwing power of the bath. When the contributions of both resistors to the total resistance are of the same order (i.e. $R_F \cong R_{sol}$), the faradaic

13.3 PROPERTIES OF PLATING BATH

resistance R_F causes poor throwing power since this resistance always decreases with increasing current density. Thus, in a recessed area of the surface, where the current density will be lower as a result of high solution resistance, it is further lowered by a higher faradaic resistance. The throwing power can be increased by operating the bath at current densities close to the diffusion-limited current density. In this region, a self-regulating mechanism is operative. If the current density is higher, say, at a protrusion on the surface, the concentration of reactant at the same part of the interphase is diminished and this limits further increase of the current density.

Consider now the effect of faradaic efficiency on the uniformity of the current distribution. Better throwing power will be observed if the faradaic efficiency decreases with increasing current density; if the opposite is the case, the result will be an enhancement of the irregularities already existing on the surface.

The throwing power of a bath can be improved by the use of suitable additives which have been developed by trial and error. The theory of operation of these materials is only partly understood and will not be discussed here.

(iii) <u>The Nature of the Species in Solution</u>

It was previously pointed out that metal ions are usually hydrated and should be written as, e.g., $[Cu(H_2O)_n]^{2+}$. In many cases of practical importance, the situation is even more complicated by the use of complexing agents which form stable complexes with the metal ions, for example,

$$[Cu(H_2O)_n]^{2+} + 4(CN)^- \rightleftharpoons [Cu(CN)_4]^{2-} + nH_2O \qquad [II.198]$$

Formation of the negative complex ions serves several purposes. Both the exchange-current density and the limiting-current density are diminished if the complex is sufficiently stable. This increases the rate of nucleation and leads to the formation of denser and more uniform electrodeposits. Copper cyanide baths, for example, are often operated at a current density close to the limiting-current density and in this way good throwing power is obtained. Movement of the ions in the electric field further improves the smoothness of the deposit. Thus, at protrusions, where the electric field is highest, the negative ions are driven away and the deposition rate is decreased. In recessed areas the concentration of negative ions is higher and the deposition rate increases. Thus, dendrite formation is inhibited and the surface tends to become uniform during electrodeposition.

There are two disadvantages in plating baths containing the metal ions as negative complexes. One is that the maximum rate of deposition is limited because of the slow supply of free metal ions to be discharged on the surface and the second is the fact that the potential of deposition is made more negative thus lowering the faradaic efficiency and increasing the danger of hydrogen embrittlement.

13.4 Plating of Active Metals

Active metals such as aluminum or magnesium cannot be plated from aqueous solutions because the solvent is decomposed before the metal can be deposited. The obvious solution to this

13.4 PLATING OF ACTIVE METALS

problem would seem to be the use of molten-salt baths. This is indeed the practice in electrowinning technology (aluminum is produced by electrolysis of molten cryolites, magnesium by electrolysis of molten $MgCl_2$) but the quality of electroplating obtained from molten-salt baths is invariably very poor. The next reasonable step was to look for a suitable aprotic solvent such as acetonitrile, liquid SO_2 or SO_3 etc. For high solubility of an inorganic salt one seeks a polar solvent with a relatively high dielectric constant. On the other hand, ions like Al^{3+} or Mg^{2+} will have a high solvation energy in polar solvents and will not be deposited easily. A bath consisting of $AlCl_3$ and $LiAlH_4$ in ether was developed in 1952. Good deposits of aluminum could be obtained from this bath but it has not become commercially useful because of its high sensitivity to moisture and air, and its dangerous nature. Recently, it has been realized that aluminum and some other active metals can be deposited from a bath containing $AlBr_3$ in toluene or some other aromatic hydrocarbons, with small amounts of bromide added.

The development of new plating baths for active metals such as aluminum, titanium and tantalum is of great technological importance. Such baths could also be used to deposit less active metals, like cadmium and chromium in better form, by avoiding the problems of hydrogen embrittlement inherent in the deposition of these metals from aqueous baths.

BIBLIOGRAPHY

1. Electrochemical Kinetics,

    K. J. Vetter, Academic Press 1967.

2. Modern Electrochemistry, Vol. II,

    J. O'M. Bockris and A. K. N. Reddy,
    Plenum Press 1970.

3. Double Layer and Electrode Kinetics,

    P. Delahay, Interscience 1966.

4. Theory and Principles of Electrode Processes,

    B. E. Conway, The Ronald Press 1965.

5. New Instrumental Methods in Electrochemistry,

    P. Delahay, Interscience 1954.

6. Electrochemistry at Solid Electrodes,

    R. N. Adams, Marcel Dekker 1969.

7. Principles of Current Methods for Study of Electrochemical Reactions,

    B. B. Damaskin, McGraw Hill 1967.

8. Adsorption of Organic Compounds on Electrodes,

    B. B. Damaskin, O. A. Petrii and V. Bartakov,
    Plenum Press 1971.

9. Physicochemical Hydrodynamics,

    V. G. Levich, Prentice-Hall 1962.

10. Principles of Polarography,

    J. Heyrovsky and J. Kuta, Academic Press 1966.

BIBLIOGRAPHY (cont'd)

11. The Elucidation of Organic Electrode Processes,
    P. Zuman, Academic Press 1969.

12. Electrosorption,
    E. Gileadi, Editor, Plenum Press 1967.

13. Fuel Cells: Their Electrochemistry,
    J. O'M. Bockris and S. Srinivasan, McGraw-Hill 1969.

14. Electrochemical Science,
    J. O'M. Bockris and D. M. Drazic,
    Taylor & Francis Ltd. London 1972.

15. A Guide to the Study of Electrode Kinetics,
    H. R. Thirsk and J. A. Harrison,
    Academic Press 1972.

16. Reference Electrodes,
    D. J. G. Ives and G. J. Janz, Editors,
    Academic Press 1961.

17. Atlas of Electrochemical Equilibria,
    M. Pourbaix, Pergamon Press 1966.

18. The Encyclopedia of Electrochemistry,
    C. A. Hampel, Editor, Reinhold 1964.

19. Handbook of Electrochemical Constants,
    R. Parsons, Butterworths 1959.

20. Modern Aspects of Electrochemistry,
    J. O'M. Bockris and B. E. Conway, Editors,
    Butterworths (Vol. 1-3) and
    Plenum Press (Vol. 4 and on).

BIBLIOGRAPHY (cont'd)

21. Advances in Electrochemistry and Electrochemical Engineering,

    P. Delahay and C. Tobias, Editors, Interscience.

22. Electroanalytical Chemistry - A Series of Advances,

    A. J. Bard, Editor, Marcel Dekker.

23. Techniques of Electrochemistry,

    E. Yeager and A. J. Salkind, Editors
    Wiley-Interscience.

24. Electrochemistry,

    J. G. Hills, Editor, The Chemical Society.

25. Electrochemistry,

    J. S. Mattson, H. B. Mark, Jr. and H. C. MacDonald, Jr., Editors, Marcel Dekker, Inc., 1972.

26. Encyclopedia of Electrochemistry of the Elements,

    A.J. Bard, Editor, Marcel Dekker, Inc.

1. EXPERIMENTS IN ELECTRONICS RELATED TO ELECTROCHEMISTRY

1.1 THE OPERATIONAL AMPLIFIER

1.1.1 General Properties of an Ideal Operational Amplifier

1.1.2 Basic Circuits Employing the Operational Amplifier

1.1.3 The Real Operational Amplifier

1.2 EXPERIMENTAL

1.2.1 Determination of the Characteristics of an Operational Amplifier

1.3 BASIC CIRCUITS IN ELECTROCHEMICAL MEASUREMENTS

1.3.1 The Galvanostat

1.3.2 The Potentiostat

1.3.3 Experiments with a Potentiostat with Positive Feedback Employing a Dummy Cell

1.3.4 Experiments with a Potentiostat with Positive Feedback Employing an Electrochemical System

1. EXPERIMENTS IN ELECTRONICS RELATED TO ELECTROCHEMISTRY*

1.1 THE OPERATIONAL AMPLIFIER

1.1.1 General Properties of an Ideal Operational Amplifier

The operational amplifier is a high-gain electronic amplifier having two inputs (+IN and -IN) and one output terminal, as shown schematically in Fig. 1.1. It is convenient, for the purpose of illustration, to describe some basic circuits employing "ideal" operational amplifiers. (The departure of real devices from ideal behavior will be discussed in section 1.1.3.) The properties of an ideal operational amplifier may be specified as follows:

(i) When the voltage difference ΔV between the two inputs is zero, the output voltage is also zero, independent of temperature and of the voltage of the two inputs with respect to ground (i.e. the operational amplifier acts strictly as a differential amplifier).

(ii) When a small voltage difference ΔV is applied between the two inputs, the output voltage tends to plus or minus infinity, depending on the polarity of ΔV. This is equivalent to the statement that the gain of an ideal operational amplifier tends to infinity.

(iii) The input impedance at both inputs is infinitely large and hence the input current is zero, irrespective of the input voltage and of temperature.

---

* This chapter was prepared by Dr. Ch. Yarnitzky of the TECHNION, Israel Institute of Technology.

1.1 OPERATIONAL AMPLIFIER

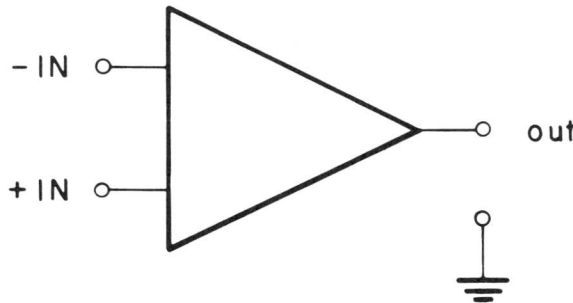

FIG. 1.1   THE OPERATIONAL AMPLIFIER

(iv) The output current depends only on the output voltage and the impedance of the load; in other words, the output impedance is zero.

(v) When the input voltage difference ΔV undergoes rapid changes, the output voltage follows these changes exactly according to the properties of the operational amplifier as specified above. Thus the frequency response or the band width of the ideal operational amplifier is unlimited.

1.1.2   Basic Circuits Employing the Operational Amplifier

(i) Feedback Networks

A common feature of simple circuits employing operational amplifiers is the existence of a feedback loop or network through which all or part of the signal appearing at the output of the amplifier is fed back to one of the inputs (usually the negative input). The output voltage is calculated in a very simple manner from the most fundamental laws of electricity (e.g., Ohm's law, Kirchhoff's law, etc.)

The simple network shown in Fig. 1.2 may serve to show how the output voltage V_{out} is calculated. Let the voltage with respect to ground at the positive input be V_{in}. It follows from the properties of the ideal operational amplifier that the voltage at the negative input is the same as V_{in}, since a finite voltage difference between the inputs would generate an infinitely large output voltage. The voltage V_{in} appears across the resistance R_{in} connected to ground, hence a current $i = V_{in}/R_{in}$ flows. This current originates from the output terminal of the operational amplifier and passes through a total resistance of $(R_{in} + R_f)$. Thus the output voltage is given by

$$V_{out} = V_{in}\left(1 + \frac{R_f}{R_{in}}\right) \qquad [1.1]$$

Assume now that for some reason the output voltage increases beyond the value calculated above. The current i will increase and this will increase the voltage at the negative input of the amplifier. Thus part of the positive change in V_{out} will be fed back and will cause a negative change in V_{out} and vice versa. In this way the value of V_{out} is stabilized.

(ii) The Voltage Follower

The voltage follower shown in Fig. 1.3 is an impedance-matching unit which has wide applications in electrochemistry and in other fields. Since the potential difference between the inputs is virtually zero, it is obvious that $V_{in} = V_{out}$; that is, the voltage follower is an amplifier with unit gain. On the other hand its input impedance is infinite and its output impedance is zero. Thus it presents no load to the system being measured

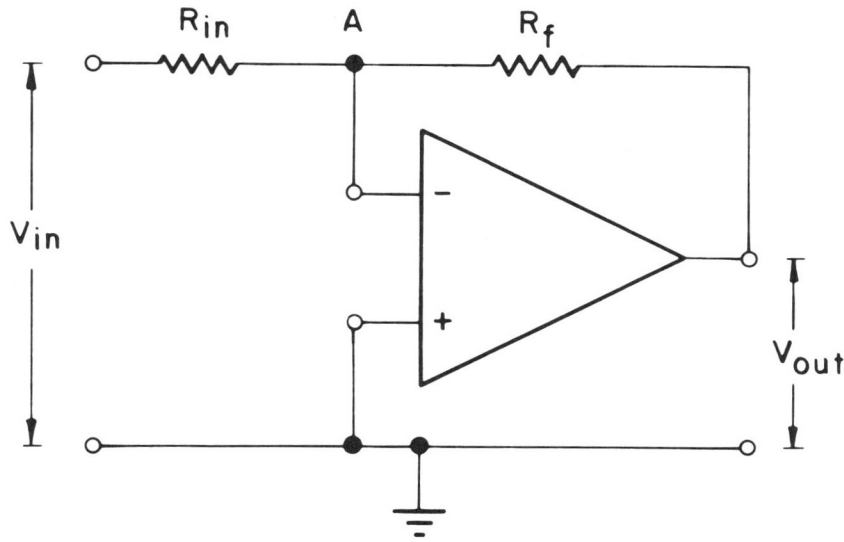

FIG. 1.5   THE INVERTING VOLTAGE AMPLIFIER

$$i_{in} = \frac{V_{in}}{R_{in}} \qquad [1.2]$$

This is also equal to the current through R_f, since the input current of the ideal operational amplifier is zero. Thus

$$i_{in} = \frac{V_A - V_{out}}{R_f} \cong -\frac{V_{out}}{R_f} = \frac{V_{in}}{R_{in}} \qquad [1.3]$$

where V_A, the voltage at the point A, is virtually equal to zero. Hence

$$V_{out} = -V_{in}\left(\frac{R_f}{R_{in}}\right) \qquad [1.4]$$

1.1 OPERATIONAL AMPLIFIER

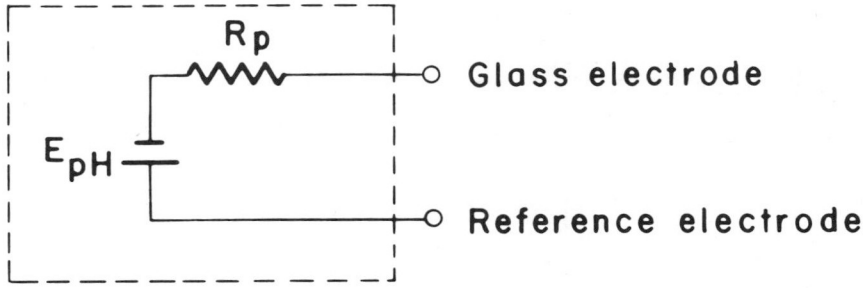

FIG. 1.4   EQUIVALENT CIRCUIT FOR pH MEASUREMENT

When the feedback network is not purely resistive (i.e., it includes capacitors and/or inductors) the complete mathematical treatment must include the reactive components of the impedances. This generally causes a delay or phase shift between the input and the output signals. When the phase shift is large enough the circuit may become unstable and start to oscillate. In certain cases this may be an undesirable phenomenon which must be eliminated (for example, when a potentiostat with resistance correction is used). In other cases the same phenomenon is used to advantage to construct generators for alternating current or voltage.

(iii) The Inverting Voltage Amplifier

The circuit shown in Fig. 1.5 serves to invert the sign of a signal and amplify it as required. The input signal is imposed between ground and a resistor R_{in} which is connected to the negative input of the operational amplifier. The positive input is grounded. Since ΔV between the two inputs is virtually zero, the current through R_{in} is given by yet is able to drive any regular measuring device, like a voltmeter or recorder.

One application of the voltage follower is found in the measurement of pH with a glass electrode. The electrochemical cell involved in pH measurement consists of a glass electrode and a suitable reference electrode. The equivalent circuit is shown in Fig. 1.4. The electromotive force of the cell depends on the pH of the solution and usually does not exceed 0.5 volt. The internal resistance R_p is of the order of 10^9 ohms. Thus, if the error due to the potential drop across the resistance R_p is to be less than 1 mV, a galvanometer connected directly to the cell should have a sensitivity of 10^{-12} Amp. for full-scale deflection. This is not practical, and a voltage follower between the cell and the measuring device must be used instead. (A remark may be in place here about the properties of a real operational amplifier. The requirement that the input current should be less than 10^{-12} Amp. is met only in special devices called "electrometer amplifiers". The input current in other operational amplifiers may be in the range of 10^{-7}-10^{-10} Amp., the lower value being achieved by having a FET (field effect transistor) input. The latter devices are satisfactory as voltage followers for most applications in electrochemistry where a glass electrode is not used.)

Looking back at Fig. 1.2 we note that it can serve in addition as a voltage follower in which the voltage is amplified to an extent depending on the relative value of the two resistors. Such a device is called an "amplifier follower".

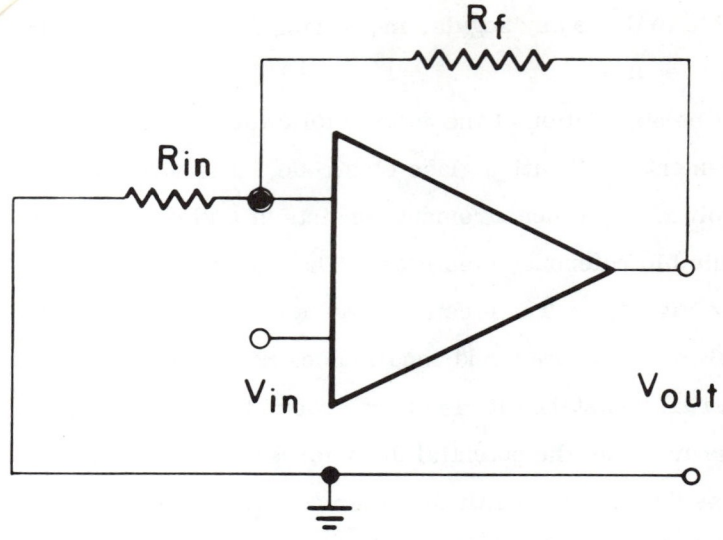

FIG. 1.2   FEEDBACK NETWORK WITH RESISTORS

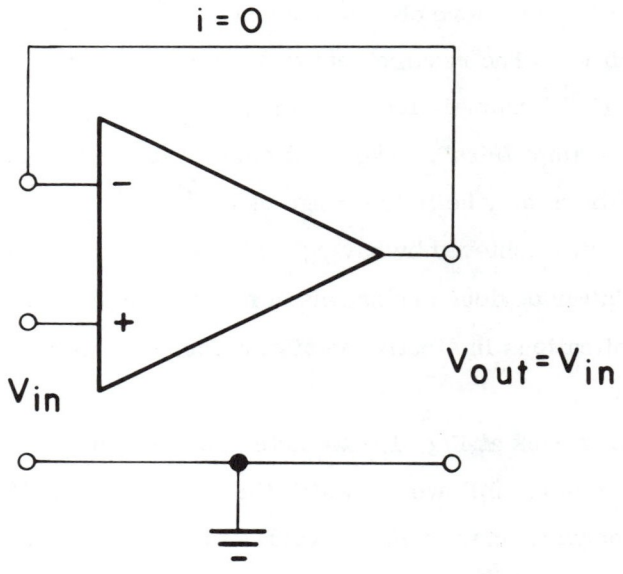

FIG. 1.3   THE VOLTAGE FOLLOWER

1.1 OPERATIONAL AMPLIFIER

This device is often used as a unit-gain inverting amplifier ($R_{in} = R_f$), to achieve a 180° phase shift of the signal, without changing its magnitude.

A similar circuit, shown in Fig. 1.6, is referred to as the "summing inverting amplifier". Following the same argument as above, we find for this circuit

$$V_{out} = -R_f \left(\frac{V_1}{R_1} + \frac{V_2}{R_2} + \frac{V_3}{R_3} \right) \qquad [1.5]$$

In the simple case when

$$R_1 = R_2 = R_3 = \frac{R_f}{n} \qquad [1.6]$$

the output voltage is simply the sum of the input voltages multiplied by (-n). By using different values of the input resistors, a sum is obtained to which the different inputs contribute to different degrees, as desired.

(iv) <u>The Voltage Subtractor</u>

The circuit shown in Fig. 1.7 can be used to measure the difference between two voltages. The potential at both inputs of the operational amplifier is given by

$$V_+ = V_- = \frac{V_1}{2} \qquad [1.7]$$

Thus the current in the feedback loop is

$$i = \frac{V_2 - V_1/2}{R} = \frac{V_1/2 - V_{out}}{R} \qquad [1.8]$$

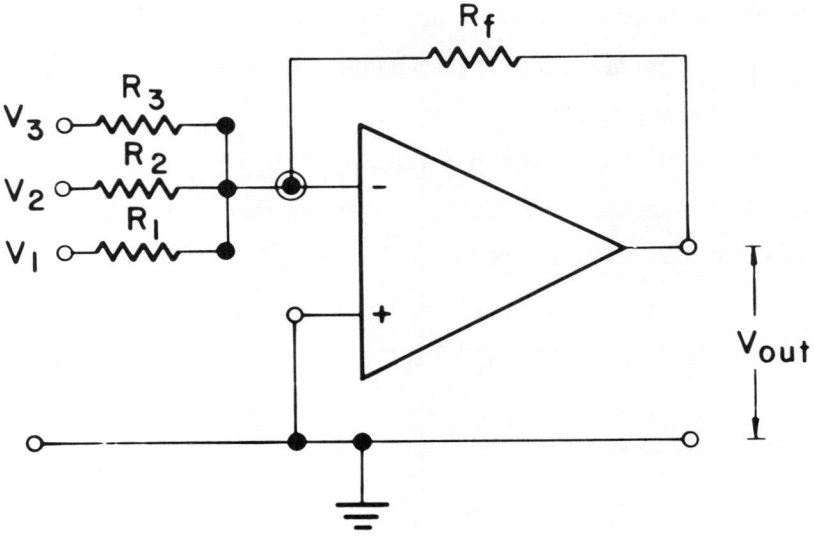

FIG. 1.6   THE SUMMING INVERTING AMPLIFIER

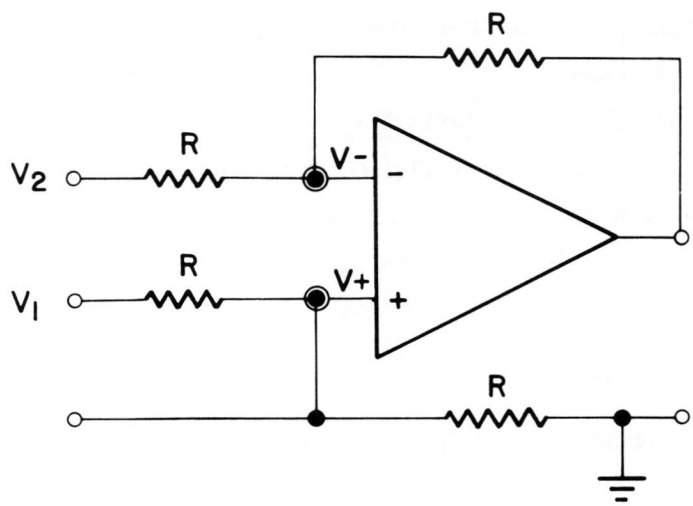

FIG. 1.7   THE VOLTAGE SUBTRACTOR

1.1 OPERATIONAL AMPLIFIER

and

$$V_{out} = V_1 - V_2 \qquad [1.9]$$

(v) <u>The Integrator</u>

A resistor in the input and a capacitor in the feedback loop (Fig. 1.8) give rise to electronic integration. The current is determined by the input voltage and resistance ($i = V_{in}/R_{in}$) and the change of voltage across the terminals of the capacitor is related to this current.

$$-\frac{dV_{out}}{dt} = \frac{i}{C_f} = \frac{V_{in}}{R_{in}C_f} \qquad [1.10]$$

hence

$$V_{out} = -\frac{1}{R_{in}C_f} \int_{t_1}^{t_2} V_{in} dt \qquad [1.11]$$

where V_{in} may be a function of time.

(vi) <u>The Differentiator</u>

A simple differentiating circuit is obtained by reversing the positions of the resistor and capacitor in the integrating circuit. Thus, referring to Fig. 1.9, we note that the current is given by

$$i = C_{in}\frac{dV_{in}}{dt} \qquad [1.12]$$

and hence the output voltage is

162   EXPERIMENTS IN ELECTRONICS

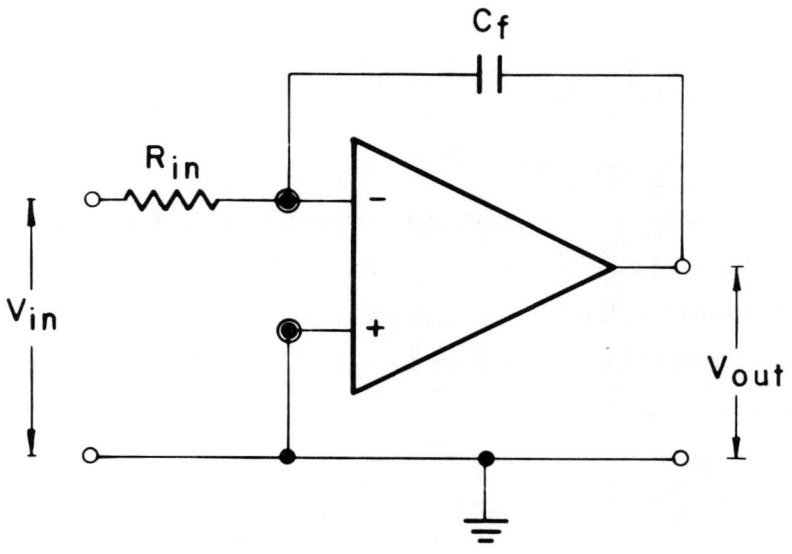

FIG. 1.8   THE INTEGRATOR

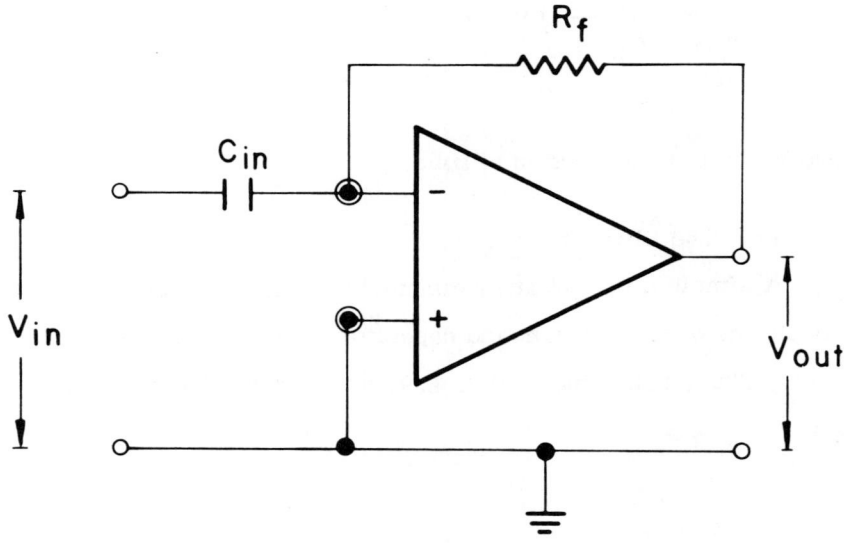

FIG. 1.9   SIMPLE DIFFERENTIATING CIRCUIT

1.1 OPERATIONAL AMPLIFIER

$$V_{out} = -R_f C_{in} \frac{dV_{in}}{dt} \qquad [1.13]$$

The differentiating circuit in its simple form is not very useful practically since it is very sensitive to noise, which is often of higher frequency than the signal itself, and therefore tends to be amplified selectively by the differentiating circuit. An improved circuit is shown in Fig. 1.10, in which $C_f \ll C_{in}$ and $R_{in} \ll R_f$. Considering the frequency response of this circuit, it is easy to see that at sufficiently low frequencies the circuit behaves like the simple differentiating circuit of Fig. 1.9, while at very high frequencies it becomes, in fact, an integrating circuit. In this manner, the high-frequency noise is filtered out while at lower frequencies the circuit operates as a proper differentiating circuit. The actual values of the capacitors and resistors in this circuit depend on the range of frequency of signals which are to be differentiated. If we choose, for simplicity, the conditions

$$R_f C_f = R_{in} C_{in} \qquad [1.14]$$

then the circuit operates as a differentiator for frequencies well below $(R_{in} C_{in})^{-1}$ and it operates as an integrator for frequencies well above this value.

(vii) <u>Current-to-Voltage Converter</u>

The measurement of the current through a cell is best performed with the current-to-voltage converter shown in Fig. 1.11. If the operational amplifier behaves ideally, this is equivalent to measurement with an ammeter with zero internal resistance. For a real operational amplifier, the effective resistance in series

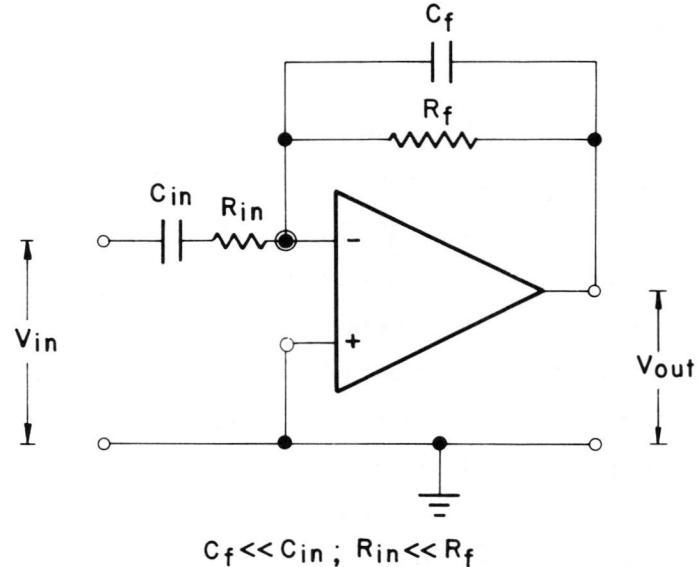

FIG. 1.10   LOW-NOISE DIFFERENTIATING CIRCUIT

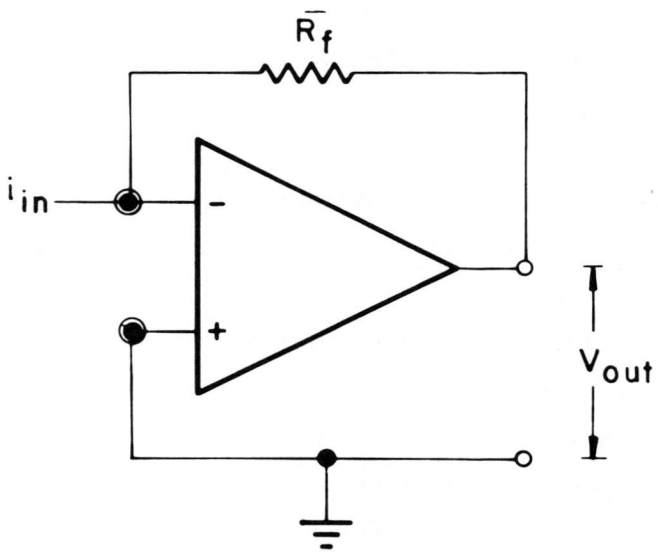

FIG. 1.11   CURRENT-TO-VOLTAGE CONVERTER

1.1 OPERATIONAL AMPLIFIER

presented by the current-to-voltage converter is given to a good approximation by Rf/A_o where A_o is the open-loop gain of the operational amplifier, and is typically in the range of 10^4-10^6.

1.1.3 The Real Operational Amplifier

The properties of the real operational amplifier deviate from ideal behavior in varying degrees. Commercial devices are priced according to performance* and it is important in any circuit to use the operational amplifier best suited to its purpose. In the following sections the important characteristics of operational amplifiers will be discussed, and methods will be outlined for testing the performance of instruments in which operational amplifiers are employed and which are used in electrochemistry.

(i) Gain

When a small voltage difference ΔV appears between the two input terminals of an operational amplifier, the potential at the output will be

$$V_{out} = A_o \Delta V \qquad [1.15]$$

---

* At present the price of an operational amplifier ranges from a few dollars to a few hundred dollars, depending on specifications. Thus, overdesign of electronic circuits (which often contain several operational amplifiers) can be a very expensive proposition.

where A_o is termed the "open-loop gain".[*] Consider now the error introduced due to this finite value of ΔV when a signal is amplified by an inverting amplifier (cf. Fig. 1.5). Assume that $R_{in} = R$ and $R_f = KR$. We can then write

$$i_{in} = \frac{V_{in} - \Delta V}{R} = \frac{\Delta V - V_{out}}{KR} \qquad [1.16]$$

Combining the last two equations, we find

$$V_{out} = -V_{in} K \left[\frac{1}{1 + (K+1)/A_o} \right] \qquad [1.17]$$

Thus the gain will be equal to K only if $A_o \gg K$. Moreover, the open-loop gain A_o is defined for d.c. operation. Its value remains constant up to a certain frequency beyond which it varies with frequency at a given rate, usually in proportion to the reciprocal frequency.[**] In this range there is also a phase shift between V_{out} and V_{in}, in addition to the $180°$ shift characteristic of the inverting amplifier and this may cause oscillations at high frequencies.

(ii) <u>Input Offset Voltage</u>

Often a finite output voltage is measured even when the

---

[*] This equation shows at once why the assumption that the two inputs are virtually at the same potential, used throughout the previous section, is justified. Thus for $V_{out} = 1.0$ volt and A_o having a typical value of 10^4 we have $\Delta V = 0.1$ mV.

[**] In the nomenclature usually used this is defined as a slope of 6 dB/octave, meaning that A_o decreases by a factor of two when the frequency is doubled. This also equals 20 dB/decade.

1.1 OPERATIONAL AMPLIFIER

inputs of the operational amplifier are shorted ($\Delta V = 0$). The voltage difference at the input needed to bring the output voltage to zero is defined as the input offset voltage. Often a small variable resistor is included in the operational amplifier, which can be adjusted so that the offset voltage is reduced to zero. The usefulness of this procedure is, however, limited by the drift of the offset voltage with time (up to 1 mV/day) and its dependence on temperature, both of which are specified by the manufacturer. A special type of operational amplifier is the chopper-stabilized amplifier in which the offset voltage is controlled and can be stabilized to $0.1 \, \mu\text{volt/day}$.

(iii) <u>Input Bias Current</u>

Of all the properties in which a real operational amplifier departs from ideal behavior, the input bias current i_b is perhaps the most troublesome. Except in very high-priced units, i_b is of the order of 10^{-7} Amp. and it substantially affects the performance of the circuit. A simple way of compensating for the finite value of i_b is shown in Fig. 1.12.

The voltage at the positive input is given by

$$\Delta V_b = -R_1 i_b \quad [1.18]$$

Since this is virtually equal to the voltage at the negative input, we can write for the total input current,

$$i_{in} = \frac{V_{in} + i_b R_1}{R} = \frac{-i_b R_1 - V_{out}}{KR} + i_b \quad [1.19]$$

hence

$$V_{out} = -V_{in} K - i_b [R_1 (1 + K) - KR] \quad [1.20]$$

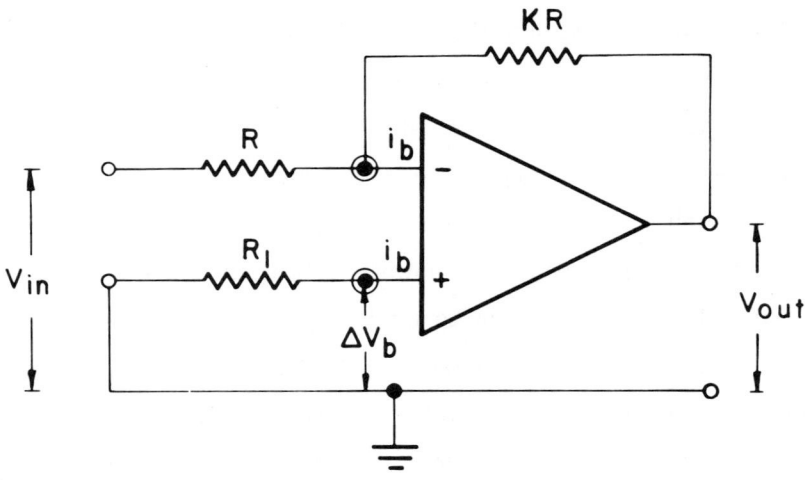

FIG. 1.12   INPUT-BIAS-CURRENT COMPENSATION

The input bias current is correctly compensated when the second term on the right-hand side of this equation is equal to zero, hence*

$$R_1 = \frac{K}{1+K} R \qquad [1.21]$$

The relative error due to the finite value of i_b in an uncompensated circuit is $i_b R / V_{in}$ and is obtained by setting $R_1 = 0$ in the above equation for V_{out}.

While the network shown in Fig. 1.12 completely eliminates this error, the same current i_b is still drawn from a

---

* Note that for an ideal operational amplifier this term is always zero since $i_b = 0$, and thus no compensation is required.

1.1 OPERATIONAL AMPLIFIER

circuit connected directly to one of the terminals of an operational amplifier. Thus, for example, for the measurement of the potential of a reference electrode, an operational amplifier with a low input bias should be used.

When i_b at the two inputs is not the same, the difference Δi_b is termed the "input offset current". Its value as well as its variation with time and temperature are usually specified by the manufacturer. If the calculation for the compensated circuit shown in Fig. 1.12 is carried out, it is easy to show that the relative error will be given by $R_1 \Delta i_b / V_{in}$. Thus if Δi_b is of the same order of magnitude as i_b itself, little is to be gained by employing the compensation network shown in Fig. 1.12 and the error can be kept low only by working in the range where $i_b R \ll V_{in}$.

(iv) Common-Mode Rejection Ratio

Among the properties of the ideal operational amplifier that were mentioned, was its function as a differential amplifier. This means that V_{out} depends only on ΔV_{in} and is independent of the absolute value of the voltage with respect to ground at either input. This does not hold true for real amplifiers. The common-mode rejection ratio is given by

$$\text{CMRR} = \frac{A_{CM}}{A_o} \qquad [1.22]$$

where A_{CM} is the common-mode gain, that is, the ratio between the output voltage and the voltage V_{CM}, with respect to ground, of each input when $\Delta V = 0$. The output voltage of a real operational amplifier is given by

$$V_{out} = A_{CM}V_{CM} + A_o\Delta V \qquad [1.23]$$

This property of the operational amplifier is of great importance when the device is used as a floating-point differential amplifier, that is, when neither input is grounded. A potential V_{in} of the inputs with respect to ground will cause an error in ΔV equivalent to V_{in}/CMRR.

(v) <u>Input Impedance and Maximum Input Voltage</u>

The input impedance between the two inputs of an operational amplifier is specified by the manufacturer. In addition to the resistive part of this impedance, a capacitance of the order of 10 pF exists between the two inputs. This may become of major importance when the network is operated at high frequencies or for fast transients.

At the input stages of an operational amplifier there are transistors which are protected up to a certain voltage. This is specified in terms of the maximum voltage between the two inputs and the maximum voltage allowed with respect to ground.

(vi) <u>Output Swing Voltage and Output Current</u>

The maximum output voltage and current of an operational amplifier are among its important characteristics. Both are symmetrical with respect to zero. When the output current (which is usually in the range of a few mA) is not sufficient for the purpose intended, a current amplifier or current booster can be used. The booster is always operated in conjunction with an amplifier and it can usually increase the output current capability of the network by up to three orders of magnitude. The specifica-

1.2 EXPERIMENTAL

tions given for a booster include the input and output currents, the voltage and the response time.

(vii) Slewing Rate

It is common to define the response time of a network in terms of the rise time, which is the time required by the signal to go from 10% to 90% of its final value. This definition is not suitable in the case of operational amplifiers and the slewing rate, defined as the rate of change of potential with time when a step-function is applied to the amplifier, is used instead. The slewing rate gives the minimum time required for the output potential to move from one value to another. The actual time is always longer and depends, among other things, on the parasitic capacity between the inputs and the output of the amplifier.

1.2 EXPERIMENTAL

1.2.1 Determination of the Characteristics of an Operational Amplifier

In this section a number of experiments will be outlined which permit the determination of the major properties of an operational amplifier. The amplifier tested will be the Fairchild type μA 741 or its equivalent, which requires a $\pm$ 15 volt power supply. Additional equipment needed for these tests are: square-wave generator, oscilloscope, and digital multimeter. It is imperative that the student perform these experiments carefully, since the very same approach will be employed in testing the potentiostat and other equipment used in electrochemistry.

(i) The Amplification (Open-Loop Gain)

The open-loop gain A_o of the operational amplifier will be determined with the aid of the simple circuit shown in Fig. 1.13. A square wave of 10 volt peak-to-peak amplitude is applied at the input and the signal at the output is measured between the points A and B at a low sensitivity on the oscilloscope (2 V/cm). Next, the oscilloscope is connected between the two inputs of the amplifier (points B and C) and the amplitude of the square wave is measured at a high sensitivity (1 mV/cm). For the circuit shown in Fig. 1.13 we have

$$V_{out} = -V_{in} \qquad [1.24]$$

and

$$A_o = -\frac{V_{out}}{\Delta V_{in}} = \frac{V_{in}}{\Delta V_{in}} \qquad [1.25]$$

The frequency of the square wave used in this measurement should be low (not higher than 1 kHz). It is convenient to use the calibrated square wave of the oscilloscope for this purpose.

An alternative method of measuring A_o is by the use of equation [1.17].

$$V_{out} = -V_{in} K \left[\frac{1}{1 + (1+K)/A_o} \right] \qquad [1.17]$$

which applies in the case of a real operational amplifier having a finite value of A_o. K is the ratio between the feedback and input resistances in a regular inverting voltage amplifier (Fig. 1.5). To evaluate A_o with reasonable accuracy, one should choose a

1.2 EXPERIMENTAL

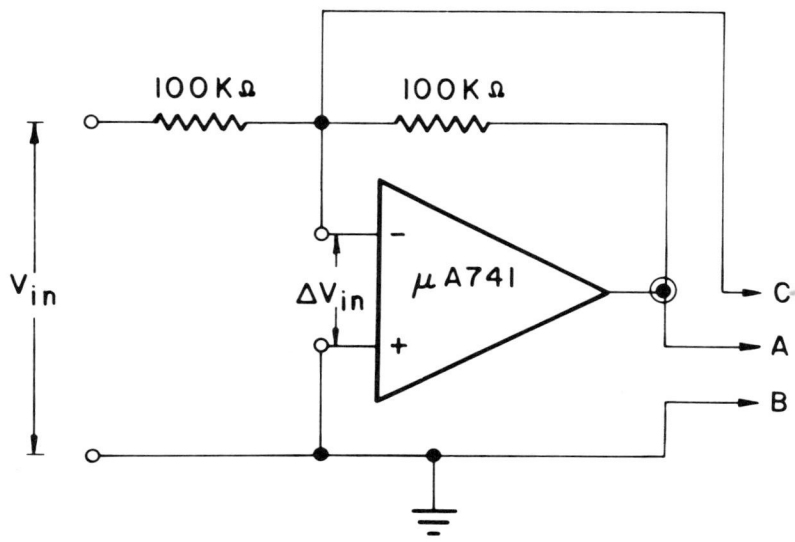

FIG. 1.13   CIRCUIT FOR DETERMINING THE GAIN

value of K of the order of magnitude of A_o. Since both quantities in this case are very large compared to unity, the above equation can be written in the simplified form as:

$$V_{out} = -V_{in} K \left(\frac{A_o}{K + A_o} \right) \qquad [1.26]$$

A numerical example shows the difficulties which may arise if this method is used. Assume that the nominal value of A_o, given by the manufacturer is 10^4. Let us use $R_{in} = 10 \, \Omega$ and $R_f = 10^5 \, \Omega$ to obtain $K = 10^4$ and assume further that the output impedance of the square-wave generator is zero. If the input offset voltage is, say, 10 mV, then the output voltage expected exceeds 50 Volts and the amplifier will be driven into saturation. (The maximum output voltage of the μA 741 is ±10 Volts). As a result, the square-wave signal applied at the input will not be detected at the output.

A second error may arise if the output impedance of the square-wave generator is not zero and cannot be neglected with respect to R_{in}. Both problems will be solved by employing the circuit shown in Fig. 1.14. The square wave V_{sq} is impressed between ground and the positive input of the amplifier, which acts as a voltage follower and does not load the square-wave generator. It can be shown that equation [1.26] is applicable in this case if V_{in} is replaced by V_{sq}.

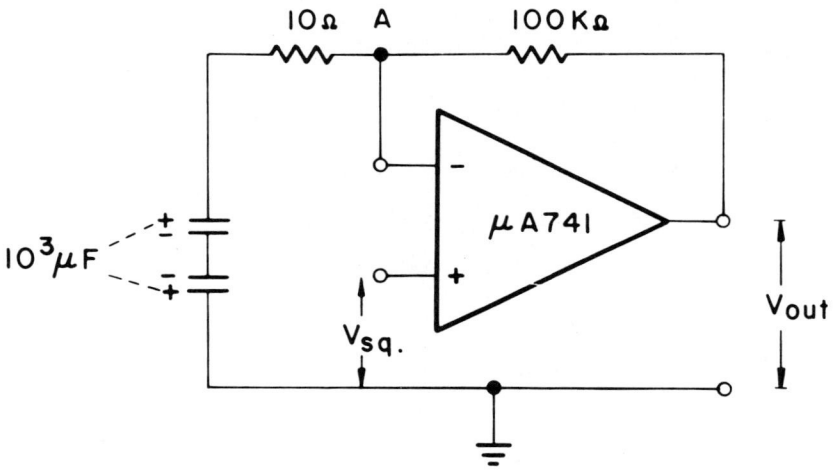

FIG. 1.14 VOLTAGE-FOLLOWER INPUT CIRCUIT FOR DETERMINING THE GAIN

1.2 EXPERIMENTAL

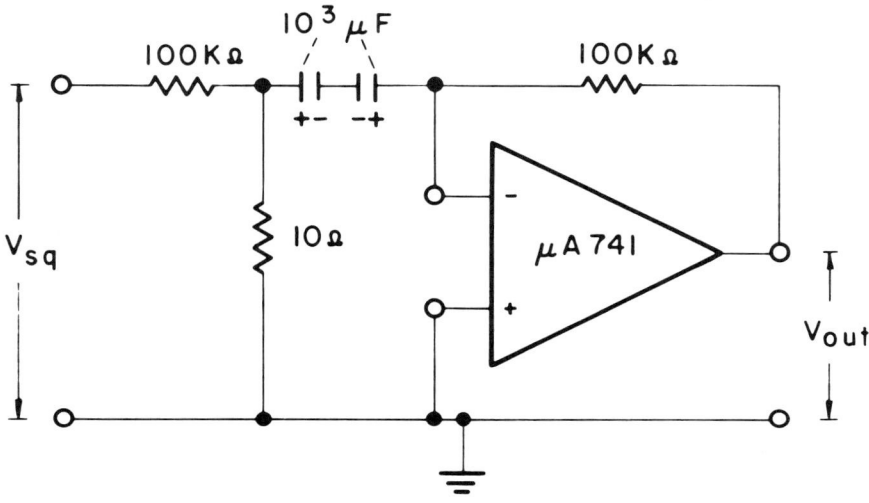

FIG. 1.15 ATTENUATOR INPUT CIRCUIT FOR DETERMINING THE GAIN

The circuit shown in Fig. 1.15, in which the input resistance serves also as the attenuator for the square wave applied, may also be used to measure A_o. Simple analysis along the lines followed above shows that

$$V_{out} \cong -V_{sq}\left(\frac{A_o}{K + A_o}\right) \qquad [1.27]$$

The advantage of this method over the previous one is that measurement of a very small voltage difference ΔV_{in} is not necessary. The two capacitors are included in the circuit to eliminate the direct current due to the input offset voltage.

Measurements should be performed with a low-frequency square wave, having an amplitude of 0.1-1.0 volt. If the last equation is rearranged it can be easily seen that a plot of V_{sq}/V_{out} vs K should yield a straight line with a slope of $1/A_o$ and an intercept of unity.

(ii) Slewing Rate

The circuit shown in Fig. 1.13 can be used to determine the slewing rate of an operational amplifier. A square wave of relatively low frequency (say 1000 Hz) having an amplitude larger than the maximum output voltage of the amplifier, is applied at the input and the rate of change of potential with time (i.e. the slewing rate) is measured on an oscilloscope connected between the output terminal of the amplifier and ground (points A and B in Fig. 1.13). The quality of the square wave used for these measurements is important and should be checked separately on the oscilloscope. Its rise-time should be small compared to the time taken by V_{out} to pass from its most negative to its most positive value. The calibrated square-wave output of a good oscilloscope is usually satisfactory for this purpose.

The response-time for the current is measured by placing a resistance at the output (between points A and B in Fig. 1.13) and measuring the voltage developed across it. A resistor of about 1000 Ω should be used in this experiment. This overloads the circuit and the current measured will be the maximum current

1.2 EXPERIMENTAL

which a μA741 type amplifier can deliver.

(iii) Input Bias Current

The input bias current at each of the two inputs can be measured with a voltage follower in which the relevant input is connected to ground through a high input resistance, as shown in Fig. 1.16a and 1.16b. Since i_b is small (generally 10^{-7} to 10^{-9} Amp.), a high resistance has to be used at the input. This should be shielded to minimize noise pickup which may be partially rectified at the input to give an erroneous value of i_b. A small error may also be introduced by the input offset voltage ΔV since

$$V_{out} = i_b R + \Delta V \qquad [1.28]$$

However if R is chosen sufficiently large to give an output voltage of several hundred millivolts, then ΔV may be neglected.

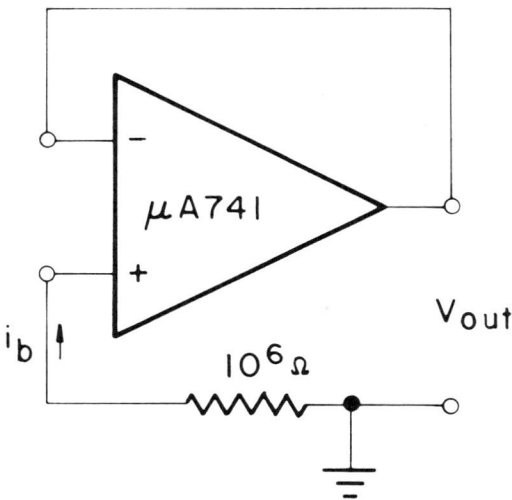

FIG. 1.16a CIRCUIT FOR MEASUREMENT OF i_b AT THE POSITIVE INPUT

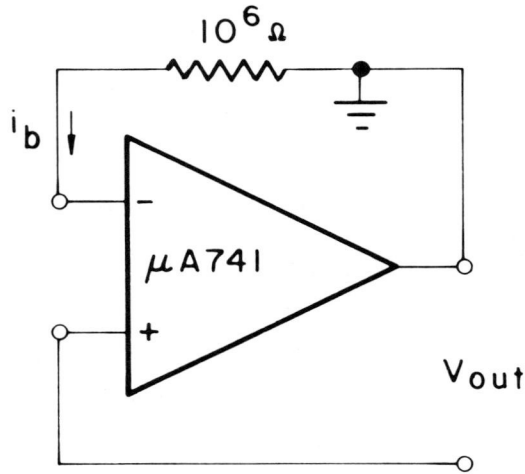

FIG. 1.16b CIRCUIT FOR MEASUREMENT OF i_b AT THE NEGATIVE INPUT

(iv) <u>Input Offset Voltage</u>

The input offset voltage is measured with the aid of an inverting amplifier network (Fig. 1.5) in which the input resistor is connected to ground. Since $V_{in} = 0$ in this configuration, the output can only be due to the small input offset voltage, ΔV, existing between the two input terminals of the amplifier. Following the usual reasoning we find (for $R_{in} = R$ and $R_f = KR$)

$$i_{in} = -\frac{\Delta V}{R} = i_b + \frac{\Delta V - V_{out}}{KR} \qquad [1.29]$$

hence

$$V_{out} = \Delta V(1 + K) + KRi_b \qquad [1.30]$$

This equation may be used directly to determine the offset voltage

ΔV from the measurement of V_{out}, with the use of the value of i_b determined in the previous section. Alternatively, the value of R may be chosen small enough to make $Ri_b \ll \Delta V$.

For the experimental determination of ΔV, use $R_{in} = 100$ Ω and $R_f = 10$ k Ω ($K = 10^2$) and measure the output voltage V_{out} with a digital multimeter. Increase R_{in} to 1 k Ω and then to 10 k Ω, maintaining $K = 10^2$, and calculate the apparent value of ΔV. Assuming that the first measurement (with $R_{in} = 100$ Ω) yields the correct value of ΔV, calculate i_b from equation [1.30] and compare with the value measured in the previous section.

1.3 BASIC CIRCUITS IN ELECTROCHEMICAL MEASUREMENTS

1.3.1 The Galvanostat

The basic circuit of an electronic galvanostat is shown in Fig. 1.17. In view of the virtual equality of the voltage at the two inputs of the amplifier, the current through the variable resistor R is V_{ref}/R. The same current also flows through the cell, which is connected in the feedback loop of the amplifier. Thus i_{cell} is independent of the cell resistance and can be controlled by varying R or V_{ref} or both as long as the current through the cell and the voltage across it do not exceed the output current and voltage of the amplifier used. It should be noted that the source of reference voltage is not loaded (beyond the very small input bias current) since it is connected between the positive input of the amplifier and ground. Although the potential drop across R is

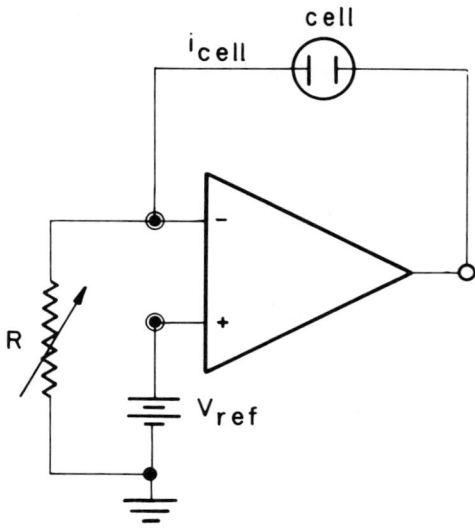

FIG. 1.17   SIMPLE CIRCUIT FOR GALVANOSTAT

generated by V_{ref}, the current through it (and through the cell) originates from the output of the amplifier.

To test the galvanostat, one replaces the cell by a variable resistor and accurately measures the current as a function of the load resistance. This should remain constant to within 0.1% throughout the range of operation. When the voltage drop across the cell exceeds the maximum voltage output of the galvanostat, the amplifier goes into saturation and the current is simply determined by $V_{out}(max)/R_{load}$. (This, in fact, is the method of determining $V_{out}(max)$).

To determine $i_{out}(max)$, decrease the load resistance nearly to zero, then decrease the variable resistance until no further change in output current is observed.

1.3 BASIC CIRCUITS

Some commercial galvanostats cannot be used to deliver very small currents (less than 5 μA) since at zero setting of the instrument a current of this magnitude still flows. Fortunately, in these cases an electronic galvanostat can be safely replaced by a flashlight battery and a suitable set of high resistances in series with the cell.

As with any instrument, stability of the current with respect to line voltage, temperature and time may be tested.

1.3.2 The Potentiostat

(i) General Properties

The potentiostat is perhaps the most widely used instrument in electrochemical-kinetics studies. It always operates with a three-electrode cell and its function is to maintain the potential of the working electrode at a preset level with respect to a fixed reference electrode. This it does by passing the necessary current between the working and a third electrode called the counter, or auxiliary electrode. The basic circuit of the potentiostat is shown in Fig. 1.18. Since the two inputs of the operational amplifier are virtually at the same potential, the potential difference between the working and reference electrodes is equal to V_{in}, which can be controlled externally. Note that the current through the reference electrode is equal to the input bias current at the negative input. In good commercial potentiostats this is kept as low as $10^{-9} - 10^{-12}$ Amp. The calibrated voltage source supplying V_{in} is also not loaded by the potentiostat. The current driven by the potentiostat (between the working and counter electrodes) can be determined by measuring the voltage

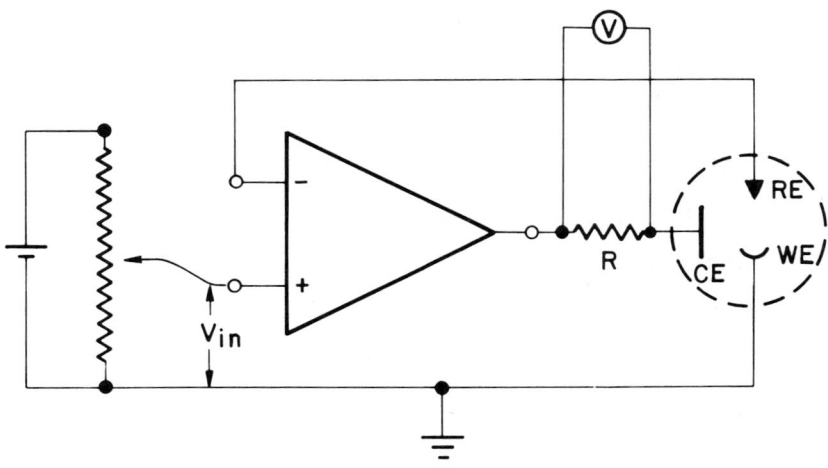

FIG. 1.18   BASIC CIRCUIT OF POTENTIOSTAT

drop across a small resistance R connected to the counter electrode.* This current is determined by the requirement that $V_{in} = V_{RE} - V_{WE}$ and is independent of the resistance in the cell within the operating range of the potentiostat, as will be shown below. In most commercial potentiostats the working electrode is connected to ground. This, however, is not a fundamental feature of the potentiostatic circuit and is usually done to minimize stray noise.

(ii) <u>The Uncompensated Resistance</u>

Ideally, the current driven by the potentiostat through the cell should be independent of the cell resistance. In practice, it

---

* Note that a floating-input voltmeter must be used here otherwise the output terminal of the amplifier will be shorted to ground and no current will flow through the cell.

1.3 BASIC CIRCUITS

must be remembered that the potentiostat can only control the potential difference between the metallic phases of the working and reference electrodes and not directly the metal/solution potential differences. To clarify this point we shall use the equivalent circuit shown in Fig. 1.19, which describes the cell under steady-state conditions. The resistance inside the reference electrode is not represented since the current through it is essentially zero. Thus the potential between the working and reference electrodes is made up of three components:

$$V_{in} = V_{RE} - V_{WE} = E_{RE} - E_{WE} \pm iR_{AB} \qquad [1.31]$$

where E_{RE} and E_{WE} are the metal/solution potential differences* at the reference and working electrodes, respectively. R_{AB} is a fraction of the total cell resistance, which depends on the geometry of the cell and particularly on the distance of the reference electrode from the surface of the working electrode. The resistance R_{AB} should not be confused with the measured resistance between the terminals of the working and reference electrodes. It is the <u>effective</u> resistance of the solution between the surface of the working electrode and the point at which electrolyte contact is made with the reference electrode or with the Luggin capillary connected to it.

---

* Since the working electrode is polarizable, E_{WE} should be regarded as the potential across the double-layer capacitor, which can be altered by passing a current across the interface.

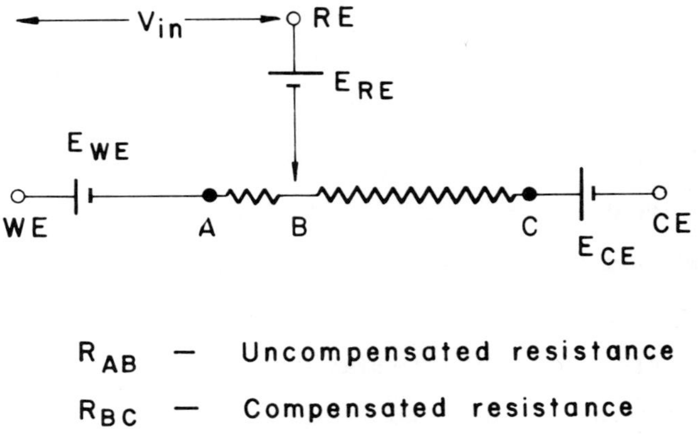

```
      ←——— V_in ———→ ○ RE
                    ⊤
                    ⊥ E_RE
                    |
    E_WE            |
  ○—|⊢—●—⋀⋀⋀—⋀⋀⋀⋀⋀—●—⊢|—○
   WE     A    B        C      CE
                              E_CE
```

R_{AB} — Uncompensated resistance

R_{BC} — Compensated resistance

FIG. 1.19 SIMPLIFIED EQUIVALENT CIRCUIT FOR THREE-ELECTRODE CELL

In potentiostatic measurements, E_{RE} is constant and it is desired to control E_{WE} and measure the corresponding current. Thus the quantity iR_{AB} represents an error due to the uncompensated part of the solution resistance. It will be demonstrated below that the best way to eliminate most of the uncompensated resistance is by employing a potentiostat with positive feedback. This, however, has become commercially available only very recently. Earlier, a Luggin capillary, placed very close (less than 1 mm) to the surface of the working electrode, was used. A discussion of the effectiveness of a Luggin capillary in reducing the value of the uncompensated solution resistance R_{AB} for different cell geometries is given in the "Introduction" section of Experiment 2. When the current density to be measured is high and/or the conductivity of the solution is low, a substantial error is introduced in the measurement of the potential as a result of the iR drop even with the best design of cell geometry. A

1.3 BASIC CIRCUITS

potentiostat with positive feedback which automatically corrects for most of the iR drop then becomes useful.

(iii) Testing the Potentiostat with a Dummy Cell

It is important to test the characteristics of a new potentiostat in order to differentiate between response arising from the electrochemistry of the system and artifacts caused by limitations of the instrument. This is best done with a dummy cell of the type shown in Fig. 1.20. The numerical values of the resistors and the capacitor depend on the nominal performance of the instrument and may be chosen by the following considerations. The faradaic resistance of an electrochemical reaction at equilibrium is given by (cf. equation [II. 70]).

$$R_F = \frac{RT}{F} \cdot \frac{\nu}{n} \cdot \frac{1}{i_o} \qquad [1.32]$$

Assuming for simplicity that $n = \nu = 1$ we have for, say, $i_o = 10^{-3}$ Amp/cm^2, $R_F = 25$ Ω cm^2. The double-layer capacity may be taken, to a first approximation, as $C_{dl} = 20$ μF/cm^2. Considering that a kinetic measurement will be typically performed with an electrode of about 0.1 cm^2 area, (except when the DME is used), a reasonable set of values for testing the potentiostat will be:

$$R_F = 0.25k\ \Omega \quad ; \quad C_{dl} = 2\mu F$$

For the solution resistance one should use two variable resistors $R_{AB} = 0-1$ k Ω , $R_{BC} = 0-10$ k Ω . Commercial potentiostats usually have a calibrated source of potential V_{in} and a terminal for an additional input, to which a function generator will be

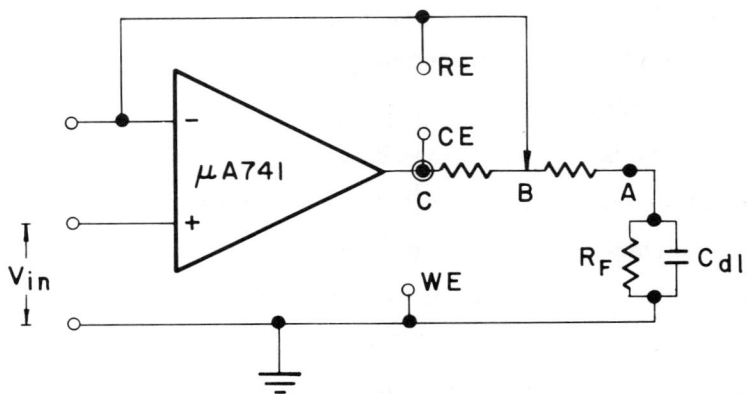

FIG. 1.20 DUMMY CELL CONNECTED TO BASIC POTENTIOSTATIC CIRCUIT

connected.

Two kinds of tests are required: those relating to low-frequency (or d.c.) behavior and those relating to high-frequency (or fast-transient) characteristics. In the former, internal calibration and long-term stability are of importance; in the latter, response-time and stability with respect to oscillations are a major factor. A commercial potentiostat and a very simple, home-made circuit shown in Fig. 1.20 will be used for these tests.

A. Low-frequency characteristics

(a) Calibration of the reference potential

When an internal potential source is included in the instrument, its calibration throughout the range should be checked. For this purpose, connect an accurate voltmeter (a four-digit multimeter is suitable) between the RE and the WE outputs of the

1.3 BASIC CIRCUITS

potentiostat. Set $R_{AB} = R_{BC} = 100$ Ω, disconnect R_F and measure the voltage V_{WE-RE} at different settings of the reference potential. Measure also at $V_{in} = 0$, since in many commercial units a large offset voltage may be found.

(b) Maximum output voltage and current

In the circuit shown in Fig. 1.20 replace R_F with a smaller, 100 Ω resistor. Set $R_{AB} = R_{BC} = 0$ and increase V_{in} in increments of 50mV. Determine the current by measuring the voltage drop across R_F. Note the maximum current which the amplifier can deliver. Repeat the experiment with reversed polarity of V_{in}.

To measure the maximum value of V_{out}, set V_{in} to deliver a current some 20% below the maximum current which the potentiostat can deliver. With this constant setting of V_{in}, increase the compensated resistance R_{BC} and measure the voltage across it. Note the maximum voltage which can develop across the resistor R_{BC}. This, with the addition of the small voltage drop across R_F in the same experiment, is the maximum output voltage of the potentiostat. The experiment should be repeated with the polarity of V_{in} reversed.

The same test can be performed with a commercial potentiostat. The resistance R_F should be chosen sufficiently small, that the maximum current output of the potentiostat can be exceeded when $V_{in} = 1.0$ Volt. Typically, for a potentiostat in which V_{out} and i_{out} have nominal values of ± 30 V and ± 0.5 A, one should choose $R_F = 1-2$ Ω and $R_{BC} = 0-1$ k Ω. (A power resistor should be taken for R_{BC}, which can stand the maximum power output of the potentiostat, in the above case 15W).*

(c) Input bias voltage and current

The input bias voltage, ΔV, represents the error of the potentiostat, since it is the difference between V_{in} and V_{RE}. In most cases this is less than one mV and is of little consequence. However, in some commercial tube-driven potentiostats ΔV may be as large as 0.1 Volt and should be taken into account or corrected. In the circuit shown in Fig. 1.20, $\Delta V = V_{RE-WE} - V_{in}$. In a commercial potentiostat one should set $V_{in} = 0$ and measure V_{RE-WE} which is then equal to ΔV.**

The input bias current, which is the current passing through the reference electrode, can be measured by placing a large resistance in the reference-electrode circuit and measuring the voltage drop across it, as discussed above (cf. Fig. 1.16b). When the circuit in Fig. 1.20 is used, the large resistance is connected between the negative input of the operational amplifier and the point B, the resistors R_{AB}, R_{BC} are set equal to 100 Ω

---

* If a large error in the potential of the working electrode is observed, repeat the above measurement, with an oscilloscope connected to the output of the operational amplifier. The error may be due to high-frequency oscillations.

** There may be some difficulty in setting V_{in} exactly at zero. To overcome this, V_{RE-WE} should be measured with V_{in} as close as possible to zero and with the polarity switch set at each position in turn. If V_{RE-WE} changes in magnitude when the polarity is changed, its average should be taken equal to ΔV.

1.3 BASIC CIRCUITS

and V_{RE-WE} is then given by

$$V_{RE-WE} = i_b R + \Delta V \qquad [1.33]$$

A sufficiently large value of R is chosen, so that $i_b R > 1$ Volt and ΔV is negligible. For a commercial potentiostat, R should be connected to the reference-electrode terminal and V_{in} set equal to zero for the above equation to hold. In most instances, i_b is sufficiently small (less than 10^{-9} Amp.) as long as the instrument is in proper operating condition.

(d) <u>Stability of reference voltage</u>

The stability of the reference voltage V_{in} is usually specified with respect to time (mV/24h), line voltage (mV/±10% change) and temperature (mV/10°C). These can be checked in a straightforward manner, if desired. In operation, a potentiostat must, in addition, be stable with respect to its output voltage and current. Specifically, V_{in} must remain constant (usually within 1 mV) as V_{out} and i_{out} are changed from zero to their maximum values.

To test the dependence of V_{in} on V_{out} use the circuit shown in Fig. 1.20. Set the resistors $R_{BC} = R_{AB} = 0$ and choose a value of R_F which will allow the passage of about half of the maximum output current at $V_{in} = 1.000$ Volt. Increase R_{BC} in steps until the maximum value of V_{out} is reached and measure V_{in} accurately for each setting (preferably with an accuracy of 0.1 mV). Note also the changes, if any, in the current.

Repeat the experiment but now keep $R_{BC} = 0$ and

increase R_{AB} in small steps (the total voltage drop across R_{AB} should be less than 1.0 Volt). Note the effect of the voltage drop across R_{AB} on the value of V_{in} and on the current. Discuss.

To test the effect of output current on V_{in}, set $V_{in} = 1.000$ Volt, $R_{AB} = R_{BC} = 100\ \Omega$. Use a variable resistor for R_F and change it stepwise from a value which allows passage of only one percent of the maximum current to a value which would allow passage of twice the maximum current. Measure V_{in} accurately (0.1 mV) for each value of the current.

B. High-frequency characteristics

(a) Rise time

The rise time is often quoted among the specifications of commercial potentiostats. This is not a clear indication of the true response time of the instrument, as previously pointed out, since the rise time depends on the magnitude of the step function applied, and on the nature of the load.

(b) Slewing rate

The slewing rate is the maximum rate of change of potential with time, when an ideal step function is applied at the input. The slewing rate can be determined with the aid of the dummy cell shown in Fig. 1.20, in which $C_{dl} = 0$, $R_{CB} = R_{AB} = 0$ and R_F has been chosen large enough to maintain the current below its maximum value. A square wave of 1.0 Volt amplitude is applied at the input (V_{in}) and the shape of the output potential pulse (V_{RE-WE}) is observed on the oscilloscope.

1.3 BASIC CIRCUITS

(c) Shape of the current transient

It is best to observe the effect of capacitors and resistors in the cell by following the current pulse when a square wave or triangular wave is applied at the input of the potentiostat. To do this, connect a current-to-voltage converter at the output terminal of the potentiostat, as shown in Fig. 1.21. Use this circuit with $R_{AB} = R_{BC} = 0$, $C_{dl} = 0$ and R_F small enough to allow the maximum output current to flow. R_F should be calculated to give $V_{out} = 1.0$ Volt for maximum current output of the potentiostat. Apply a square wave at the input (1.0 Volt amplitude) and observe the current transient on the oscilloscope. Repeat with a triangular-wave signal of the same amplitude and a frequency of 1000 Hz (or a sweep rate of 2000 Volt/sec). Reduce the frequency to 1 Hz (or 2 Volt/sec), and repeat the experiment. Repeat all three measurements with the capacitor C_{dl} having a value of 10 μF per 100 mA maximum current of the potentiostat.

Repeat once more with $R_{BC} = 1/2 \, V_{out}(max)/i_{out}(max)$ and R_{AB} changing in steps from zero to $0.1/i_{out}(max)$, maintaining C_{dl} at the value set above. Discuss the effect of the double-layer capacitance, the compensated and the uncompensated resistance on the shape of the current/time transient caused by a square-wave or a triangular-wave input signal.

(iv) Potentiostat with Positive Feedback

We have seen above that proper design of the electrolytic cell can reduce the effect of solution resistance but does not eliminate it completely, particularly when solutions of low specific

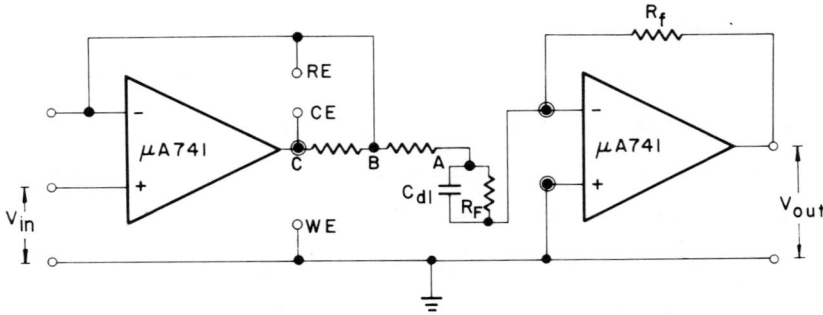

FIG. 1.21 CIRCUIT FOR TESTING A POTENTIOSTAT IN THE HIGHER FREQUENCY RANGE

conductance are used. The problem may be clarified by a numerical example. Assume that the total cell resistance is 20 Ω and 95% of this resistance is compensated by bringing the tip of the Luggin capillary close to the surface of the working electrode. Assume further that an experiment in cyclic voltammetry at relatively high sweep rates is conducted and the current goes from 1 mA at some point at the foot of the wave to 100 mA at the peak. The error in potential due to 1 Ω of uncompensated resistance will rise from 1 mV at the foot of the wave to 100 mV at the peak, rendering meaningless an analysis of the peak potential or its variation with any experimental parameter. In polarography, the error is smaller if relatively concentrated aqueous electrolytes are used because of the much lower currents (in the range of 10-100 μA) but becomes important if nonaqueous or dilute aqueous solutions are studied. On the assumption that optimum cell design has been achieved, further reduction of the error can be

1.3 BASIC CIRCUITS   193

effected by reducing the total current (e.g. by the use of a more dilute solution of the reactant or slower sweep rates in cyclic voltammetry) or by employing a potentiostat with positive feedback shown schematically in Fig. 1.22. Close examination of this figure reveals that operational amplifiers 2 and 3 function as the potentiostat proper and as a current-to-voltage converter, respectively, as shown in Fig. 1.21. The output of operational amplifier 3 can be measured in the usual way to determine the total current through the cell. Part of the potential appearing at this output is fed to the summing point (i.e. the negative input) of operational amplifier 1, where V_{in} is applied. This amplifier serves as a unit-gain inverting amplifier and its output signal feeds the input of amplifier 2, which serves as the potentiostat proper. By suitable choice of the fraction γ of V_{out} at amplifier 3 which is fed back to the input of amplifier 1, it is possible to correct for the uncompensated solution resistance up to a maximum value of R_f, the optimum value of which can be chosen for the system studied.

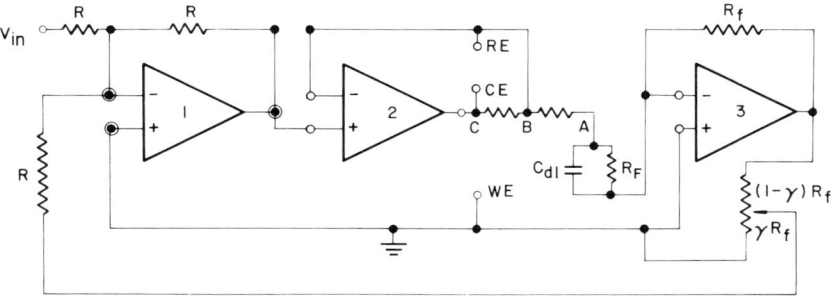

FIG. 1.22   POTENTIOSTAT WITH POSITIVE FEEDBACK

To understand better the operation of the potentiostat with positive feedback, suppose that for a given setting of V_{in}, a current i_{cell} flows through the cell. The potential at the point B is $-V_{in}$ since operational amplifier 1 acts as a unit-gain inverting amplifier. At the point A, the potential will be $-V_{in} - R_{AB} i_{cell}$. This is the potential across the capacitor, or in a real cell the actual potential applied to the interphase between the solution and the working electrode.* The potential at the output of amplifier 3 will be $-R_f i_{cell}$ and a fraction γ of this potential, (γ depends on the setting of the variable resistance connected between this output and ground), is added to V_{in}. Thus, at the output of amplifier 1, the potential will be $-(V_{in} - \gamma R_f i_{cell})$ and the net potential applied to the interphase (or to the capacitor in the dummy cell) will be $-(V_{in} - \gamma R_f i_{cell}) - R_{AB} i_{cell}$. The uncompensated part of the cell resistance will be accurately corrected when $\gamma R_f = R_{AB}$.

The usefulness of the potentiostat with positive feedback shown schematically in Fig. 1.21 is somewhat limited by the tendency of this type of circuit to go into oscillations due to phase shifts in the three amplifiers and in the cell and due to possible overcompensation (e.g. if γ is set for very nearly full compensation, a small decrease in the resistance near the electrode due to the reaction taking place there will cause oscillation). This may account for the fact that this type of device is not yet widely available commercially, although there are many home-made

---

* There is an additional constant which depends on the reference electrode, but this we shall consider for the moment equal to zero, for convenience.

units in research laboratories. The risk of oscillations is diminished if γ is set so that only, say, 90-95% of the total uncompensated resistance is corrected. This can still be very useful since the uncompensated and uncorrected resistance is reduced to 0.5% or less of the total cell resistance which, in most cases, should reduce the error introduced by this factor to a tolerable level.

1.3.3. Experiments with a Potentiostat with Positive Feedback Employing a Dummy Cell

In the previous section, the basic circuit was described for a potentiostat with positive feedback, which can correct for the uncompensated part of the solution resistance. Here, a few experiments in which such a potentiostat is employed are suggested. As pointed out above, this device tends to be unstable and will readily go into oscillations. It is thus recommended that a commercial instrument, or one built by an electronics expert be used, rather than attempt to build the circuit shown in Fig. 1.22.

(i) The Uncompensated Solution Resistance

Connect a dummy cell to the potentiostat as shown in Fig. 1.23. Apply a small (20 mV peak-to-peak) triangular wave at the external-modulation input of the potentiostat and record on an oscilloscope the current transient produced by the potentiostat. If an oscilloscope with two input channels is available, record simultaneously the triangular wave at the input of the potentiostat. The slope of the triangular wave should be chosen so that it is small compared to the slewing rate of the potentiostat and so that

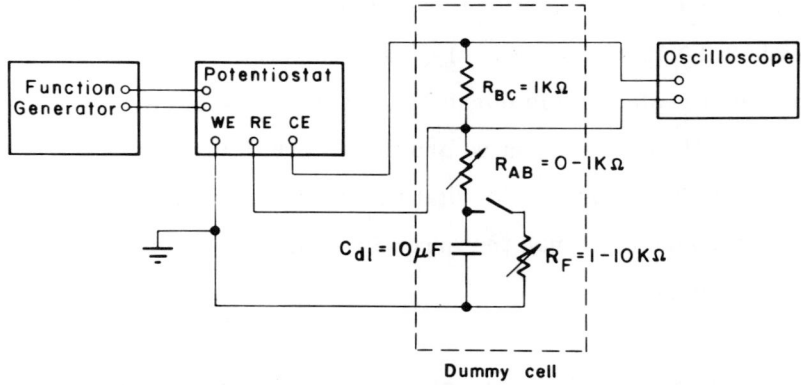

FIG. 1.23   POTENTIOSTAT WITH POSITIVE FEEDBACK EMPLOYING A DUMMY CELL

the resulting charging current C_{dl} (dV/dt) will be small compared to the maximum output current of the potentiostat used. The lower limit of dV/dt is set by the requirement that the resulting charging current should be large compared to the noise and to the limit of sensitivity of the measurement.

Disconnect R_F from the circuit and set $R_{AB} = 0$. The dummy cell is now equivalent to an ideal polarized electrode with all of the solution resistance compensated. Set the "iR correction" on the potentiostat to zero, apply the triangular wave at the input and observe the square-wave current at the output. Increase R_{AB} in steps to its maximum value of 1 k Ω and observe the resulting changes in the shape of the square wave. Photograph the square wave at $R_{AB} = 0, 0.1, 1.0$ k Ω .

Repeat the above experiment with values of dV/dt five times and one fifth of the first value.

Return to the value of dV/dt used in the first measurement

1.3 BASIC CIRCUITS

and repeat the experiment with two other values of C_{dl}, say 5 and 20 μF.

(ii) Correction for the Uncompensated Solution Resistance

Connect the dummy cell as in the previous section (R_F disconnected), set the uncompensated solution resistance R_{AB} to 1.0 k Ω and observe the shape of the current pulse at the output. Increase the "iR correction" of the potentiostat until the current pulse at the output closely resembles the shape which was observed in the previous section for $R_{AB} = 0$. Note the resistance required to make this correction and compare it to R_{AB}. Increase the "iR correction" further and note on the oscilloscope the beginnings of oscillation. Turn the resistance down again until oscillations cease. Note the value of the resistance at which oscillations just stop, and compare to R_{AB}.

(iii) iR Correction When Faradaic Reaction Takes Place

Connect R_F to the circuit, set $R_{AB} = 0$ and observe the changes in the shape of the square-wave current as R_F is decreased from 10 k Ω to 1.0 k Ω. Photograph at $R_F = 10, 5.0$ and 1.0 k Ω and compare to the picture obtained for $R_F = \infty$ (i.e. when the resistor was disconnected) in section (i) above. Repeat at two other frequencies as in section (i).

Set $R_{AB} = 1.0$ k Ω and repeat the above experiments at one frequency. The shapes of the current transients for different combinations of the uncompensated part of the solution resistance R_A and the faradaic resistance R_F are shown schematically[2] in Fig. 1.24. Increase the "iR correction" of the potentiostat until the shape of the current transient changes from the form shown in

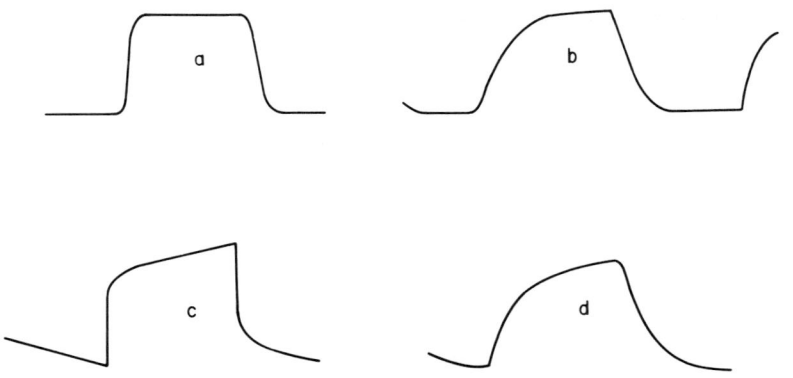

FIG. 1.24 CURRENT/TIME TRANSIENTS OBTAINED WHEN A TRIANGULAR WAVE IS APPLIED TO A POTENTIOSTAT EMPLOYING A DUMMY CELL.
(a) $R_{AB} = 0$, $R_F = \infty$ (b) $R_{AB} \neq 0$, $R_F = \infty$
(c) $R_{AB} = 0$, R_F finite (d) $R_{AB} \neq 0$, R_F finite

Fig. 1.24d to that shown in Fig. 1.24c. Note the value of the resistance required to reach this situation and compare it with R_{AB}. Increase the resistance further until the circuit starts to oscillate, then decrease until oscillation just stops and note the corresponding value of the resistance.

Comment on the different methods of finding the appropriate setting of the "iR correction" on the potentiostat in the presence

1.3 BASIC CIRCUITS   199

and absence of faradaic reactions.*

1.3.4. Experiments with a Potentiostat with Positive Feedback Employing an Electrochemical System

The effect of a finite value of R_{AB} on the observed current/potential behavior in cyclic voltammetry (cf. Experiment No. 9) has been discussed by Delahay[6] and is shown schematically in Figs. 1.25 and 1.26. The potential appearing across the interphase differs from the potential applied by the function generator by an amount iR_{AB}. This departure is negligible at the foot of the wave, where the current is small, but reaches a maximum at the potential corresponding to the peak current, as seen in Fig. 1.25 (curve b). As a result, the basic condition of cyclic voltammetry, namely that dV/dt = constant, is no longer maintained and, as seen in Fig. 1.26, the voltammogram deviates from that expected for ideal compensation of the solution resistance (R_{AB} = 0).

---

* In recent years [3,4], several potentiostats have been designed which measure the uncompensated part of the solution resistance R_{AB} and automatically set the appropriate correction. This is particularly useful when the dropping-mercury electrode is employed, since in this case R_{AB} changes continuously and a single setting of the correction cannot be made. At the present state of development, commercial instruments have a relatively slow response time (of the order of milliseconds) and can only be used for steady-state measurements or for slow transients. This shortcoming will probably be overcome within a few years. An instrument with a reponse time of $<10\,\mu$ sec has already been constructed [5].

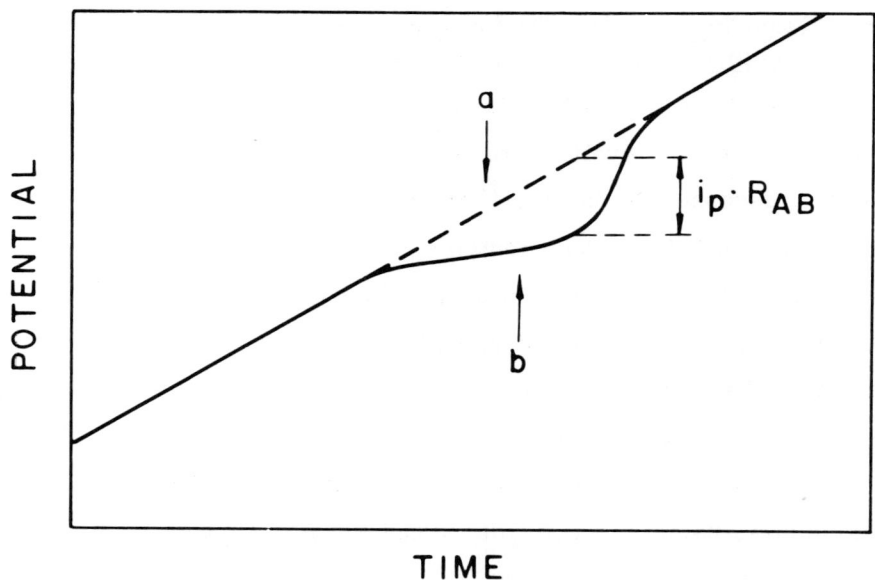

FIG. 1.25 THE VARIATION OF THE METAL/SOLUTION POTENTIAL DIFFERENCE WITH TIME DURING LINEAR POTENTIAL SWEEP

(a) WITH ZERO SOLUTION RESISTANCE

(b) WITH FINITE SOLUTION RESISTANCE

1.3 BASIC CIRCUITS   201

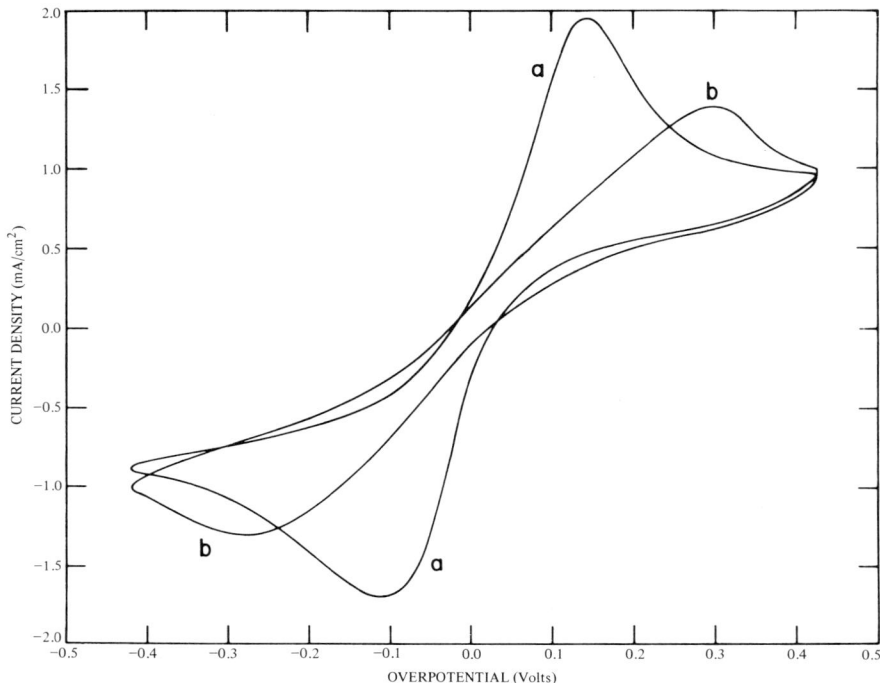

FIG. 1.26   CYCLIC VOLTAMMOGRAM FOR 5 mM
            QUINHYDRONE, 1 mM H_2SO_4. v = 75 mV/sec

    (a) WITH iR CORRECTION
    (b) WITHOUT iR CORRECTION

The Electrochemical Setup

The cell and electrode configuration will be as shown in Fig. 8.4 except that the auxiliary electrode compartment will contain a commercial reference electrode instead of the auxiliary electrode. It is advisable to increase the distance between the parallel working and reference electrodes to about 4 mm and/or to use thicker wires (ϕ = 0.05 cm) to increase the effect of the iR drop. A cell with this configuration is used to permit calculation of the resistance of the solution between the working and reference electrodes. The electrical setup shown in Fig. 9.5 will be used. The solutions will be 5 m$\underline{M}$ in quinhydrone and 1 m$\underline{M}$, 10 m$\underline{M}$ and 500 m$\underline{M}$ in H_2SO_4. The quinhydrone solution will be prepared on the day of the experiment and the required volumes of 0.5 $\underline{M}$ H_2SO_4 solution will be added to reach the concentrations required.

Determination of the Uncompensated Solution Resistance

Add to the cell 0.5 $\underline{M}$ H_2SO_4. Carry out the electrochemical pretreatment on each of the two platinum-wire electrodes (working and reference) according to the procedure outlined in Experiment 6.1, using the commercial electrode as reference. Rinse the cell and the electrodes thoroughly with distilled water until the conductivity of the rinse water falls below 10 μmho cm^{-1}.

To determine the uncompensated solution resistance, R_{AB}, add 5 m$\underline{M}$ quinhydrone in 1 m$\underline{M}$ H_2SO_4 to the cell. Insert the electrodes, remove the commercial reference electrode, bubble purified nitrogen through the solution for 10 minutes and pass it over the solution during the course of the experiment.

Adjust the sweep rate to 40 mV/sec and the voltage range to −0.5 to +0.5 V vs the platinum reference electrode. Set the Y-scale of the X-Y recorder at about 100 μA/cm for 1 cm^2 of electrode area and the X-scale at 50 mV/cm. With the "iR correction" on the potentiostat set to zero, record the voltammogram.

Choose a potential at which the faradaic current is very low ($R_F \to \infty$) and determine the best setting of the "iR correction" by observing the shape of the square wave on the oscilloscope, as

1.3 BASIC CIRCUITS

described in section 1.33 (ii)*.

Voltammograms of Quinhydrone, with and without iR Correction

Add to the cell the solution which is 5 m$\underline{M}$ quinhydrone, 1 m$\underline{M}$ H_2SO_4 and deaerate as before. Use the platinum-wire electrode as reference. Record, on the same paper, cyclic voltammograms at 40 mV/sec, with and without iR correction.

Compare the values of the peak current i_p and peak potential E_p in the anodic and cathodic directions as obtained with and without correction. Repeat at sweep rates of 10 and 160 mV/sec. Adjust the sensitivity of the current axis on the recorder, keeping in mind that the peak current is proportional to $(dV/dt)^{1/2}$.

Repeat the experiment with 10 m$\underline{M}$ H_2SO_4 and with 0.5 $\underline{M}$ H_2SO_4.

Treatment of Results

Collect the results in the form of the table given below for the three quinhydrone solutions.

---

* It is not advisable to find this setting by allowing the system to oscillate and then turning the setting down until oscillation just stops. During oscillations the potential across the interphase can reach extreme values and the surface of the electrode may change irreversibly.

Sweep Rate mV/sec	Uncorrected			Corrected		
	$i_p(a)$	$i_p(c)$	ΔE_p	$i_p(a)$	$i_p(c)$	ΔE_p
10						
40						
160						

Calculate the resistance V_{iR}/i between the reference platinum electrode and the working electrode for the cell configuration employed in the experiment (using equation 2.3) and compare with the "iR correction" setting.

Comment on the reversibility of the Q/QH_2 system in the three solutions of H_2SO_4 employed in this experiment.

REFERENCES

1. H. V. Malmstadt and C. G. Enke, "Electronics for Scientists", W. A. Benjamin Co., New York, 1963

2. N. Tshernikovski and E. Gileadi, Electrochim. Acta, <u>16</u>, 579 (1971)

3. Y. Nemirovsky, Ch. Yarnitzky and M. Ariel, Israel J. Chem. <u>6</u>, 41p (1968)

4. J. Devay, B. Lengyel, Jr. and L. Meszaros Acta Chim. Acad. Sci. Hung., <u>66</u> (3), 269 (1970)

5. Ch. Yarnitzky, Anal. Chem. March, (1975)

6. P. Delahay "New Instrumental Methods in Electrochemistry" p. 133, Interscience 1954.

2. REFERENCE ELECTRODES

Introduction

2.1 The Dynamic Hydrogen Electrode
2.1.1 General
2.1.2 The Cell, Solutions, Electrodes and Electrical Setup
2.1.3 Preparation of Platinized-Platinum Electrodes
2.1.4 The Potential of the Dynamic Hydrogen Electrode as a Function of Current Density
2.1.5 Stability of the Potential of the Dynamic Hydrogen Electrode
2.1.6 Treatment of Results

2.2 The Palladium/Hydrogen Electrode
2.2.1 General
2.2.2 The Cell, Solutions, Electrodes and Electrical Setup
2.2.3 Charging Curves for Palladium
2.2.4 Preparation of the Palladium/Hydrogen Electrode
2.2.5 Stability of the Palladium/Hydrogen Electrode
2.2.6 Treatment of Results

2.3 The $Ag/AgClO_4$ Electrode in Propylene Carbonate
2.3.1 General
2.3.2 The Cell, Solutions and Electrical Setup
2.3.3 Construction of the Reference Electrode
2.3.4 Reproducibility and Stability of the Potential
2.3.5 Thermodynamic Behavior of $Ag/AgClO_4$ in Propylene Carbonate
2.3.6 Polarization Curves of $Ag/AgClO_4$ in Propylene Carbonate near Equilibrium
2.3.7 Treatment of Results

INTRODUCTION

2. REFERENCE ELECTRODES

Introduction

The characteristics of various types of reference electrodes have been discussed in the literature in great detail.[1] Highest accuracy is achieved for electrodes of the second kind (e.g. calomel or Ag/AgCl), the potentials of which can be reproduced to within 0.01 mV, if proper precautions are taken in construction. This high accuracy is not required in the study of electrode kinetics, first, because of other sources of error involved in the measurement and second, since it is usually the difference in potential, rather than its absolute value, which is of importance. Thus, the value of the potential is usually reported to the nearest millivolt and a reproducibility of one or two millivolts is considered satisfactory for most purposes. As a result, some of the stringent requirements set for the construction of reference electrodes used in thermodynamic studies may be relaxed. Commercial reference electrodes are of the "fixed potential" type, that is, they constitute a complete half-cell (e.g. $Hg/Hg_2Cl_2/KCl$ (sat)) the potential of which does not depend on the composition of the solution. These electrodes are usually constructed so that they have a very low liquid-junction potential. In certain cases it is convenient to have a reference electrode which is reversible to one of the ions in solution (e.g. a silver wire coated with AgBr in a solution of KBr). These are sometimes called "indicator" electrodes. The advantages of indicator electrodes are that they can often be prepared <u>in situ</u> and there is no liquid junction, thus the addition of a foreign ion or a different concentration of the same ion is avoided. Indicator electrodes are particularly convenient for electrocapillary studies (cf. Experiment 3) and in studies of electrosorption (cf. Experiments 11, 12) and of organic reactions. In the last case the potential of an indicator electrode reversible to H_3O^+ ions in solution often varies with pH in the same way as does the reversible potential for the process being studied, thus providing a convenient scale for overpotential, independent of pH.

In recent years the electrochemistry of nonaqueous solvents has gained impetus and this has required the development of new types of reference electrodes suitable for each new solvent system examined.

In this group of experiments three kinds of reference electrodes will be tested. The dynamic hydrogen electrode and the palladium/hydrogen electrode, both of which can serve as indicator electrodes reversible with respect to H_3O^+ ions in aqueous solutions; and the $Ag/AgClO_4$ electrode for use as a fixed-potential reference in propylene carbonate. The choice of these electrodes is rather arbitrary and they do not have any special merit among the many indicator or fixed-potential electrodes which could have been chosen. Indeed, some of the best reference electrodes (e.g. calomel) have such high exchange-current densities that polarization measurements required to determine their characteristics are very difficult. In the following experiments the emphasis will be on the methods of testing the suitability of reference electrodes for a given system.

A good reference electrode should have the following characteristics: (a) high exchange-current density; (b) variation

$$R_{AB} = 1.6 \times 10^{-2}/\kappa \qquad [2.5]$$

The position of the Luggin capillary with respect to the working electrode is less important in this case than for planar electrodes since V_{iR} varies logarithmically rather than linearly with δ. Thus, if we position the capillary as far as 1 cm from the working electrode, the uncompensated part of the solution resistance will rise to $R_{AB} = 1/\kappa$ for the planar configuration while for the cylindrical electrode it will increase only by a factor of about three to $R_{AB} = 4.6 \times 10^{-2}/\kappa$. Thus it may often be more expedient to position the Luggin capillary relatively far from a thin-wire working electrode (making cell construction much simpler and avoiding shielding effects). If cylindrical symmetry is maintained, V_{iR} can then be calculated and corrected for.

A similar situation arises when the working electrode has the shape of a small drop and the counter electrode is a sphere concentric with it. In this case,

$$V_{iR} = \frac{i\delta}{\kappa}\left(\frac{r}{r+\delta}\right) \qquad [2.6]$$

where r is the radius of the drop. Taking $r = 0.01$ cm and $\delta = 5d/3$ with $d = 0.02$ cm as above, we have $R_{AB} = 0.78 \times 10^{-2}/\kappa$. There will clearly be some difficulty in placing the Luggin capillary as close as 0.04 cm from, say, a dropping-mercury electrode. However, this is not critical as can be seen by substituting $\delta = 1$ cm in equation [2.6]. The uncompensated solution resistance will increase by less than 30% to $R_{AB} = 1 \times 10^{-2}/\kappa$. As in the case of cylindrical symmetry, it is often better to

INTRODUCTION

of potential with concentration (activity) according to the Nernst equation; (c) reproducibility and stability of the potential; (d) fast response. When a fixed-potential reference electrode is used, a medium value of i_o (say, in the range of $10^{-5} - 10^{-7}$ A/cm^2) may be suitable, provided that impurities are rigorously excluded from the reference-electrode compartment. The term "impurity" is used here in a very broad sense and may include the reactant or product of the reaction being studied. This may be the case, for example, when a hydrogen electrode is used in the study of organic reactions.

In polarography and in certain studies of electrode kinetics where low currents are passed and a very low level of noise is essential, a two-electrode system may be employed. The same electrode then serves as both counter and reference and its total resistance should be kept to a minimum. The resistance of the reference should also be kept low in three-electrode systems when fast transients are measured, but in steady-state or slow-transient measurements a resistance of the order of $(1-5) \times 10^4$ Ω will not interfere with the measurements. On the other hand, the glass electrode cannot be used with presently available potentiostats because its very high internal resistance (ca 2×10^8 Ω), and its relatively slow response makes it unsuitable for any but steady-state measurements.

The position of the reference electrode with respect to the working electrode is of importance since it determines which part of the potential drop between the working and counter electrodes will be included in the measurement; in other words, it determines the uncompensated part of the solution resistance

INTRODUCTION

place the Luggin capillary further away from the working electrode and make corrections, if necessary, by calculating V_{iR} from equation [2.6].

It should be noted that the error V_{iR} in the measured potential is proportional in each case to the current <u>density</u>. It can be decreased by decreasing the radius of the spherical or cylindrical electrode. Plots of V_{iR} against the distance of the tip of the capillary from the electrode surface are shown in Fig. 2.2 for different geometries.

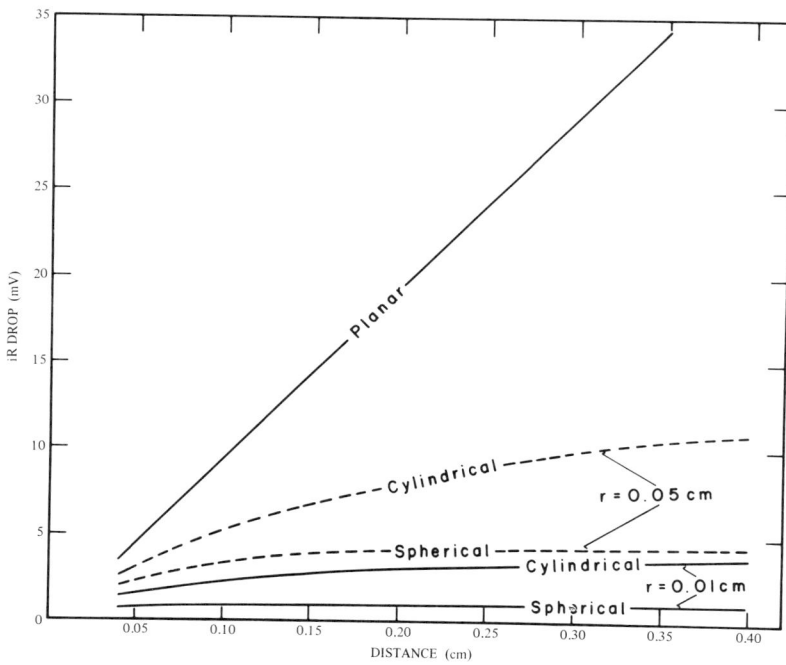

FIG. 2.2 DEPENDENCE OF THE iR DROP ON THE DISTANCE OF THE TIP OF THE LUGGIN CAPILLARY FROM THE SURFACE OF THE WORKING ELECTRODE. COUNTER ELECTRODE SYMMETRICAL.
$i = 1 \text{ mA/cm}^2$; $\kappa = 0.01 \text{ } \Omega^{-1} \text{ cm}^{-1}$; $d = 0.02 \text{ cm}$.

Several conclusions may be drawn from this figure. First, the increase in V_{iR} is much less for spherical and cylindrical electrodes than for planar electrodes. Second, for cylindrical electrodes, there is a continual small increase in V_{iR} with distance; for spherical electrodes, V_{iR} reaches a constant value after a relatively short distance. For the planar configuration V_{iR} increases almost linearly with distance.

2.1   The Dynamic Hydrogen Electrode

2.1.1   General

The reversible hydrogen electrode is an indicator-type reference electrode which is useful in the study of different electrode processes. This is particularly so for reactions of the type

$$RH \rightleftharpoons R + H^+ + e_M \qquad [2.7]$$

since the reversible potential for this reaction varies with pH in the same way as does the potential of the reference electrode and the overpotential at a given measured potential is thus independent of pH. Specific examples of reactions of the above type are to be found in fuel-cell processes, for example,

$$CH_3OH + H_2O \longrightarrow CO_2 + 6H^+ + 6e_M \qquad [2.8]$$

organic redox reactions such as

$$\phi\overset{O}{\overset{\|}{C}}CH_3 + 2H^+ + 2e_M \longrightarrow \phi\overset{OH}{\overset{|}{C}}HCH_3 \qquad [2.9]$$

2.1 DYNAMIC HYDROGEN ELECTRODE

or simply adsorption of oxygen or hydrogen

$$H_2O \longrightarrow OH_{ads} + H^+ + e_M \qquad [2.10]$$

A disadvantage of the reversible hydrogen electrode is that it requires a constant supply of purified hydrogen, which may be undesirable under certain circumstances and is generally inconvenient. This difficulty has been overcome by Giner[4] who proposed the use of a dynamic hydrogen electrode. In this system, a small cathodic current is passed through the reference electrode and the solution in its vicinity is thus saturated with hydrogen. The potential of the dynamic hydrogen electrode is slightly more cathodic than that of the reversible hydrogen electrode, but it is stable with time and the difference between the two potentials does not depend on pH.

In this experiment the properties of the dynamic hydrogen electrode will be studied in a number of acid and alkaline solutions. The electrode operates best in unstirred solutions, where the layer of solution adhering to it can be saturated with molecular hydrogen. This, however, precludes its use in unbuffered solutions, since the change in pH in the same thin layer of solution as a result of the current passing (due to a reaction such as $H_2O + e_M \longrightarrow 1/2\ H_2 + OH^-$) may become significant even when the pH of the bulk of the solution is unaffected.

Note that although the dynamic hydrogen electrode is used as a reference, its potential is determined by both thermodynamic and kinetic factors. Thus, for example, the addition of an inert electrolyte may affect the potential both by changing the activity coefficients of the species in solution and by changing the potential of the diffuse double layer, ϕ_2. In practice, the potential of the dynamic hydrogen electrode should be measured in each system against a conventional reference electrode.

2.1.2 The Cell, Solutions, Electrodes and Electrical Setup

The cell shown in Fig. 2.3 will be used in this experiment. The following solutions are used: H_2SO_4, 0.5, 0.05, 0.005 $\underline{M}$; NaOH, 1, 0.1, 0.01 $\underline{M}$ (with no inert electrolyte present). The reagents used must be of analytical grade and dilutions must be made with triply distilled water. Platinized-platinum electrodes[*] of about 1 cm^2 area, arranged as shown in Fig. 2.3 will serve as working (lower) and counter (upper) electrodes. This pair of electrodes constitutes the dynamic hydrogen electrode. Two reference electrodes will be used: a commercial calomel electrode and a platinized-platinum reversible hydrogen electrode.

The galvanostatic circuit shown in Fig. 2.4 will be used in this experiment. The change of potential with time will be measured with a digital voltmeter and a strip-chart recorder connected in parallel. The impedance-matching unit in Fig. 2.4 should have an input impedance of at least 10^7 Ω.

2.1.3 Preparation of Platinized-Platinum Electrodes

<u>Surface preparation</u>: To strip the platinization from old electrodes, immerse briefly in a 3:1:4 mixture of conc. HCl,

---

[*] See following section for details of preparation.

2.1 DYNAMIC HYDROGEN ELECTRODE

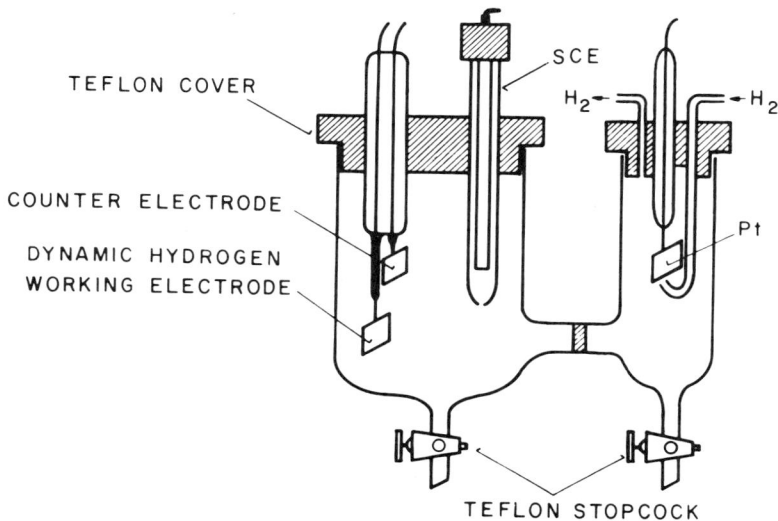

FIG. 2.3 CELL FOR TESTING THE DYNAMIC HYDROGEN ELECTRODE

conc. HNO_3 and water. Transfer to warm conc. HNO_3 and rinse with distilled water. Immediately before platinization, polarize cathodically for five minutes in 0.1 $\underline{M}$ H_2SO_4 at 0.50 A/cm^2 and wash with distilled water.

Platinization: Polarize cathodically for two minutes in a 2% solution of chloroplatinic acid at 100 mA/cm^2.

Activation: The platinized-platinum electrode is activated by polarizing it cathodically and anodically (five cycles) in 0.10 $\underline{M}$ H_2SO_4 at 0.5 A/cm^2 for about 15 seconds in each direction. (For best results, this step should be carried out before each set of experiments.)

The electrode should be stored in distilled water or in dilute sulfuric acid and should not be allowed to dry. Electrodes

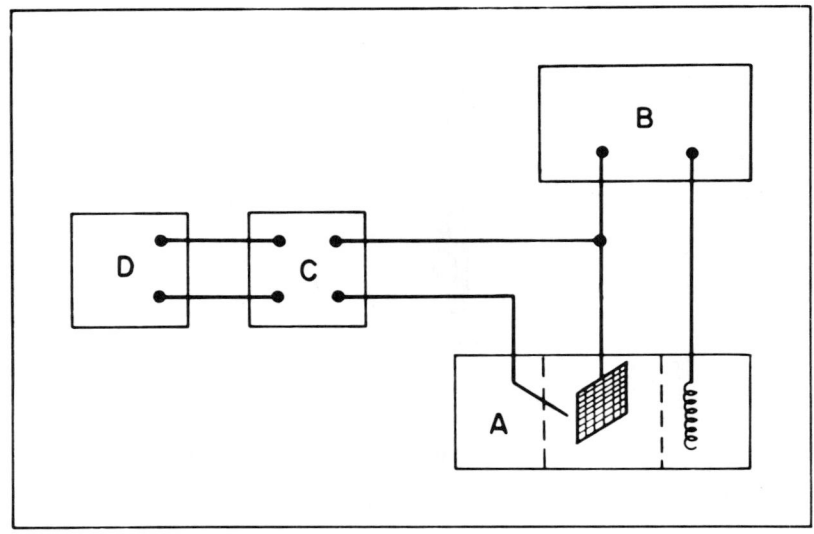

FIG. 2.4   BLOCK DIAGRAM FOR GALVANOSTATIC CIRCUIT
A - CELL; B - GALVANOSTAT: C - IMPEDANCE-
MATCHING UNIT; D - OSCILLOSCOPE OR STRIP-
CHART RECORDER

more than a week old should not be used, but should be stripped and re-platinized.

2.1.4 The Potential of the Dynamic Hydrogen Electrode as a Function of Current Density

Set up the cell as shown in Fig. 2.3 but without the SCE. Add to both compartments 0.5 $\underline{M}$ H_2SO_4 and bubble purified hydrogen. Note that the platinized-platinum reference electrode should be positioned in the cell so that its surface is swept by the hydrogen bubbles. Pretreat the working electrode by polarizing cathodically at 10 mA/cm^2 for two minutes in the unstirred solution. Note the potential at the end of this pretreatment.

2.1 DYNAMIC HYDROGEN ELECTRODE

Determine the potential as a function of current density from 4.0 to 0.1 mA/cm^2, taking about ten measurements in the order of decreasing current density. Hold the current constant at each value for about two minutes or until the potential has become constant. Note the time required to reach constant potentials.

Repeat the above measurements in all the solutions of H_2SO_4 and NaOH prepared for this experiment.

2.1.5 Stability of the Potential of the Dynamic Hydrogen Electrode

Set up the dynamic hydrogen electrode together with a saturated-calomel reference electrode in 0.5 $\underline{M}$ H_2SO_4. (The reversible hydrogen reference electrode will not be used in this part of the experiment and hydrogen bubbling may be discontinued.) Measure the potential simultaneously with a recorder and digital voltmeter by connecting these two instruments in parallel (cf. Fig. 2.4). Set the current density to 0.1 mA/cm^2 and follow the change of potential with time at a sensitivity setting of 1 Volt full scale and a time base of about 3 min./cm. When steady-state has almost been reached, mark on the recorder chart paper the values of the potential read on the digital voltmeter. After a constant potential has been reached, continue the measurement for ten more minutes. Stop the current and follow the change of potential with time on open circuit.

Set $i = 10$ mA/cm^2 and follow the change of potential with time. When steady-state is reached, switch the current to 0.1 mA/cm^2 and again follow the change of potential with time. Compare the steady-state potential at this current with the value obtained in the previous experiment. Compare the rate of change of potential with time at 0.1 mA/cm^2 with and without pretreatment at 10 mA/cm^2.

2.1.6 Treatment of Results

Plot the current density against potential. Use one sheet of graph paper to plot all the results obtained in acid solution and another sheet to plot the results obtained in alkaline solution.

Plot the potential difference between the reversible and the dynamic hydrogen electrodes as a function of H_2SO_4 or NaOH concentrations at three different current densities (0.1, 1, 3mA/cm^2).

Discuss the differences between the E/i curves obtained in different solutions taking into consideration factors such as depletion of the solution in the vicinity of the electrode and the effect of composition on the exchange-current density and on the diffuse-double-layer potential, ϕ_2.

Is the dynamic hydrogen electrode suitable for use in any aqueous solution at any pH? Why is it preferable to work in unstirred or in gently stirred solutions? At what current density does the electrode operate best?

2.2 The Palladium/Hydrogen Electrode

2.2.1 General

The absorption of hydrogen by palladium was discovered over 100 years ago.[5] Later studies[6,7] showed that two phases can coexist in palladium when hydrogen is absorbed. The α-phase is formed first and remains the only one as long as the concen-

2.2 PALLADIUM/HYDROGEN ELECTRODE

tration of hydrogen is below 0.3 atom percent. At higher concentrations the β-phase appears; this corresponds to the non-stoichiometric formula $PdH_{0.6}$. X-ray analysis shows no change in the lattice parameter for the α-phase while a 3% increase in the lattice parameter is observed in the β-phase. Thus in the former, a small amount of hydrogen is absorbed interstitially, while the latter may be considered as an alloy which has mechanical and electrical properties which are quite different from those of the metal.[7]

Isotherms for the absorption of hydrogen in palladium have been measured.[7] It was found that in the region of coexistence of the two phases, the equilibrium pressure of hydrogen was independent of the total amount of hydrogen absorbed, as was expected.

The electrochemical behavior of a palladium electrode in equilibrium with H_3O^+ ions in solution parallels the behavior of palladium in equilibrium with molecular hydrogen. This may be seen by writing the respective equilibria

$$1/2\ H_2 \rightleftharpoons H_{ads} \rightleftharpoons H_{abs} \qquad [2.11]$$

$$H_3O^+ + e_M \xrightleftharpoons{-H_2O} H_{ads} \rightleftharpoons H_{abs} \qquad [2.12]$$

Thus the equilibrium is controlled in one case by the partial pressure or fugacity of molecular hydrogen and in the other case by the activity of H^+ ions in solution and by the applied potential. The relationship between these two quantities is expressed by the Nernst equation

$$E = E^o + \frac{RT}{F} \ln \frac{a_{H^+}}{P_{H_2}^{1/2}} \qquad [2.13]$$

This relationship makes it very convenient to study the palladium/hydrogen system electrochemically, since the potential, which can be easily and accurately controlled, determines the equivalent pressure. Thus, a measured potential of +90 mV vs a reversible hydrogen electrode in the same solution corresponds to $P_{H_2} = 10^{-3}$ atm., while -90 mV RHE corresponds to $P_{H_2} = 10^3$ atm. When palladium is electrochemically charged with hydrogen, the existence of two phases is manifested by a constant potential of approximately +50 mV vs RHE. This potential is independent of the amount of hydrogen absorbed over a wide range. This property has been utilized in the construction of a palladium/hydrogen reference electrode. It is similar to the dynamic hydrogen electrode in that it is an indicator-type reference electrode, reversible to H_3O^+ ions in solution, although its potential differs from that of a reversible hydrogen electrode in the same solution.

2.2.2 The Cell, Solutions, Electrodes and Electrical Setup

The cell shown in Fig. 2.5 will be used in these experiments. Solutions will be 0.5 M H_2SO_4 and 1 M NaOH, both prepared from analytical-grade reagents and triply distilled water. A commercial calomel reference and a platinum-wire counter electrode will be used. The palladium working electrode will be constructed from a thin wire (about 0.2 mn diameter and 1-2 cm length) sealed in glass.

2.2 PALLADIUM/HYDROGEN ELECTRODE

The galvanostatic circuit shown in Fig. 2.4 will be used.

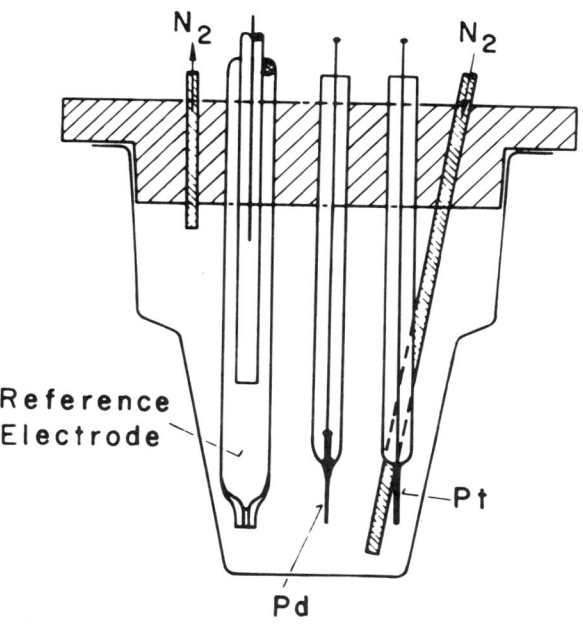

FIG. 2.5   CELL FOR PREPARATION AND TESTING OF THE PALLADIUM/HYDROGEN ELECTRODE

2.2.3 Charging Curves for Palladium

Add 0.5 $\underline{M}$ H_2SO_4 to the cell, introduce the electrodes and connect to the galvanostat and measuring instruments as shown in Fig. 2.4. Set the galvanostat in the "standby" position. Set the sensitivity of the potential scale on the recorder to 2 Volts full scale and the chart speed to 0.5 cm/min and follow the change of potential with time during the pretreatment and preparation of the Pd/H_2 electrode which are outlined below.

Pretreat the electrode by alternately passing through it

anodic and cathodic currents of about 100 mA/cm^2 until the evolution of oxygen and hydrogen, respectively, is observed. Do this several times. At the end of the final cycle, when hydrogen just starts to bubble rapidly from the palladium (the wire at this point is filled with hydrogen) stop the current and note the total cathodic charge passed in the last cycle. Leave on open circuit for 5 minutes and note the potential. Reverse the current and polarize anodically at the same current density for 20 sec. Turn off the current and follow the potential until it reaches a steady-state value. Repeat for 8-10 similar anodic polarization periods.

2.2.4 Preparation of the Palladium/Hydrogen Electrode

The electrode is prepared by pretreating the palladium wire as in the previous experiment. At the end of the last cycle, when hydrogen is bubbling rapidly from the wire, reverse the current until about one-quarter of the charge measured in the last cathodic cycle has been removed. Rinse in distilled water. The electrode is now ready for use.

2.2.5 Stability of the Palladium/Hydrogen Electrode

Effect of time

Follow the potential of the palladium/hydrogen electrode just prepared for 10 minutes, in both acid and alkaline solution.

Effect of mechanical motion

Check the effect of the following modes of mechanical motion on the potential of the Pd/H_2 electrode in acid solution:

2.2 PALLADIUM/HYDROGEN ELECTRODE

    a) Move the electrode back and forth in the solution.

    b) Stir the solution with a magnetic stirrer.

    c) Bubble nitrogen but do not allow the bubbles to sweep the electrode itself.

Effect of oxygen

Use a Pd/H_2 electrode in 0.5 $\underline{M}$ H_2SO_4. Bubble oxygen or compressed air through the solution for several minutes. Stop the bubbling and follow the potential change. Note the steady-state potential.

Effect of organic materials

Introduce the Pd/H_2 electrode in a solution which is 10^{-5} $\underline{M}$ in benzene and 0.5 $\underline{M}$ in H_2SO_4. Measure the potential. Increase the benzene concentration to 10^{-4} $\underline{M}$ and measure again.

Long-term stability

Prepare a Pd/H_2 electrode and measure the potential with a digital voltmeter for several hours.

2.2.6 Treatment of Results

Discuss the shape of the potential/time curves during the charging and discharging processes.

Tabulate the potential of the Pd/H_2 electrode at different quantities of anodic charge passed (i.e. at different atomic ratios of palladium to hydrogen) beginning from the fully charged electrode, and discuss the observed dependence.

Discuss the effect of mechanical motion, presence of

oxygen and presence of benzene in the solution.

Use the Einstein-Smoluchowski relationship ($x = (2Dt)^{1/2}$) to estimate the depth of penetration of hydrogen into the palladium wire during the last stage of cathodic pretreatment.

2.3 The Ag/AgClO$_4$ Electrode in Propylene Carbonate

2.3.1 General

Propylene carbonate is a polar nonaqueous solvent which has attracted increasing interest in recent years.[10] It is a stable, non-toxic compound which can be easily purified. Because of its high dielectric constant, it can serve as a good solvent for different inorganic salts. Several reference electrodes have been developed for use in propylene carbonate. The Li/LiClO$_4$ system[11] and the Tl(Hg)/TlCl/LiCl system[12] have been employed successfully, but the system Ag/AgCl/LiCl was found unsuitable because of the enhanced solubility of AgCl in propylene carbonate in the presence of chloride ions.[13]

In this experiment the properties of the Ag/AgClO$_4$ reference electrode in propylene carbonate will be tested. This system was studied by Kirowa-Eisner and Gileadi[14] and was found to have suitable properties as a reference electrode. The experiment will largely follow the work of these authors.

2.3.2 The Cell, Solutions and Electrical Setup

A cell such as that shown in Fig. 2.5, well sealed and of small (5-10 ml.) volume will be used in this experiment. The

2.3 Ag/AgClO$_4$ ELECTRODE IN PROPYLENE CARBONATE

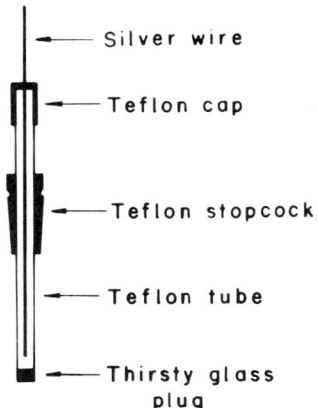

FIG. 2.6   THE Ag/AgClO$_4$ REFERENCE ELECTRODE IN PROPYLENE CARBONATE

galvanostatic circuit shown in Fig. 2.4 will be used with only the strip-chart recorder for measuring the potential. The impedance-matching unit is very important in this system (and in most other nonaqueous systems) and its input impedance should be 10^9 Ω at least.

High purity of the solvent is essential for satisfactory operation of this reference electrode. This is easily achieved for propylene carbonate by distilling it at a reduced pressure of 1 mm Hg in an adiabatic fractional-distillation column.[15] Reagent-grade LiClO$_4$ and AgClO$_4$ should be dried overnight in a vacuum oven at a pressure of 1 mm Hg at 100°C and 50°C respectively. A 0.5 $\underline{M}$ AgClO$_4$, 1 $\underline{M}$ LiClO$_4$ stock solution should be prepared before the experiment from the purified solvent and the dried salts.

2.3.3 Construction of the Reference Electrode

The structure of the reference electrode is shown in Fig. 2.6. A Teflon tube is fitted at the top with a Teflon cup, and a silver wire is inserted through the cup. At the bottom, the tube is plugged with a "thirsty glass" rod (Corning type 7930 glass). This arrangement provides for effective separation between the solution inside the tube and the solution in the rest of the cell while at the same time giving rise to a relatively low electrical resistance (in the range of 5-10 kΩ for a rod thickness of 3-5 mm).

When a new electrode is constructed, the "thirsty glass" plug must be equilibrated with the solution for about one hour before use. It is not necessary to protect the reference electrode from light. This construction provides a "fixed-potential" reference electrode, in contrast to the "indicator-type" reference electrodes used in Experiments 2.1 and 2.2. The potential of the electrode depends on the concentration of $AgClO_4$ inside the teflon tube. In this experiment, the solution inside the reference electrode will be 1 $\underline{M}$ $LiClO_4$, 0.02 $\underline{M}$ $AgClO_4$. The solution in the cell will be 1 $\underline{M}$ $LiClO_4$ with $AgClO_4$ at concentrations varying between 1 m$\underline{M}$ and 0.1 $\underline{M}$. In this way, the liquid-junction potential is small and its variation with the concentration of Ag^+ in solution is negligible.

For polarization measurements, a silver wire wrapped tightly with Teflon tape and inserted directly into the cell will be used as the working electrode. A platinum wire sealed through glass will serve as the counter electrode.

All silver wires employed for the construction of

2.3 Ag/AgClO$_4$ ELECTRODE IN PROPYLENE CARBONATE

electrodes should be abraded with fine emery paper, degreased in acetone, rinsed thoroughly with triply distilled water and dried.

2.3.4 Reproducibility and Stability of the Potential

Prepare three to five reference electrodes and place them in a cell containing a 1 $\underline{M}$ LiClO$_4$ solution in propylene carbonate. Measure the potential between all pairs of electrodes 10 minutes after they have been introduced into the cell. Repeat these measurements about every half-hour during the course of the experiment.

2.3.5 Thermodynamic Behavior of Ag/AgClO$_4$ in Propylene Carbonate

Measure into the cell 5.00 ml. of 1 $\underline{M}$ LiClO$_4$ solution and introduce the reference electrode and two silver wires to serve as two separate indicator electrodes. Add the required amount of the AgClO$_4$/LiClO$_4$ stock solution to reach a concentration of 1 m$\underline{M}$ Ag$^+$ and measure the potential of each of the indicator electrodes with respect to the reference. Increase the concentration of Ag$^+$ ions by adding exactly measured amounts of the stock solution in 5 to 10 steps, up to a maximum concentration of 0.1 $\underline{M}$ AgClO$_4$. Measure the potential of each of the two indicator electrodes with respect to the reference in each solution.

Estimate the time required for the electrode to reach

constant potentials.*

2.3.6 Polarization Curves of $AgClO_4$ in Propylene Carbonate near Equilibrium

Add to the cell a solution which is 0.02 $\underline{M}$ in $AgClO_4$ and 1 $\underline{M}$ in $LiClO_4$. Introduce the reference electrode ($Ag/AgClO_4$ (0.02 $\underline{M}$), $LiClO_4$ (1 $\underline{M}$)), a platinum counter electrode and a silver-wire working electrode into the cell and measure the potential of the working electrode with respect to the reference. This is the reversible potential in the system studied. Polarize the electrode galvanostatically at current densities ranging from 0.02 to 0.2 $\mu A/cm^2$ in both the anodic and cathodic directions and observe the steady-state overpotential in each case.** Repeat several times in order of increasing and decreasing current density to test the reproducibility of your results.

2.3.7 Treatment of Results

Plot the potential of the indicator electrode as a function of log C_{Ag^+} and calculate the slope. For each point mark the approximate time required to reach steady-state and the difference

---

* Note that at low concentrations of Ag^+, the potential difference between the two indicator electrodes may be a few millivolts and the response time will be of the order of several minutes.

** Some commercial galvanostats may not be suitable here because of the very low currents required. They may be replaced by a fresh flashlight battery with a set of resistors of suitable high values to control the current.

2.3 Ag/AgClO$_4$ ELECTRODE IN PROPYLENE CARBONATE 231

in potential between the two indicator electrodes. Discuss the causes of deviation, if any, from the expected Nernst slope.

Plot the current density against overpotential and calculate the exchange-current density for the Ag/Ag$^+$ couple in this system.

Choose one of the electrodes used in the reproducibility test and calculate the measured potentials of all the other electrodes with respect to it. Plot these potentials against time.

REFERENCES

1. D. J. G. Ives and G. J. Janz, "Reference Electrodes", Academic Press, New York, 1961.
2. S. Barnatt, J. Electrochem. Soc. **99**, 549 (1952).
3. S. Barnatt, J. Electrochem. Soc. **108**, 102 (1961).
4. J. Giner, J. Electrochem. Soc. **111**, 376 (1964).
5. T. Graham, Phil. Trans. Roy. Soc. 156, 415 (1866).
6. F. A. Lewis, "The Palladium/Hydrogen System", Academic Press, London, 1967.
7. Technical Bulletin VII of Engelhard Industries Inc. (1966).
8. M. Fleischmann and J. N. Hiddleston, J. Sci. Instrum. **1** (2) 667 (1968).
9. R. V. Bucur and L. Stoicoviciu, J. Electroanal. Chem. **11**, 152 (1966).
10. R. F. Nelson and R. N. Adams, J. Electroanal. Chem. **13**, 184 (1967).
11. B. Burrows and R. Jasinski, J. Electrochem. Soc. **115**, 365 (1968).
12. F. G. K. Baucke and C. W. Tobias, J. Electrochem. Soc. **116**, 34 (1969).
13. J. N. Butler, D. R. Cogley and W. Zurosky, J. Electrochem. Soc. **115**, 445 (1968).
14. E. Kirowa-Eisner and E. Gileadi, J. Electroanal. Chem. **25**, 481 (1970).
15. R. Jasinski and S. Kirkland, Anal. Chem. **39**, 1663 (1967).

3. DOUBLE-LAYER MEASUREMENTS

Introduction

DIFFERENTIAL-CAPACITY MEASUREMENTS AND THE DETERMINATION OF THE POTENTIAL OF ZERO CHARGE

General

3.1 Systems without Specific Adsorption

3.1.1 Measurement of Differential Capacity in NaF Solutions
3.1.2 Determination of the Potential of Zero Charge in NaF Solutions with a Streaming-Mercury Electrode
3.1.3 Treatment of Results

3.2 Systems with Specific Adsorption

3.2.1 General
3.2.2 Measurements of E_z - The Esin-Markov Effect
3.2.3 Effect of Addition of Foreign Ions on E_z
3.2.4 Effect of Addition of Organic Compounds
3.2.5 Double-Layer Capacitance in the Presence of Neutral Organic Molecules
3.2.6 Treatment of Results

ELECTROCAPILLARY MEASUREMENTS

General

3.3 Determination of the Relative Surface Excess

3.3.1 The Cell, Electrodes, Solutions and Electrical Setup
3.3.2 Electrocapillary Measurements in KBr Solutions
3.3.3 Treatment of Results

3.4 The Dependence of Charge q_M on Potential in the Presence of Strong Specific Adsorption

3.4.1 Experimental Setup and Procedures
3.4.2 Determination of the Electrocapillary Curve
3.4.3 Treatment of Results

DOUBLE-LAYER MEASUREMENTS

Introduction

The structure of the ionic double layer may be investigated directly by measuring the interfacial tension or the differential capacity, or indirectly by its effect on electrode processes (cf. Experiment 4). The bulk of experiments in this field have been performed on mercury electrodes (and some dilute amalgams), since in this case both the interfacial tension γ and the differential capacity C_{dl} can be measured with high accuracy. These two quantities are related through the equation

$$\gamma - \gamma_{max} = -\iint_{E_z}^{E} C_{dl}\, dE\, dE \qquad [3.1]$$

This equation is derived from the Gibbs adsorption isotherm and the equation of electrocapillarity, and is based on purely thermodynamic considerations, as applied to an ideally polarized electrode. The quantity E_z is the potential of zero charge and it coincides with the potential of the electrocapillary maximum, i.e. the potential at which γ reaches its maximum value.

The relationship between γ and C_{dl} is given here in its integral form rather than the more commonly used differential form

$$C_{dl} = -(\partial^2 \gamma / \partial E^2)_\mu \qquad [3.2]$$

because equation [3.1] can be, and is, often used to evaluate surface-tension data from capacitance measurements, while the

INTRODUCTION

reverse process of double differentiation yields less accurate values of C_{dl}, because of the errors inherent in the differentiation procedure.

The relative surface excess Γ' is related to the surface tension by an equation of the form

$$(\partial \gamma / \partial \mu_i)_{\mu_j, E} = -\Gamma'_i \qquad [3.3]$$

where all surface excesses are calculated with respect to the solvent, in other words, μ_j refers to all ionic or molecular species in the system except the solvent. Thus adsorption data can be derived from surface-tension or differential-capacity measurements. It is well to remember, though, that Γ' is an integral quantity and thus gives no information on the distribution of adsorbed species in the interphase, as a function of, say, the distance from the electrode. Such information cannot, in principle, be derived without use of a specific model or some other non-thermodynamic assumption. A pictorial representation of the various integral and differential relationships between γ, q, Γ' and C is shown in Fig. 3.1.

The literature covering double-layer thermodynamics and double-layer structure is very wide and an account of current theories will be found in almost any text dealing with interfacial electrochemistry. A list of readily available review articles is given[1-8] which may serve as a source for further references. The theoretical background for the experiments presented below may be found in concise form in Delahay's monograph[9].

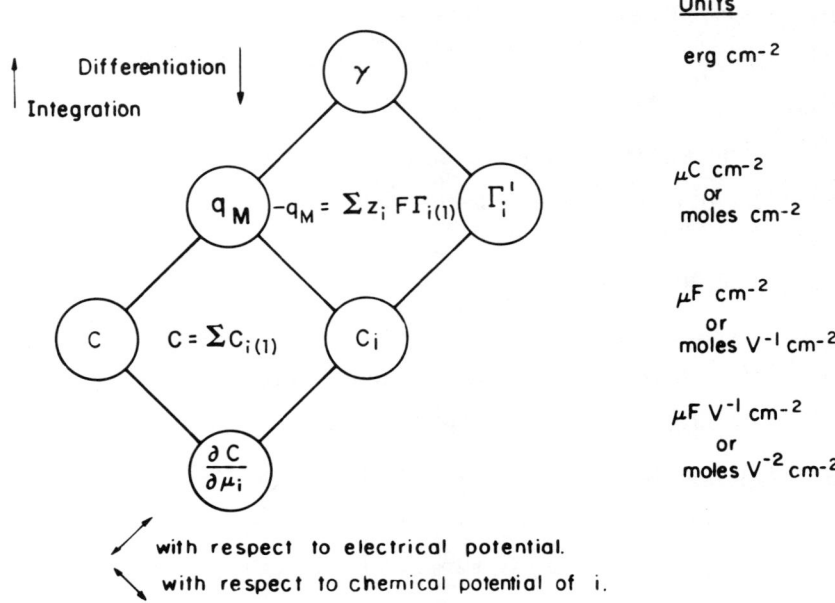

FIG. 3.1   FUNDAMENTAL RELATIONS IN ELECTROCAPILLARY THERMODYNAMICS

Ionic-double-layer studies are particularly sensitive to traces of impurities in the system. Workers in the field usually spend more time cleaning the system than performing the experiment. Although a similar degree of purity cannot be achieved in the time available for the following experiments, the effort should be made to approach high-purity conditions as far as possible, in order to obtain results comparable to those reported in the

INTRODUCTION

literature.*

DIFFERENTIAL-CAPACITY MEASUREMENTS AND DETERMINATION OF THE POTENTIAL OF ZERO CHARGE

General

The analysis of differential double-layer capacities is relatively simple in the absence of specific adsorption. The Stern model may be used (cf. section II.1.3), according to which the measured capacity C is a series combination of the capacity C_{M-2} of the Helmholtz layer and the capacity C_{2-S} of the Gouy-Chapman layer

$$1/C = 1/C_{M-2} + 1/C_{2-S} \qquad [3.4]$$

The validity of the Gouy-Chapman theory can then be tested in the following way:

(i) The dependence of charge q_M on the potential*** $\bar{E}$ is obtained by integrating the experimental $C/\bar{E}$ plot.

---

* A sensitive test for the cleanliness of the cell is to rinse it with triply distilled water. When the conductivity of the water is the same before and after rinsing, the cell is probably clean. Very efficient cleaning is attained by holding the cell in 1:1 HNO_3 in an ultrasonic cleaner for 10 minutes. If more drastic cleaning is required, the cell may be filled with a 5% solution of HF for a few minutes, rinsed with distilled water and then cleaned with HNO_3 as above.

*** $\bar{E}$ is the rational potential, i.e. the potential measured with respect to the potential of zero charge.

(ii) The potential ϕ_2 of the outer Helmholtz plane is calculated as a function of $\overline{E}$ from diffuse-double-layer theory, with the equation*

$$q_M = 2A \sinh(|z| F\phi_2 / 2RT) \qquad [3.5]$$

in which

$$A = (RT\varepsilon C^o / 2\pi)^{1/2} \qquad [3.6]$$

where C^o is the electrolyte concentration in the bulk of the solution and ε is the dielectric constant, taken equal to its value in the bulk.

(iii) The diffuse-double-layer capacitance C_{2-S} is calculated as a function of ϕ_2 (and hence of $\overline{E}$ or q_M) from the equation

$$C_{2-S} = \frac{|z| FA}{RT} \cosh(|z| F\phi_2 / 2RT) \qquad [3.7]$$

(iv) The capacity C_{M-2} of the Helmholtz layer is calculated with the use of equation [3.4] from the experimental value of C and the value of C_{2-S} computed above.

(v) In the absence of specific adsorption, C_{M-2} is independent of the solution concentration and varies only with q_M. The graph of C_{M-2} <u>vs</u> q_M obtained in one solution can hence be used at all other concentrations of the same electrolyte, to compare the measured

---

* This equation is only correct for a symmetrical electrolyte. The calculation, however, can be performed without difficulty for unsymmetrical electrolytes with the use of the more general relationship between q_M and ϕ_2. (cf. page 35 of ref. 9.)

INTRODUCTION

capacitance C with that obtained from equation [3.4] in which C_{2-S} is evaluated by the use of diffuse-double-layer theory.

The above procedure was followed by Grahame[10] who employed NaF as the electrolyte. The variation of C_{M-2} with q_M was calculated from measurements in the most concentrated solution (ca 1 M), since in this case C_{2-S} is very large (cf. equation [3.7]) and its contribution to the measured capacity is small.

A more sensitive method, based on equation [3.4], was proposed by Parsons[11] to verify the diffuse-double-layer theory and detect specific adsorption. According to Parsons' method, C_{2-S} is calculated as a function of q_M as outlined above for a series of solutions of varying concentration. The reciprocal of the capacity $(1/C)q_M$ is then plotted as a function of $(1/C_{2-S})q_M$, both taken at a constant value of the charge. In the absence of specific adsorption this should give a straight line of unit slope, with $1/C_{M-2}$ as the intercept.

Another distinct criterion for specific adsorption is the shift of the potential of zero charge with concentration of electrolyte, known as the Esin-Markov effect. Following the treatment given by Parsons[12,13] we can modify the electrocapillary equation

$$-d\gamma = q_M dE_\pm + \Gamma_\mp d\mu \qquad [3.8]$$

by subtracting $d(q_M E_\pm)$ from both sides, to obtain

$$-d\gamma' = \Gamma_\mp d\mu - E_\pm dq_M \qquad [3.9]$$

where $d\gamma'$, defined by

$$d\gamma' \equiv d\gamma + d(q_M E_\pm) \qquad [3.10]$$

is a complete differential. Cross-differentiating equation [3.9] we have

$$-(\partial \Gamma_\mp / \partial q_M)_\mu = (\partial E_\pm / \partial \mu)_{q_M} \qquad [3.11]$$

which can also be written in the form

$$(\partial E_\pm / \partial \ln a)_{q_M=0} = -\frac{RT}{z_\mp F}(\partial q_\mp / \partial q_M)_\mu \qquad [3.12]$$

where the surface excess has been expressed in units of charge

$$q_\mp = z_\mp F \Gamma_\mp \qquad [3.13]$$

The potential E with respect to a fixed reference electrode may be written as

$$E = E_\pm \pm \frac{RT}{|z|F} \ln a_\pm + \text{const.}$$

$$= E_\pm \pm \frac{RT}{2|z|F} \ln a + \text{const.} \qquad [3.13]$$

since $a_\pm = a^{1/2}$, thus

$$\left(\frac{\partial E}{\partial \ln a_\pm^2}\right)_{q_M=0} = \pm \frac{RT}{|z|F}\left[\left(\frac{\partial q_\mp}{\partial q_M}\right)_\mu + \frac{1}{2}\right] \qquad [3.14]$$

In the absence of specific adsorption

3.1 SYSTEMS WITHOUT SPECIFIC ADSORPTION

$$q_\pm = q_\pm^{2-S} = \pm A[\exp(z_\pm F\phi_2/2RT) - 1] \qquad [3.15]$$

hence at $q_M = 0$ where $\phi_2 = 0$, one has, by differentiating equations [3.5] and [3.15],

$$(\partial q_\mp / \partial q_M) = \mp 1/2 \qquad [3.16]$$

and the potential of zero charge is independent of the activity of the electrolyte.

In general, specific adsorption of anions causes a shift of the potential of zero charge (E_z) in the cathodic direction and <u>vice versa</u> in the case of cations. When both anions and cations are specifically adsorbed (e.g. for $(CH_3)_4NI$) the effects of the two ions may compensate for each other and the shift of E_z can be rather small. Specific adsorption in such cases is easily detected from the strong suppression of the whole electrocapillary curve and from major changes in the C/E curve.

Neutral organic species are adsorbed over a limited range of potential, near the potential of zero charge. The capacitance in the region of adsorption has a low, roughly constant value, bounded by two capacitance peaks (known as "adsorption-desorption" peaks) beyond which no significant adsorption occurs (cf. Fig. II 12b).

3.1   Systems Without Specific Adsorption

3.1.1 Measurement of Differential Capacitance in NaF Solutions
       <u>The capacitance meter</u>

The capacitance meter[*] to be used in this experiment has been developed recently[14,15]. It is based on the technique of applying a small triangular voltage wave to the double layer and detecting the square current wave which results if the interphase acts as a pure capacitor. The amplitude of the square-wave current pulse is proportional to the capacitance, since

$$i = C(dE/dt) \qquad [3.17]$$

The square wave is rectified in a suitable manner and a recorder output signal results, which can be calibrated in terms of known capacitors. In the advanced version of this device[15], it is possible to compensate automatically for the solution resistance, which acts in series with the double-layer capacitance in the equivalent circuit representation. Correction for the solution resistance is achieved by a positive-feedback system similar to that employed in a potentiostat with IR correction (cf. Fig. 1.22). The signal observed on an oscilloscope (before rectification) is shown in Fig. 1.24 for four cases: (a) pure capacitor or capacitor with series resistance after compensation, (b) capacitor with uncompensated series resistance, (c) capacitor with parallel resistance, (d) capacitor with both series and parallel resistances. The last two cases occur when the interphase departs from ideal polarizability, i.e. when a faradaic current flows. It is evident from Fig. 1.24c that in this case erroneous readings of the capacity will result.

---

[*] If such an instrument is not available, this experiment can be performed with any other device available for the measurement of double-layer capacitance.

3.1 SYSTEMS WITHOUT SPECIFIC ADSORPTION

Calibration of the capacitance meter

In order to begin familiarizing yourself with the capacitance meter, repeat sections 1.3.3 and 1.3.4 of experiment 1 which deal with the potentiostat with positive feedback.

To calibrate the instrument for measurement of a capacitance of about 0.1 μF, first determine the zero by shorting the inputs of the recorder. Then set the instrument to the internal dummy cell with a standard capacitor of 0.1 μF, observe the square wave on the oscilloscope and set the series-resistance compensation to obtain the best square wave.* Set the controls of the capacitance meter and the recorder to obtain highest sensitivity (80-90% of full-scale deflection). Now connect the dummy cell with the capacitor to be measured as shown in Fig. 1.23, with $R_{sol} \equiv R_{AB} = 0$ and R_F disconnected. Readjust the resistance compensation to obtain again the best square-wave form, without changing the sensitivity of either instrument. Read the capacitance on the recorder. Add a series resistance (R_{AB} in Fig. 1.23) of about 100 Ω, readjust the resistance compensation and measure the capacitance. Repeat for series resistances having the approximate values 0.5, 2.0, 5.0, 10 and 20 kΩ and note the error, if any, in the measured capacitance due to incomplete compensation of the series resistance.

Double-layer capacitance in NaF solutions

The cell shown in Fig. 3.2 will be used in this experiment.

---

* Overcompensation at this point will drive the circuit into oscillations.

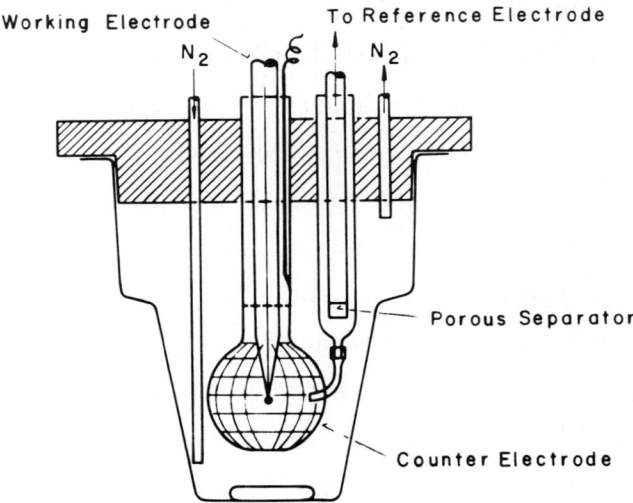

FIG. 3.2 CELL FOR MEASUREMENTS WITH A SMALL SPHERICAL ELECTRODE

Solutions which are 0.01$\underline{M}$, 0.1$\underline{M}$ and 0.9$\underline{M}$ in NaF should be prepared from highest purity NaF and triply distilled water. Purified nitrogen should be bubbled through the solution for 10 min. before each experiment, and over the solution during the experiment. Very careful cleaning of all parts of the system is required in this experiment, and in particular, traces of organic matter must be excluded.

Add 0.9$\underline{M}$ NaF to the cell and bubble nitrogen for 10 min. before introducing the electrodes into the cell.* Introduce the

---

* It is preferable to perform measurements of the potential of zero charge (by the streaming-electrode method described below) in the same solution before measurement of the capacitance is begun.

3.1 SYSTEMS WITHOUT SPECIFIC ADSORPTION

dropping-mercury working electrode, the calomel reference electrode and a platinum-gauze counter electrode into the deaerated solution and continue passing nitrogen over the solution. (Contact between mercury and a solution which contains dissolved oxygen may cause the formation of traces of mercuric salts which interfere with subsequent measurements.) The counter electrode should be positioned symmetrically with respect to the working electrode as far as possible.

Calibrate the measuring circuit (meter and recorder) with the accurate internal capacitors. Connect the electrochemical cell and set the potential to a value in the range of zero to -1.8 volt vs SCE. Observe the square wave on the oscilloscope and adjust for best compensation of the series resistance at an accurately determined time near the end of the life of the drop.*
Set the sensitivity of the system to give about 50-90% of full-scale deflection at the end of the drop-life. Adjust the recorder speed to 0.5-1.0 cm/sec.

The shape of the trace obtained on the recorder is shown schematically in Fig. 3.3. Measurement should be made after a fixed time τ which is about 80% of the drop-time, since the rate of change of surface area is proportional to $t^{-1/3}$ and the error in measuring C decreases with increasing t.

Measure the capacitance at intervals of 50 mV between

---

* As the drop grows, its capacitance increases while the resistance becomes smaller. Thus, resistance compensation should be adjusted for a particular instant in the life of the drop at which the capacity is measured.

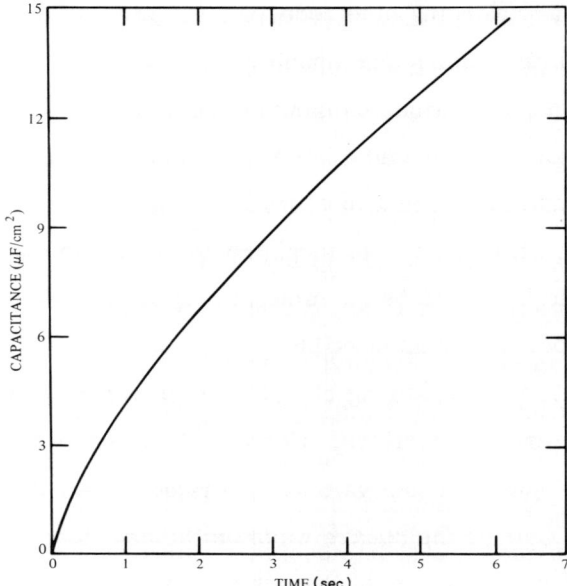

FIG. 3.3 THE VARIATION OF CAPACITANCE WITH TIME DURING GROWTH OF A MERCURY DROP

zero and -1.8 volt vs SCE. Repeat in solutions of 0.1M and 0.01M NaF. In the latter, measure at closer intervals near the minimum in the curves. Remember to deaerate each solution before the electrodes are introduced into it.

The area of the mercury drop

The area A of a drop at time t is related to the rate of flow m of mercury by

$$A = 4\pi \left(\frac{3}{4\pi\rho}\right)^{2/3} m^{2/3} t^{2/3} \qquad [3.18]$$

in which ρ = 13.53 g/cm^3 at 25°C, is the density of mercury.

To determine A at the time of measurement τ of the capacity, place a 5 ml beaker into a 250 ml beaker and introduce

3.1 SYSTEMS WITHOUT SPECIFIC ADSORPTION 247

into the beakers about 100 ml of 0.01 M NaF solution. Allow the mercury to drop for a while into the larger beaker, then move the small beaker into place under the capillary with the aid of a bent glass rod and collect the mercury for 3-5 min. Measure this time accurately with a stopwatch. Transfer the mercury from the beaker to a filter paper folded into a cone and rinse with distilled water. Transfer to another similar filter paper and rinse with acetone. Transfer to a dry beaker and weigh accurately. Calculate m and from it the surface area of the drop at the time τ.

3.1.2 Determination of the Potential of Zero Charge of NaF Solutions With a Streaming-Mercury Electrode

The experimental method

The streaming-mercury electrode provides one of the simplest and fastest experimental methods for the determination of the potential of zero charge. Its operation is based on the idea that when mercury is allowed to stream from an electrode which is not connected to any source of potential, the excess charge on the surface of the metal is carried away and the electrode is left at its potential of zero charge.

The capillary to be used for the streaming electrode is similar to that employed for the classical dropping-mercury electrode (cf. Fig. 3.4). This can be prepared by heating and drawing a clean capillary of 1-2 mm inner diameter.* The

---

* Best results are obtained if the capillary is drawn out in two stages. First it is made narrow by heating and pulling slowly. The thin part is then heated again and pulled rapidly to form the capillary.

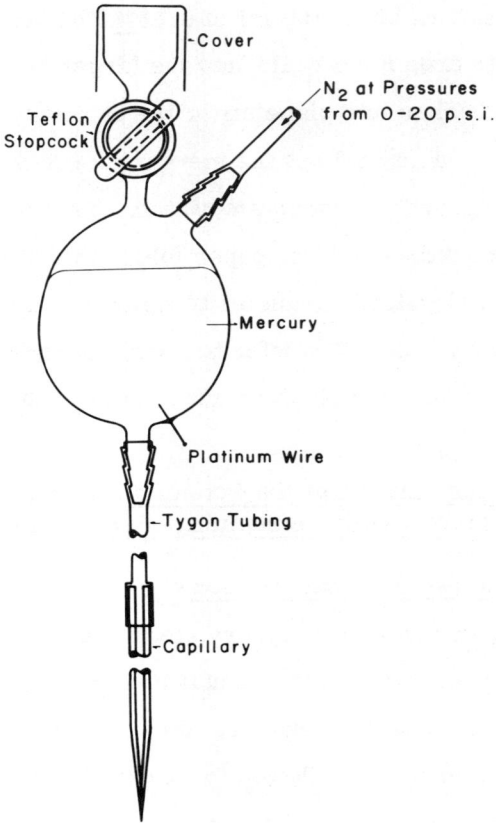

FIG. 3.4 STREAMING-MERCURY ELECTRODE ASSEMBLY

drawn-out capillary is broken about 3 cm from its wider part and examined under a microscope to see if it was not sealed accidentally during heating and drawing. It is then connected to the mercury reservoir with tygon tubing and secured with a metal wire. The reservoir is filled with mercury and raised to a height of about 30 cm while the capillary is immersed in distilled water. The capillary must be shortened (or the reservoir raised) if the

3.1 SYSTEMS WITHOUT SPECIFIC ADSORPTION

drop-time exceeds 15 seconds and it must be replaced if the drop-time is less than 3-5 seconds.

Measurement with the streaming electrode

Add about 100 ml of 0.01$\underline{M}$ NaF into the clean cell and deaerate by bubbling purified nitrogen for 10 min. Rinse the clean capillary and reference electrodes with a small amount of the deaerated solution in the cell (by opening the stopcock at the bottom of the cell) and then introduce the electrodes into the cell while mercury is not flowing from the capillary. Bubble nitrogen for an additional two minutes, then start the flow of mercury from the capillary. Measure the potential with respect to a suitable reference (a commercial calomel electrode may be used here) with a high-input-impedance voltmeter (at least 10^9 Ω). The potential will be seen to change during the life of the drop. Carefully* apply nitrogen pressure on the mercury reservoir (cf. Fig. 3.4) and measure the change of potential with pressure as the mercury now flows continuously from the capillary. Beyond a certain flow rate the potential becomes constant, independent of further increase in pressure. This is the potential of zero charge,** E_z, for the system studied. Release the pressure carefully, then apply it again and measure the potential. Repeat

---

* Do not exceed a pressure of 20 psi. The mercury reservoir should be built to withstand this pressure and should be protected against shattering.

** A correction for liquid-junction potential should be applied to obtain the accurate value of E_z.

several times to test the reproducibility of the measurement.

Repeat in solutions of 0.1M and 0.9M NaF.

3.1.3 Treatment of Results

Tabulate and plot the $C/\bar{E}$ data and compare with literature data, for the three solutions studied.

Evaluate q_M, ϕ_2 and C_{2-S} from the experimental data following the procedure outlined in the "General" section above. Tabulate these quantities together with C and $\bar{E}$. Plot q_M and ϕ_2 against $\bar{E}$ and plot C_{2-S} against q_M.

Calculate C_{M-2} in 0.9M NaF as a function of q_M, using equation [3.4]. Plot C_{M-2} and C for this solution as a function of q_M on the same graph.

Use the values of C_{M-2} obtained for 0.9M NaF and the calculated values of C_{2-S} to compute the C/q_M curves for 0.1M and 0.01M NaF. Compare with experiment.

Plot $(1/C)_{q_M}$ against $(1/C_{2-S})_{q_M}$ and evaluate $(1/C_{M-2})_{q_M}$.
Discuss the shape of the $\phi_2/\bar{E}$ plots in the three solutions studied with respect to their effect on the elucidation of the mechanism of electrode reactions.

Plot ϕ_2 against $\log C_{NaF}$ at three potentials: 0, −0.50, −1.40 volt vs SCE. Discuss the shape of these lines.

Plot the capacity against time during the growth of the drop on a log-log scale and determine the slope.

Correct the values obtained for the potential of zero charge for the liquid-junction potential according to the equation of Henderson[16] and compare with results reported in the literature.

3.2 Systems with Specific Adsorption

3.2.1 General

The two experimental techniques used in Experiment 3.1 will be employed here to determine E_z and C in solutions of electrolytes in which specific adsorption is known to occur. All experimental techniques are identical to those employed in Experiment 3.1.

Particular care must be taken in cleaning all parts of the system when the solutions are changed.

3.2.2 Measurement of E_z - The Esin-Markov Effect

Prepare solutions of KBr of the following concentrations: 0.001, 0.003, 0.01, 0.03, 0.1, 0.3, 1.0$\underline{M}$. Determine the potential of zero charge in each solution using the streaming-mercury electrode (cf. Experiment 3.1).

3.2.3 Effect of Addition of Foreign Ions on E_z

Prepare a solution of 0.1$\underline{M}$ KCl and determine E_z. Add, in turn, the required amounts of 1.0$\underline{M}$ KCNS* to reach final concentrations of 10^{-4}, 10^{-3} and $10^{-2}\underline{M}$ in this salt. Measure the potential of zero charge in each solution.

Repeat this experiment using tetrabutyl ammonium perchlorate (TBAP) as the additive. The stock solution of this salt will be $10^{-3}\underline{M}$ (and 0.1$\underline{M}$ with respect to KCl, as above) and the concentrations of TBAP in the solutions tested will be 1×10^{-6}.

---

* The stock solution of KCNS should also be 0.1$\underline{M}$ in KCl so that the concentration of KCl will be the same in all four solutions.

1×10^{-5}, 6×10^{-5} and 1×10^{-4} M. Measure E_z in each solution.

3.2.4 Effect of Addition of Organic Compounds

Prepare two solutions of 0.1M KCl, one of which is also 1 mM in benzene. From these, prepare three solutions having benzene concentrations of 1×10^{-5}, 1×10^{-4} and 1×10^{-3} M. Measure the potential of zero charge in each solution.

3.2.5 Double-Layer Capacitance in the Presence of Neutral Organic Molecules

Prepare a solution of 0.1M KCl and measure the capacitance as a function of the rational potential $\overline{E}$ at 50 mV intervals, as in Experiment 3.1.

Prepare a solution which is 10^{-3} M in n-amyl alcohol and 0.1M in KCl. Measure the potential of zero charge with the streaming electrode, then measure the capacitance in the same solution at 50mV intervals from -0.20 to -1.20 volt vs SCE. Plot your data while measurements proceed and take more closely spaced measurements in regions where the capacitance varies rapidly with potential.

Repeat in a solution of 10^{-2} M amyl alcohol, 0.1M KCl.

3.2.6 Treatment of Results

Plot the values of E_z for the KBr solutions against the logarithm of the concentration. Evaluate the differential parameter $(\partial q_-/\partial q_M)$ appearing in equation [3.14]. Note the concentration above which specific adsorption becomes significant at E_z.

On a single sheet of graph paper plot E_z against $\log C_{KBr}$, $\log C_{KCNS}$ and $\log C_{TBAP}$.

3.2 SYSTEMS WITH SPECIFIC ADSORPTION

Discuss the relative adsorption of the various ions in the system studied.

Compare graphically and discuss the effect of benzene on the potential of zero charge.

Discuss the limitation of the streaming electrode for measurements of the potential of zero charge in very dilute solutions.

Discuss the shape of the C/E curves in the presence of n-amyl alcohol. What is the origin of the capacitance peaks observed? How would you expect them to depend on the frequency of the applied signal? Why is the capacitance lowered in the presence of neutral organic molecules? What is the shape (schematically) of the electrocapillary curve corresponding to the C/E curve found for 0.01$\underline{M}$ n-amyl alcohol?

ELECTROCAPILLARY MEASUREMENTS

General

Electrocapillary measurements involve the experimental determination of the metal/solution interfacial tension γ as a function of electrode potential E and the chemical potential μ of all the species in solution. Accurate measurement of interfacial tension is possible only if the metal involved is in the liquid state, as, for example, in the case of mercury, dilute amalgams, gallium and some soft metals in molten-salt electrolytes.

The most accurate measurements of interfacial tension are obtained by the Lippmann electrocapillary electrometer and by the method of maximum bubble pressure. These, however, are time-consuming and a faster, somewhat less accurate method of drop-time measurement will be adopted here.* This method depends on the assumption that a slowly growing drop will be detached from the end of a fine capillary at the moment when the gravitational force Wg acting on the drop equals the force due to interfacial tension $2\pi r\gamma$. Thus the weight W of the drop is proportional to the interfacial tension. If conditions are set so that the flow-rate is constant, then, for a given capillary, the drop-time becomes proportional to the interfacial tension.

It has been pointed out[18] that the way in which the drop becomes detached is rather complex and a correction factor should be applied to the simple equation relating γ to W[19, 20].

$$2\pi r\gamma = Wg\chi(r/V^{1/3}) \qquad [3.19]$$

where $\chi(r/V^{1/3})$ is a known function of the radius of the capillary and the volume of the drop. This, however, is a small correction and the variation of the function $\chi(r/V^{1/3})$ over the whole electrocapillary curve is of the order of 0.5% and may be neglected in the following experiments.**

The electrocapillary equation for a mercury electrode in contact with an aqueous solution of KBr may be written in the form

---

* The maximum-bubble-pressure method has been developed recently by Mohilner[17] into an automatic system with an "on-line" computer which both carries out the experiment and evaluates the experimental data.

** When this method is used in research to obtain electrocapillary data, the correction factor is quite important since γ is then measured with an accuracy of better than 0.05%.

3.3 DETERMINATION OF RELATIVE SURFACE EXCESS

$$-d\gamma = q_M dE_- + \Gamma'_K + d\mu_{KBr} \qquad [3.20]$$

or

$$-d\gamma = q_M dE_+ + \Gamma'_{Br^-} d\mu_{KBr} \qquad [3.21]$$

depending on whether the potential is measured with respect to an electrode reversible to the anion or the cation, respectively. From these equations, the charge density on the metal as well as the relative surface excess of both ions can be determined as a function of potential. The extent of specific adsorption can then be evaluated by application of diffuse-double-layer theory. Thus, one can write

$$F\Gamma'_{Br^-} = q_- = q_-^{M-2} + q_-^{2-S} \qquad [3.22]$$

Assuming no specific adsorption of the cation one can write $F\Gamma'_+ = q_+^{2-S}$. Equation [3.15] is then used twice, first to calculate ϕ_2 from q_+^{2-S} and then to calculate q_-^{2-S} from ϕ_2. The amount of specific adsorption of the anions q_-^{M-2} is then obtained from equation [3.22]. Thus, although the relative surface excess is an integral quantity and gives no information on the distribution of ions in the interphase, it can be combined with calculations based on a model to provide an insight into the concentration profile of ions in this region.

3.3 Determination of the Relative Surface Excess

3.3.1 The Cell, Electrodes, Solutions and Electrical Setup

The same cell as used in Experiment 3.1 (cf. Fig. 3.2)

will be used here. The dropping-mercury electrode may be constructed with a commercial capillary or by drawing out a piece of glass tubing of 1-2 mm inner diameter (cf. Experiment 3.1). The height of the mercury column above the capillary is held constant by means of a side arm and levelling bulb. Care should be taken to eliminate mechanical disturbances as much as possible in order to avoid premature detachment of the drops.

A commercial SCE will be used as reference.* A platinum electrode, separated from the bulk of the solution by means of a fritted disk will be used as the counter electrode.

Prepare solutions of KBr of the following concentrations in triply distilled water**: 0.001, 0.005, 0.01, 0.05, 0.1, 0.5, 1.0 molal.***

The potential of the dropping-mercury electrode can be

---

* The potential on the E_+ and E_- scale (i.e. the potential with respect to an electrode reversible to the cation or anion in solution, respectively) is calculated from the Nernst equation. Alternatively, these potentials may be obtained directly from measurements with a K^+ specific-ion electrode and a Ag/AgBr electrode in each solution.

** Purity of materials, thorough purification of all parts of the apparatus which come into contact with solutions and avoidance of de-ionized water are particularly important in this experiment.

*** The solutions are prepared at rounded values of molality rather than molarity since the molal activity coefficients are well established over the whole range of concentrations[21], thus permitting calculation of the corresponding activities (see appendix).

3.3 DETERMINATION OF RELATIVE SURFACE EXCESS

controlled by a constant-voltage supply consisting of a battery and voltage divider. The charging current during growth of the drop is recorded to simplify counting the drops.

The simple circuit shown in Fig. 4.3 may be used as a low-noise device for following the current.

The chart speed of the recorder should be set so that over a distance of 20 to 30 cm on the chart paper at least 10 drops are recorded.

Calibrate the chart speed employed with a stop watch.

In order to save paper, trace several sets of drops over the same portion of the chart, adjusting the zero in order to distinguish between the different traces.

3.3.2 Electrocapillary Measurements in KBr Solutions

Add to the clean cell a solution of 0.001 m KBr and de-aerate for 10 minutes by bubbling purified nitrogen. Rinse the capillary and reference electrodes with a small amount of this solution and introduce them into the cell. Bubble nitrogen for an additional 5 minutes, then allow the mercury to begin dropping and continue to pass nitrogen over the solution. Adjust the height of the mercury column above the capillary by means of the levelling bulb to produce a drop time of 5-15 seconds. Adjust the potential of the mercury electrode to -0.10 Volt vs SCE and record a series of drops as described in the previous section.

Repeat at 50 mV intervals up to -1.00 Volt vs SCE.[*]

---

[*] The measurements could be extended to -1.5 V, but for considerations of time, this has not been required here.

Check the reproducibility by carrying out two successive measurements at a number of potentials and by repeating some of the measurements after going through the whole voltage range.

Repeat the measurements in the other KBr solutions in order of increasing concentration.

3.3.3 Treatment of Results

Plot the drop time τ as a function of potential for all solutions on a single graph paper.* Determine the coordinates of the maximum in each solution and plot the potential of zero charge as a function of log a_{KBr} (the Esin–Markov effect).

The value of γ_{max} for KBr solutions is given [22] as 426.3, 424.5 and 417.4 dyne/cm for 0.01, 0.1 and 1.0 $\underline{M}$ solutions, respectively. Use these values and the experimental maximum drop-times in the corresponding solutions to calculate the average numerical factor relating τ to γ. Use this factor for all further calculations.

Make plots of γ <u>vs</u> log a_{KBr} at fixed values of E_- and E_+ at about 0.10 Volt intervals, and determine the relative surface excesses of both ions.

Use the values of Γ_+' to calculate q_+^{2-S}, assuming no specific adsorption and from it calculate q_-^{2-S}. Compare with the values of Γ_-' obtained above for 1 $\underline{m}$ KBr solution and determine the extent of specific adsorption of Br^- ions as a function of q_M.

---

* It is important that great care be taken at this point to obtain a smooth curve. The validity of the derivatives obtained depends on the effort expended in drawing this curve.

3.4 The Dependence of Charge q_M on Potential in the Presence of Strong Specific Adsorption

3.4.1 Experimental Setup and Procedures

The experimental setup and all procedures are identical to those employed in the previous experiment, except that different solutions will be used.

3.4.2 Determination of the Electrocapillary Curve

Prepare the following four solutions: 0.1 $\underline{M}$ NaF, 0.1 $\underline{M}$ NaI, 0.1 $\underline{M}$ $[(CH_3)_4N]_2SO_4$ and 0.01 $\underline{M}$ n-amyl alcohol in 0.1 $\underline{M}$ NaF. Determine the electrocapillary curve in all four solutions, at 50 mV intervals from −0.10 Volt to −1.4 Volt $\underline{vs}$ SCE, (except in NaF, where points should be taken every 20 mV).

3.4.3 Treatment of Results

Plot all four electrocapillary curves on the same graph paper. Use the factor calculated in Experiment 3.3 to convert drop-time to interfacial tension (provided that the same capillary and same mercury-column height are used).

Determine the region of potential in which I^- and $(CH_3)_4N^+$ ions and n-amyl alcohol molecules are adsorbed in these solutions.

Determine the surface charge density q_M as a function of potential in each of the solutions studied.

Use the data for NaF to calculate the differential capacity at different potentials. Compare with the data obtained in Experiment 3.1 for the same system.

REFERENCES

1. D. C. Grahame, Chem. Revs, 41, 441 (1947).
2. R. Parsons in "Modern Aspects of Electrochemistry," Vol. 1, Chap. 3. J. O'M. Bockris and B. E. Conway eds., Butterworths (London), 1954.
3. J. T. G. Overbeek in "Colloid Science", Vol. I, H. R. Kruyt, ed., Elsevier (Amsterdam), 1952, pp. 115-193.
4. R. Parsons in "Advances in Electrochemistry and Electrochemical Engineering", Vol. 1, Chap. 1, P. Delahay and C. Tobias, eds., Interscience (New York), 1961.
5. D. M. Mohilner in "Electroanalytical Chemistry" - A Series of Advances, Vol. I, A. J. Bard, ed., Marcel Dekker Inc., (New York) 1966.
6. A. N. Frumkin in "Advances in Electrochemistry and Electrochemical Engineering", Vol. I, Chap. 2 and Vol. III, Chap. 5, P. Delahay and C. Tobias, eds., Interscience (New York) 1961, 1963.
7. A. N. Frumkin and B. B. Damaskin in "Modern Aspects of Electrochemistry", Vol. III, Chap. 3, J. O'M. Bockris and B. E. Conway, eds. Butterworths (London), 1964.
8. B. B. Damaskin, O. A. Petri, and V. V. Betrakov, "Adsorption of Organic Compounds on Electrodes", Plenum Press (New York) 1971.
9. P. Delahay, "Double Layer and Electrode Kinetics", Interscience (New York), 1965.
10. D. C. Grahame, J. Amer. Chem. Soc. 76, 4819 (1954); 79, 2093 (1957).

REFERENCES

11. R. Parsons and F.G.R. Zobel, J. Electroanal. Chem. $\underline{9}$, 333 (1965).
12. R. Parsons, Proc. 2nd Int. Congress Surface Activity, Butterworths (London) $\underline{3}$, 38 (1957).
13. P. Delahay, ref. 9, pp. 53-58.
14. N. Tshernikovski and E. Gileadi, Electrochim, Acta $\underline{16}$, 579 (1971).
15. N. Tshernikovski, M. Babai and E. Gileadi, J. Electrochem. Soc. $\underline{119}$, 1018 (1972).
16. P. Henderson, Z. physik. Chem. $\underline{59}$, 118 (1907). See also R.G. Bates "Determination of pH, Theory and Practice", pp. 39, 264 J. Wiley and Sons, New York 1964.
17. J. Lawrence and D.M. Mohilner, J. Electrochem. Soc., $\underline{118}$, 1596 (1971).
18. N.K. Adam, "The Physics and Chemistry of Surfaces", Dover (New York) 1968, p. 377.
19. P. Corbusier and L. Gierst, Anal. Chim. Acta $\underline{15}$, 254 (1956).
20. R.G. Barradas and F.M. Kimmerle, Can. J. Chem. $\underline{45}$, 109 (1967).
21. R. Parsons, "Handbook of Electrochemical Constants", pp. 122, 123 Butterworths (London), 1959.
22. M.A.V. Devanathan and P. Peries, Trans. Faraday Soc. $\underline{50}$, 1236 (1954).

4. STRUCTURE OF THE IONIC DOUBLE LAYER AND ELECTRODE PROCESSES

Introduction

4.1 The Reduction of Iodate Ion on Mercury
4.1.1 General
4.1.2 The Cell, Solutions, Electrodes and Electrical Setup
4.1.3 Measurements at the Foot of the Wave
4.1.4 Kinetic Parameters from the Region of Mixed Control by Activation and Diffusion
4.1.5 Treatment of Results

4.2 The Reduction of Persulfate Ion on Mercury
4.2.1 General
4.2.2 The Cell, Solutions and Electrical Setup
4.2.3 The Effect of Monovalent Cations
4.2.4 The Effect of Polyvalent Cations
4.2.5 The Effect of the Supporting Electrolyte Concentration
4.2.6 Treatment of Results

4.3 Reduction of Nickel Ion on Mercury
4.3.1 General
4.3.2 The Cell, Solutions and Electrical Setup
4.3.3 The Effect of Ionic Strength on the Polarogram
4.3.4 Variation of Current with Time During Growth of the Drop
4.3.5 Treatment of Results

INTRODUCTION

STRUCTURE OF THE IONIC DOUBLE LAYER AND ELECTRODE PROCESSES

Introduction

The structure of the ionic double layer plays a major role in determining the course of charge-transfer processes which take place across the interphase. A detailed knowledge of the potential profile and the location of the reacting ion just before charge-transfer takes place would permit the calculation of two important parameters. One is the effective concentrations of reactants (which differ from the bulk concentrations) and the other is the part of the metal/solution potential difference which assists or retards charge-transfer. Such detailed information is not generally available, particularly when the system is complicated by specific adsorption of the reactant or other ionic species in solution. In the absence of specific adsorption, the situation is substantially simplified. The inner Helmholtz plane does not exist and the initial state for reactants may be identified to a good approximation with the outer Helmholtz plane. This approach was taken by Frumkin in his classical paper of 1933[1] for the reduction of hydrogen ion and has been followed by numerous investigators.*

The equilibrium concentration $C(s)_i$ of an ionic species at

---

* A complete list of references to papers in which the "Frumkin correction" has been used cannot be given here. A few representative papers are listed in refs. 2-8, where further references may be found.

the outer Helmholtz plane is given by

$$C(s)_i = C_i^o \exp\left(-\frac{z_i F \phi_2}{RT}\right) \qquad [4.1]$$

where C_i^o is the bulk concentration of the same ion and ϕ_2 is the potential at the outer Helmholtz plane. The rate equation for an activation-controlled cathodic process at high overpotential may be written in the form* (cf. equation [II.79])

$$i = -k_c C_i^o \exp\left[\frac{(\alpha - z_i)F \phi_2}{RT}\right] \exp\left(-\frac{\alpha F \phi_M}{RT}\right) \qquad [4.2]$$

This equation has been tested for several systems[2-9] but its most common application has been to the reduction of anions on mercury. There are two reasons for this. One is the possibility of calculating ϕ_2 as a function of applied potential and concentration in solution and the other is the lack of specific adsorption of the anion at the negative potentials employed for its reduction.

The dependence of ϕ_2 on applied potential in solutions of various concentrations of a 1:1 electrolyte is shown in Fig. 4.1, which is based on the calculations of Russell[10] (Table 4.1). These data can be used for quantitative correction for double-layer effects in kinetic studies. Two qualitative features of the curves should be noted: (a) the diffuse-double-layer potential ϕ_2

---

* The transfer coefficient α is defined here in an operational manner as $\alpha \equiv \frac{2.3RT}{F}\left(\frac{\partial \log i}{\partial E}\right)_\mu = \frac{2.3RT}{bF}$. It is thus an experimental quantity which cannot be given any <u>a priori</u> assumed value.

INTRODUCTION

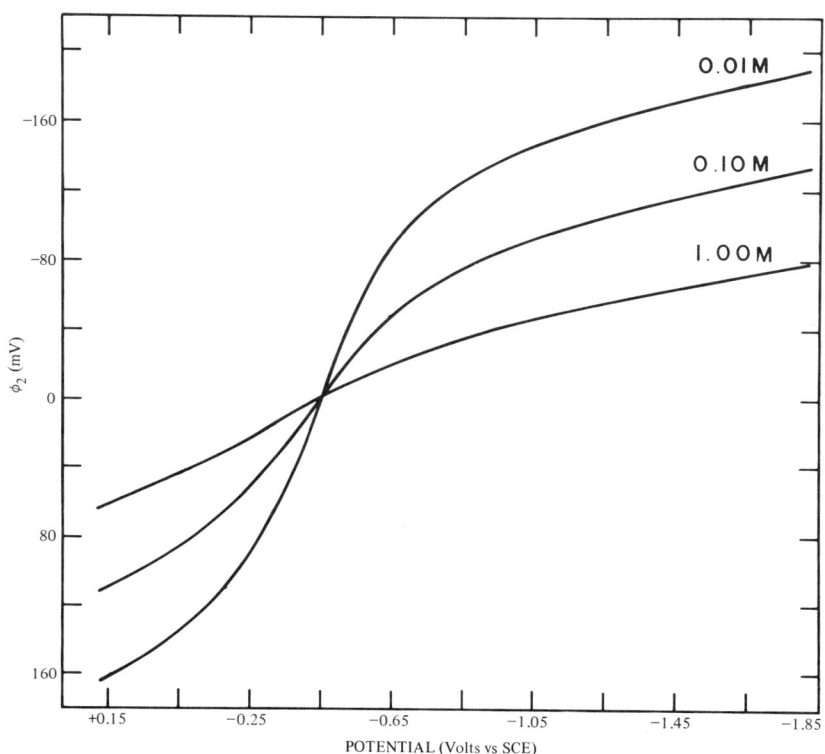

FIG. 4.1   THE POTENTIAL ϕ_2 AT THE OUTER HELMHOLTZ PLANE AS A FUNCTION OF APPLIED POTENTIAL IN AQUEOUS NaF SOLUTIONS AT 25°C

varies most in the vicinity of the potential of zero charge (pzc) and becomes almost independent of applied potential at high rational potentials,* and (b) the curves at high rational potentials become essentially parallel, and the variation of ϕ_2 with concen-

---

* The rational potential is the potential measured with respect to the potential of zero charge.

Table 4.1 Outer Helmholtz Potential for a Mercury Electrode in Aqueous NaF at 25°C (10)

E, mV vs. SCE \ C, mM	2	3	5	7.5	10	15	20	30	50	75	100	200	300	500	750	1000
-150	154_0	146_0	135_9	127_8	122_0	113_8	107_9	99_7	89_4	81_2	75_6	62_3	55_0	46_2	39_9	35_9
-250	125_6	118_7	110_0	103_0	97_9	90_8	85_7	78_5	69_5	62_5	57_5	46_4	40_4	33_4	28_5	25_4
-350	83_1	78_2	72_0	66_9	63_3	58_1	54_4	49_2	42_8	37_8	34_4	27_0	23_2	18_8	15_8	14_1
-450	15_8	15_8	15_1	14_1	13_4	12_2	11_3	10_1	8_6	7_5	6_7	5_1	4_3	3_3	2_7	2_4
-550	-57_5	-53_1	-47_9	-44_0	-41_3	-37_7	-35_1	-31_4	-26_8	-23_4	-21_0	-16_2	-13_8	-11_2	-9_5	-8_5
-650	-105_7	-99_6	-91_7	-85_3	-80_7	-74_0	-69_3	-62_7	-54_4	-48_0	-43_7	-34_3	-29_5	-24_0	-20_3	-18_1
-750	-137_3	-129_5	-119_6	-111_6	-105_9	-97_8	-92_1	-84_0	-73_9	-66_1	-60_6	-48_5	-42_0	-34_5	-29_3	-26_1
-850	-157_5	-148_8	-137_7	-128_9	-122_7	-113_9	-107_6	-98_8	-87_8	-79_2	-73_1	-59_4	-51_9	-43_0	-36_8	-33_0
-950	-171_3	-162_1	-150_5	-141_3	-134_7	-125_5	-118_9	-109_7	-98_2	-89_2	-82_9	-68_1	-60_0	-50_2	-43_3	-39_0
-1050	-181_6	-172_1	-160_2	-150_7	-144_0	-134_5	-127_8	-118_4	-106_6	-97_3	-90_8	-75_5	-66_9	-56_5	-49_0	-44_2
-1150	-189_9	-180_3	-168_2	-158_6	-151_8	-142_2	-135_4	-125_8	-113_8	-104_3	-97_6	-81_9	-73_0	-62_2	-54_2	-49_1
-1250	-197_1	-187_3	-175_1	-165_3	-158_4	-148_7	-141_8	-132_2	-120_0	-110_4	-103_7	-87_7	-78_5	-67_4	-59_1	-53_7
-1350	-203_2	-193_5	-181_1	-171_3	-164_4	-154_6	-147_6	-137_9	-125_6	-116_0	-109_2	-92_9	-83_7	-72_3	-63_8	-58_1
-1450	-208_9	-199_0	-186_6	-176_8	-169_8	-160_0	-153_0	-143_2	-130_9	-121_2	-114_3	-97_9	-88_5	-77_0	-68_2	-62_3
-1550	-214_1	-204_3	-191_8	-181_9	-174_9	-165_1	-158_1	-148_2	-135_8	-126_1	-119_2	-102_7	-93_2	-81_5	-72_5	-66_4
-1650	-219_1	-209_2	-196_7	-186_8	-179_7	-169_8	-162_8	-152_9	-140_5	-130_7	-123_8	-107_2	-97_6	-85_8	-76_7	-70_5
-1750	-223_8	-213_8	-201_3	-191_4	-184_3	-174_3	-167_4	-157_5	-145_1	-135_2	-128_3	-111_6	-102_0	-90_0	-80_8	-74_4
-1850	-228_3	-218_3	-205_8	-195_9	-188_8	-178_9	-171_9	-162_0	-149_5	-139_6	-132_7	—	—	—	—	—

INTRODUCTION

tration may be expressed approximately (cf. p. 42 in ref. 11) by the equation

$$\phi_2 = \pm \frac{2.3RT}{|z|F} \log C° + \text{const} \qquad [4.3]$$

The potential ϕ_2 decreases with increasing ionic strength. It should be noted that the product $z_i \phi_2$ appears in the exponent in the rate equation (equation [4.2]). Thus, for ions having a high charge, diffuse-double-layer effects may not be negligible even in concentrated solutions.

In the following experiments the application of Frumkin's theory to three systems will be studied. The reduction of iodate ion is taken as a case in which the reaction occurs at potentials far removed from the pzc. At these potentials the change of ϕ_2 with potential is negligible and its change with concentration causes a shift in the $E/\log i$ plots. The reduction of persulfate ion occurs not far from the pzc. Here the effect of the variation of ϕ_2 with potential is pronounced and is sufficient to cause a decrease in current with increasing cathodic potential. The reduction of Ni^{++}_{aq} is a rather more complex case in which a chemical dehydration step precedes the electrochemical reaction. In this case the chemical reaction is rather fast and occurs near the surface, hence the concentration of reacting species is affected by the diffuse-double-layer potential, ϕ_2.

For a simple charge-transfer process of the type

$$Ox + ne_M \xrightarrow{k_c} Red \qquad [4.4]$$

taking place at an overpotential sufficiently high that the reverse process may be neglected, the current i at a constant potential is related to the diffusion-limited current i_d by the equation[12]

$$\frac{i}{i_d} = \pi^{1/2} \lambda \exp \lambda^2 \, \text{erfc} \, \lambda \qquad [4.5]$$

This equation is applicable in the case of a planar electrode (assuming semi-infinite linear diffusion) when the initial concentration of the product Red is zero. The parameter λ is given by

$$\lambda = \frac{k_c t^{1/2}}{D^{1/2}} \qquad [4.6]$$

For the expanding-sphere electrode used in polarography, the ratio i/i_d has been calculated by Koutecky[13] as a function of the parameter χ, which is related to λ

$$\chi = \left(\frac{12}{7}\right)^{1/2} \lambda \qquad [4.7]$$

Thus, by measuring the potential dependence of the current i under mixed activation and diffusion control and measuring the current i_d under pure diffusion control, it is possible to evaluate λ or χ from tables of numerical values (Tables 4.2 and 4.3) and hence to obtain k_c as a function of potential.

A somewhat more complex situation arises when the electroactive species Ox is formed in solution from the reactant Y and the rate of this chemical step partially controls the overall rate of the reaction. For the reaction

INTRODUCTION

Table 4.2 Values of i/i_d vs λ for Semi-Infinite Linear Diffusion

λ	i/i_d	λ	i/i_d	λ	i/i_d	λ	i/i_d
0.01	0.0175	0.16	0.239	0.70	0.653	1.7	0.879
0.02	0.0345	0.20	0.287	0.75	0.674	1.8	0.889
0.03	0.0514	0.25	0.341	0.80	0.694	1.9	0.898
0.04	0.0678	0.30	0.391	0.90	0.728	2.0	0.905
0.05	0.0838	0.35	0.435	1.0	0.758	2.2	0.919
0.06	0.0995	0.40	0.476	1.1	0.783	2.4	0.930
0.08	0.130	0.45	0.512	1.2	0.805	2.6	0.938
0.09	0.145	0.50	0.546	1.3	0.824	2.8	0.946
0.10	0.159	0.55	0.576	1.4	0.841	3.0	0.952
0.12	0.187	0.60	0.604	1.5	0.855	3.5	0.964
0.14	0.213	0.65	0.629	1.6	0.868	4.0	0.976

$$Y \underset{k_b}{\overset{k_f}{\rightleftarrows}} Ox \xrightarrow{ne_M} Red \qquad [4.8]$$

occurring under mixed control by diffusion and by the preceding chemical step, assuming that charge transfer is fast one obtains, under the simplifying assumptions listed below, equations identical to equations [4.5] and [4.7], but in which the parameter λ is defined by[12,14]

$$\lambda = (Kk_f t)^{1/2} \qquad [4.9]$$

where K is the equilibrium constant of the chemical step preceding charge transfer.

Table 4.3 Values of i/i_d vs X for Diffusion to an Expanding-Sphere Electrode[13]

X	i/i_d	X	i/i_d	X	i/i_d	X	i/i_d
0.01	0.00880	0.3	0.219	1.4	0.597	5	0.858
0.02	0.0175	0.4	0.275	1.6	0.632	6	0.880
0.03	0.0260	0.5	0.325	1.8	0.662	7	0.895
0.04	0.0345	0.6	0.369	2.0	0.690	8	0.909
0.05	0.0428	0.7	0.409	2.5	0.740	9	0.919
0.06	0.0510	0.8	0.444	3.0	0.777	10	0.927
0.08	0.0671	0.9	0.476	3.5	0.803	12	0.939
0.1	0.0828	1.0	0.505	4.0	0.827	14	0.946
0.2	0.155	1.2	0.555	4.5	0.844	16	0.953
						20	0.963

$$K = \frac{k_f}{k_b} \qquad [4.10]$$

Equation [4.5] is applicable here only if the following requirements are fulfilled:

(i) Charge transfer is fast and does not control the current

(ii) $K \ll 1$, or in other words the bulk concentration of the electroactive species Ox is negligible. Thus, one may neglect the current resulting from the reduction of molecules of Ox which diffuse from the bulk of the solution to the surface of the electrode.

(iii) The initial concentration of Red is zero.

(iv) The species Y and Red are not electroactive in the potential

INTRODUCTION

region covered.

The factors which control various sections of the i/E curve for reactions of the type given by equation [4.4] and equation [4.8] are shown in Fig. 4.2.

For a kinetically controlled electrode process of the type given by equation [4.8] above, the concept of a reaction layer having a thickness μ is useful. A molecule of Ox formed somewhere near the electrode surface may either diffuse to the electrode and there react electrochemically, or it may be transformed back into a reactant molecule Y. It was shown by Budevski[15] that the reaction-layer thickness μ is related to the mean life time, τ, of the species in solution by the equation*

$$\mu = (D\tau)^{1/2} \qquad [4.11]$$

and since for a monomolecular reaction

$$\tau = \frac{1}{k_b} \qquad [4.12]$$

one has

$$\mu = \left(\frac{D}{k_b}\right)^{1/2} = \left(\frac{DK}{k_f}\right)^{1/2} \qquad [4.13]$$

A distinct characteristic of kinetically controlled reactions (cf.

---

* Note that equation [4.11] is very similar to the well known Einstein-Smoluchowski equation for random walk $X = (2Dt)^{1/2}$. Thus the reaction layer is approximately equal to the average distance traversed by a particle in a random fashion during its (average) life time.

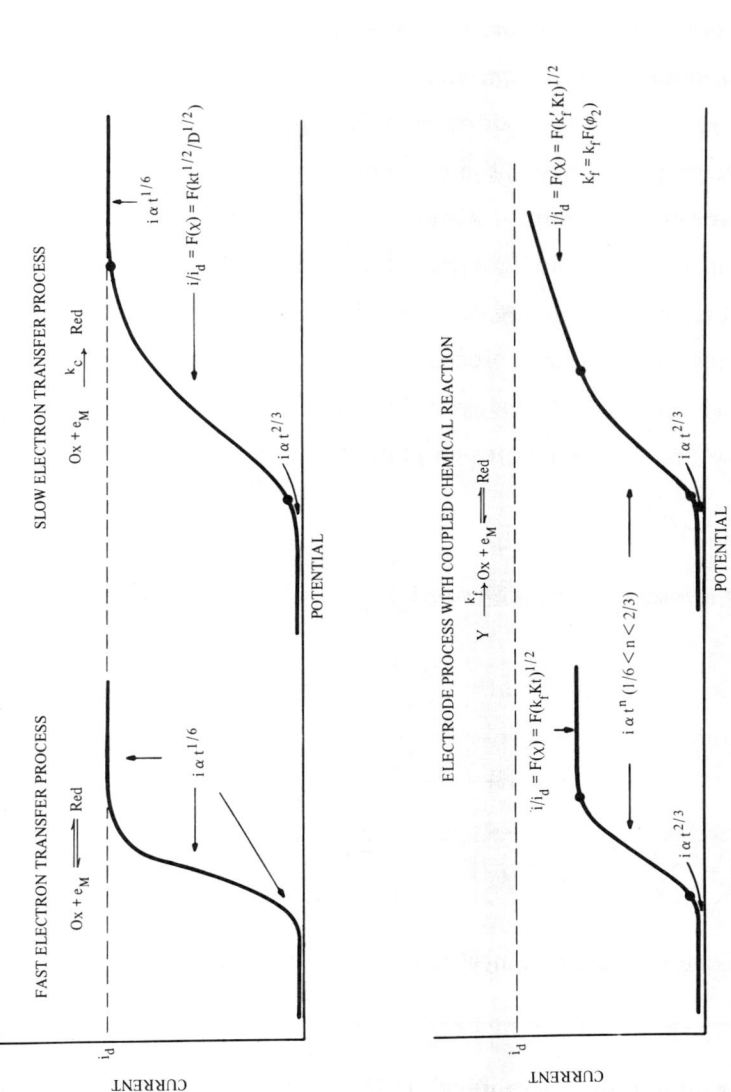

FIG. 4.2 POLAROGRAMS FOR VARIOUS ELECTRODE PROCESSES. POINTS ON THE CURVES MARK THE BOUNDARIES BETWEEN THE DIFFERENT REGIONS.

4.1 REDUCTION OF IODATE ION

equation [4.8]) is that the apparent rate constant is a function of the diffuse-double-layer potential, ϕ_2, provided that the reaction-layer thickness μ is small or comparable to the diffuse-double-layer thickness given by the reciprocal Debye length κ^{-1}. This is the situation in the case of the reduction of Ni^{++}_{aq} in noncomplexing aqueous media[7] and will be treated in Experiment 4.3 below.

4.1 The Reduction of Iodate Ion on Mercury

4.1.1 General

The reduction of iodate ion at the dropping-mercury electrode takes place in alkaline solutions at potentials of -1.0 Volt to -1.2 Volt vs SCE, where adsorption of the anions present in solution (I^-, IO_3^-, OH^-) is negligibly small. The cathodic current may be written in the form

$$i = -k_c C^o_{IO_3^-} \exp \frac{(\alpha - z) F \phi_2}{RT} \exp\left(-\frac{\alpha E F}{RT}\right) \qquad [4.14]$$

where E is the applied potential, measured with respect to some reference electrode.* From this the change of potential ΔE caused by a given change in diffuse-double-layer potential $\Delta \phi_2$ at constant current density is given by

---

* The applied potential E differs from the potential ϕ_M on the metal (with respect to the potential in the solution taken as $\phi_S = 0$) by a constant which cannot be determined (cf. section II.2).

$$\Delta E_{i=\text{const}} = \frac{\alpha - z}{\alpha} \Delta \phi_2 \qquad [4.15]$$

Since the reaction takes place at high negative rational potentials, the variation of ϕ_2 with the applied potential at constant composition may be neglected. Thus the transfer coefficient α can be calculated from measurements at the foot of the polarographic wave ($i < 0.05\ i_d$) where the electrode reaction is activation-controlled. The diffuse-double-layer potential ϕ_2 is changed by changing the composition of the solution and equation [4.15] can be used to determine the charge of the reacting particle from the measured values of α and $\Delta E_{i=\text{const}}$. The quantity $\Delta \phi_2$ may be calculated from the approximate equation [4.3] or from tables given in the literature[10] (cf. Table 4.1). Alternatively, when the charge of the reacting particle is known, it is possible to test the applicability of diffuse-double-layer theory by comparing the experimental value of $\Delta E_{i=\text{const}}$ with its value calculated from equation [4.15]. The experiment below is based in part on the papers of Delahay et al.[5,6]

4.1.2 The Cell, Solutions, Electrodes and Electrical Setup

A standard polarographic cell will be used in this experiment. The solutions will be 5 m$\underline{M}$ KIO_3, 20 m$\underline{M}$ KOH with the following concentrations of KCl: 0; 0.02; 0.06; 0.25; 0.50 $\underline{M}$. A further set of solutions, identical with the above except for KIO_3, which is 0.25 m$\underline{M}$, will also be prepared. All solutions will be prepared from analytical-grade reagents and triply distilled water.

A polarographic dropping-mercury electrode will be used

4.1 REDUCTION OF IODATE ION

as a working electrode with a low-resistance calomel electrode of large area as reference.* A commercial polarograph, designed for operation with three electrodes may also be used in this experiment. In any case, since the instantaneous rather than the average currents during drop time are of interest, measurements should be made without damping and with a recorder having sufficiently fast response (one second for full-scale deflection is suitable). The simple circuit shown in Fig. 4.3 can be used in place of a commercial polarograph. The function generator in this circuit should be able to produce a sweep rate of 1 mV/sec so that the change of potential during the life of the drop will not exceed 3 mV. It is best to use an electromechanical knocker which will ensure a constant drop-time, independent of the change of surface tension with potential.

4.1.3 Measurements at the Foot of the Wave

These experiments will be performed with solutions containing 5 m$\underline{M}$ KIO_3.

Add to the cell the solution which is 5 m$\underline{M}$ KIO_3, 20 m$\underline{M}$ KOH and 0.5 $\underline{M}$ KCl. Deaerate by bubbling purified nitrogen for ten minutes before the experiment and continue to pass the gas above the solution during the experiment. Set the drop-time at about 3 seconds. Obtain the polarogram and determine accurately the limiting current.

---

* The very small currents passing through the DME permit the use of a two-electrode system, provided that the reference/counter electrode has a large area.

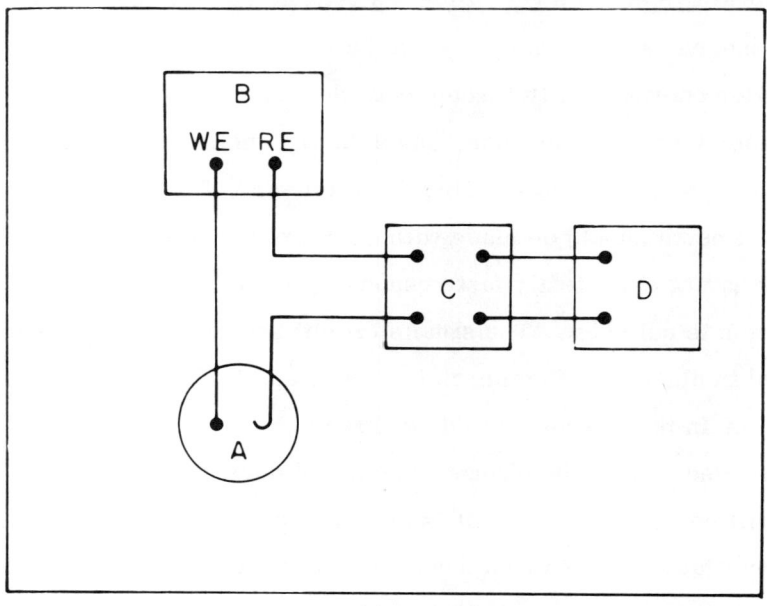

FIG. 4.3 BLOCK DIAGRAM FOR SIMPLE POLAROGRAPH
       A - CELL;
       B - POTENTIAL-STEP OR SLOW-SWEEP GENERATOR;
       C - CURRENT-TO-VOLTAGE CONVERTER;
       D - STRIP-CHART RECORDER.

Confine all measurements at the foot of the wave to currents of up to 5% of the diffusion-limited current obtained here. In this range the reaction is not influenced by mass

4.1 REDUCTION OF IODATE ION

transfer.* For measurements at the foot of the wave, the sensitivity of the potential axis of the recorder should be about 10 mV/cm. The current should be measured at a sensitivity about twenty times higher than that used to obtain the complete polarogram. Drop time should be at least 3 seconds, in order to minimize the effect of charging currents due to growth of the drop (this current is proportional to $1/t^{1/3}$). The rate of change of potential with time should be slow, (0.5-1 mV/sec) in order to ensure that there is little change in potential during the life of a single drop.

Set the initial potential at a value about 0.3 Volt less negative than the least cathodic potential at which a reduction current could be detected on the complete polarogram, and determine the current/potential curve at the foot of the polarographic wave. Repeat for solutions of all five concentrations of KCl.

4.1.4 Kinetic Parameters from the Region of Mixed Control by Activation and Diffusion

In this experiment the intermediate range of the polarogram (where $0.1\ i_d < i < 0.9\ i_d$) will be analyzed by the method of Koutecky to obtain the kinetic parameters.

---

* Remember that the concentration at the surface is related to the concentration in the bulk by $C(s) = C^o(1 - i/i_d)$, hence under the experimental conditions chosen, $C(s) \geqslant 0.95\ C^o$ and the difference between these quantities may be ignored in most cases.

The concentration of the electroactive material will be 0.25 m$\underline{M}$ as compared to 5 m$\underline{M}$ in measurements at the foot of the wave. This is done in order to minimize errors arising from the iR drop* in the solution and in the reference electrode, which serves also as the counter electrode in the present system.** It also ensures that mass-transport by migration of ions in the electric field is kept to a negligible level even when no KCl is added to the solution.

Obtain polarograms in solutions of 0.25 m$\underline{M}$ KIO$_3$, 20 m$\underline{M}$ KOH and the following concentrations of KCl: 0; 0.02; 0.06; 0.25; 0.50 $\underline{M}$.

4.1.5 Treatment of Results

Plot log i $\underline{vs}$ E at the foot of the wave for all solutions on the same graph paper. Calculate the Tafel slope and the transfer coefficient from these curves. Estimate the error introduced in the measured values of $\underline{b}$ by neglecting the change of diffuse-double-layer potential, ϕ_2, with applied potential. (Use the data given in reference 10 for this purpose.) Tabulate the results obtained at the foot of the wave in the form given below:

---

* This is a six-electron reduction, hence significant currents will flow even when the concentration of reactant in solution is quite low.

** Equipment is available which compensates automatically for resistance loss at all times during the growth of the drop.$^{(16)}$ When such instrumentation is used, higher concentrations of the electroactive material may be used without loss of accuracy in measurements.

4.1 REDUCTION OF IODATE ION

C_{KC1} M	C_{Total} M	$\Delta\phi_2$ (mV) from Table 4.1	$\Delta\phi_2$ (mV) from eq. [4.3]	ΔE experimental	ΔE calc. with $\Delta\phi_2$ from Table 4.1	ΔE calc. with $\Delta\phi_2$ from eq. [4.3]
0						
0.02						
0.06						
0.25						
0.50						

For the purpose of these calculations, use the average experimental value of b obtained at the foot of the wave and $z = -1$ to calculate ΔE from equation [4.15]. The potential of zero charge of mercury may be taken as -0.450 Volt vs SCE.

Plot ΔE_{exp} as a function of $\Delta\phi_2$ as calculated from: a) Table 4.1; b) equation [4.3]. Compare the slopes with the values calculated from equation [4.15].

Plot the envelope of each of the five complete polarograms (i.e., the line connecting the current maxima of the drops) on the same graph paper. Discuss the effect of supporting electrolyte (total ionic strength) on the diffusion-limited current, i_d, and on the half-wave potential, $E_{1/2}$. Evaluate the Tafel slope, b, and the transfer coefficient, α, from a plot of $(E-E_{1/2})$ vs log $[(i_d-i)/i]$ according to the equation for irreversible polarographic waves[17] for one of the polarograms

$$E = E_{1/2} + 0.916\, b \log \frac{i_d - i}{i} \qquad [4.16]^*$$

Plot $E_{1/2}$ as a function of $\Delta\phi_2$ and calculate the transfer coefficient α (equation [4.15] may be employed with $\Delta E_{i=const}$ replaced by $E_{1/2}$ as long as the limiting current is independent of ϕ_2).

Using the same polarogram, calculate k_c' (the charge-transfer rate constant uncorrected for double-layer effects), by Koutecky's method.

Tabulate the results in the following manner (with i and i_d measured at the end of the drop life).

E	i/i_d	X	k_c

Plot E against $\log k_c$ and calculate b.

Discuss the range of applicability of diffuse-double-layer theory to this system.

Show how the results found above would be expected to change if a dropping-thallium-amalgam electrode were used

---

* The factor 0.916 is due to the fact that the current is proportional to the drop time raised to a power which is itself potential dependent. This equation is valid only in the range, $0.15 < i/i_d < 0.95$. See section II.10.3 for a discussion of the form of this equation.

4.2 REDUCTION OF PERSULFATE ION

instead of the DME. Assume that the only change is that caused by a shift in the potential of zero charge to about -0.85 Volt <u>vs</u> SCE.

4.2 The Reduction of Persulfate Ion on Mercury

4.2.1 General

The reduction of persulfate ion on mercury, studied in detail by Frumkin and coworkers[8] is perhaps the best known example of the effect of double-layer structure on electrode kinetics. The phenomena observed are much more influenced by the value of the diffuse-double-layer potential, ϕ_2, and its dependence on total concentration and on applied potential E than is the case in the reduction of iodate ion. This is because the reaction takes place in the vicinity of the potential of zero charge, where ϕ_2 depends strongly on E and because of the higher charge of the reacting ion. It will be recalled that the ionic charge appears in the exponent of the diffuse-double-layer correction term (cf. equation [4.2]). A very pronounced dependence on the cation of the supporting electrolyte has also been observed in this case and has been explained alternatively by ion-pair or ion-bridge formation, and by specific adsorption.[8,9] A linear Tafel plot, corrected for the effect of the diffuse double layer may be obtained by rearranging equation [4.2] to the form

$$\ln i + \frac{zF\phi_2}{RT} = -\frac{\alpha F(\phi_M - \phi_2)}{RT} + \text{const} \qquad [4.17]$$

and plotting the l.h.s. of this equation against $(\phi_M - \phi_2)$ as proposed by Delahay et al.[18]

This experiment follows in part the paper by Frumkin, Petry and Nikolaeva-Fedorovich.[8]

4.2.2 The Cell, Solutions and Electrical Setup

The cell and electrical setup are identical with those used in Experiment 4.1. The solutions should all be 0.25 m$\underline{M}$ in $K_2S_2O_8$ with various concentrations and types of supporting electrolyte, as will be detailed below.

All reagents should be analytical grade and solutions should be made up with triply distilled water. It is important to minimize the time of contact of the persulfate solution with mercury, since catalytic decomposition of the $S_2O_8^=$ ion may occur. Thus, deaeration with purified nitrogen should be performed before mercury is introduced into the cell. Also when increasing the concentration of a supporting electrolyte, one should make up new solutions in each case instead of adding portions of a concentrated solution of the supporting electrolyte to the existing solutions.

4.2.3 The Effect of Monovalent Cations

Prepare five solutions, each 0.25 m$\underline{M}$ in $K_2S_2O_8$ and 10 m$\underline{M}$ in one of the following salts as supporting electrolyte: LiCl, NaCl, KCl, RbCl, CsCl. Deaerate one of the solutions by bubbling purified nitrogen for ten minutes before the experiment. Introduce the dropping-mercury electrode and obtain the complete polarogram, starting from a potential of -0.40 V $\underline{vs}$ SCE. Repeat with the other solutions. Plot the envelope of each of the five

4.2 REDUCTION OF PERSULFATE ION

polarograms (i.e. the line connecting the current maxima of the drops) on a single sheet of graph paper.

4.2.4 The Effect of Polyvalent Cations

Prepare six solutions containing 0.25 m$\underline{M}$ $K_2S_2O_8$ and one of the following supporting electrolytes: 3.0 m$\underline{M}$ NaCl, 30 m$\underline{M}$ NaCl, 0.20 m$\underline{M}$ $BaCl_2$, 2.0 m$\underline{M}$ $BaCl_2$, 6.0 $\mu\underline{M}$ $LaCl_3$, 60 $\mu\underline{M}$ $LaCl_3$. Obtain the polarograms as in the previous section and plot the envelopes of all six polarograms on the same sheet of graph paper.

4.2.5 The Effect of the Supporting Electrolyte Concentration

Prepare six solutions containing 0.25 m$\underline{M}$ $K_2S_2O_8$ and NaCl as the supporting electrolyte at the following concentrations: 3.0, 5.0, 7.5, 10, 15 and 30 m$\underline{M}$. Obtain the complete polarograms of these solutions as before. Plot the envelopes of the complete polarograms on a single graph paper. Determine the background current in 10 m$\underline{M}$ NaCl as a function of potential. This may be used as the baseline for all six polarograms.

4.2.6 Treatment of Results

Discuss the influence of the nature and valency of the cations of the supporting electrolyte on the shape of the polarogram.[9]

Plot log X (calculated from the experimental values of (i/i_d)) as a function of potential near the minimum and on the rising part of the wave, for different concentrations of NaCl. How is the slope of this plot related to the Tafel slope b?

Make a corrected Tafel plot of the above data (cf. equation [4.17]) and evaluate the Tafel parameters b and α. Values of ϕ_2 as a function of E at different concentrations are obtained from Table 4.1.

Plot log χ as a function of potential at constant ϕ_2.

Calculate $(\partial \log k_c/\partial E)_{\phi_2}$ at two or three suitable values of ϕ_2 (cf. ref. 8) and calculate b and α. Compare with the values calculated above.

4.3 The Reduction of Nickel Ion on Mercury

4.3.1 General

The reduction of Ni^{++} in aqueous solutions in the absence of complexing ions has been studied in detail by Gierst et al.[7] It was found that charge transfer is preceded by partial dehydration of the aquo-complex. Thus, the reaction may be written in the form

$$[Ni(H_2O)_6]^{2+} \underset{k_b}{\overset{k_f}{\rightleftharpoons}} [Ni(H_2O)_{6-m}]^{2+} + mH_2O \xrightarrow{2e_M} Ni + 6H_2O \qquad [4.18]$$

which is equivalent to equation [4.8]. It was further found that the lifetime of the partially dehydrated species is very short (k_b is large) so that the reaction-layer thickness μ is very small, and the dehydration step may be considered as a heterogeneous surface reaction which can be treated by the method of Koutecky,[13] who related i/i_d to the parameter χ (cf. Table 4.3), where

4.3 REDUCTION OF NICKEL ION

$$X = \left(\frac{12}{7}\right)^{1/2} k_f t^{1/2}/D^{1/2} \qquad [4.19]$$

(cf. equations [4.6] and [4.7]).*

Furthermore, since the chemical reaction occurs close to the plane where the potential is ϕ_2, the structure of the diffuse double layer has a marked effect on the rate of discharge of nickel ions on mercury.

While the heterogeneous rate constant k_f is not affected by ϕ_2, the measured rate of the chemical reaction is given by

$$v = k_f C_Y(x = x_2) = k_f' C_Y \qquad [4.20]$$

where k_f' is the apparent value of the rate constant and C_Y is the concentration of the reactant just outside the diffuse double layer. From equations [4.1] and [4.20] one has

$$k_f' = k_f \exp(-z_Y F\phi_2/RT) \qquad [4.21]$$

where Y represents the reactant $[Ni(H_2O)_6]^{2+}$ in equation [4.18]. Since X is derived from the experimental value of the ratio (i/i_d), the rate constant derived from it with the use of equation [4.19] is the apparent rate constant k_f'. Thus we may write

$$X' = X \exp(-z_Y F\phi_2/RT) \qquad [4.22]$$

---

* Note that equation [4.9] with $\lambda = (Kk_f t)^{1/2}$ or $X = (12/7)^{1/2}(Kk_f t)^{1/2}$ only applies if the preceding chemical step is a homogeneous reaction.

where χ' is the experimental (apparent) value of this parameter while χ is its true value, corrected for double-layer effects.

When the reaction layer is not quite so thin and the preceding chemical reaction (equation [4.8] or [4.18]) is homogeneous, the parameter X is given by

$$X = \left(\frac{12}{7}\right)^{1/2} (Kk_f t)^{1/2} \qquad [4.23]$$

and is related to the ratio (i/i_d) in the same way as χ given in equation [4.19]. As long as μ is small or comparable to the thickness of the diffuse double layer, κ^{-1}, an effect of ϕ_2 on the apparent rate constant can still be measured. A rigorous treatment of this case has been given by Matsuda,[14,19] who expressed the double-layer effect in terms of a multiplying factor G, which is a function of the ratio κ^{-1}/μ and of ϕ_2.

When the product $z_Y \phi_2 F$ is negative, the rate of the homogeneous reaction is enhanced by the double-layer effect and the overall rate may become diffusion-controlled. The factors which control various sections of the i/E curve for reactions of the type given by equation [4.4] and equation [4.8] are shown in Fig. 4.2.

For the experimental evaluation of the double-layer effect, the current is measured, χ' is obtained as a function of i/i_d from mathematical tables[13] (cf. Table 4.3), and its logarithm plotted against ϕ_2 to ascertain the validity of equation [4.22].

4.3 REDUCTION OF NICKEL ION

4.3.2 The Cell, Solutions and Electrical Setup

The cell and electrical setup are identical to those used in Experiment 4.1. For "first-drop" measurements the polarographic setup shown in Fig. 4.3 will be used with a step-function generator. A drop-life timer is used to produce a constant drop time, in the range of 2–4 seconds. The exact drop time should be measured with a stopwatch. The flow rate m is determined by collecting the mercury delivered from the capillary over a period of several minutes, drying and weighing (cf. Experiment 3.1).

4.3.3 The Effect of Ionic Strength on the Polarogram

Prepare 50 ml. solutions which are 1 m$\underline{M}$ in Ni^{2+} and contain the following concentrations of $NaClO_4$: 0.0500, 0.100, 0.300, 0.500, 0.750, 1.00 $\underline{M}$.* Transfer about 20 ml of one of the solutions into the polarographic cell, bubble nitrogen for ten minutes through the solution, then pass the gas above the solution during the measurement. Obtain the polarogram. Repeat with each of the six solutions.

4.3.4 Variation of Current with Time during Growth of the Drop

Measurement of the current as a function of time during the lifetime of a single drop allows a clear distinction to be made between a diffusion-limited current and one that is limited by a chemical reaction preceding charge-transfer. In the former case,

---

* These particular concentrations are used here since for them ϕ_2 is given as a function of ϕ_M in Table 4.1.

the slope of log i vs log t has a calculated value of 1/6 (in practice a slightly higher value of 0.19 - 0.20 is observed). In the latter case the current is proportional to the area of the electrode, hence it varies with $t^{2/3}$. For mixed control the slope has a value intermediate between 1/6 and 2/3.

Measurement of i/t curves should be made only on the first drop, since the depleted diffusion layer, which exists once the first drop has fallen, causes deviations from the predicted i/t relationship. For this purpose, the electrode is allowed to drop continuously, but the potential is maintained in a region where no reaction takes place and then, just as a drop has fallen, is stepped to the desired value, as shown in Fig. 4.4. The electrical circuit used for this operation is shown in Fig. 4.3.

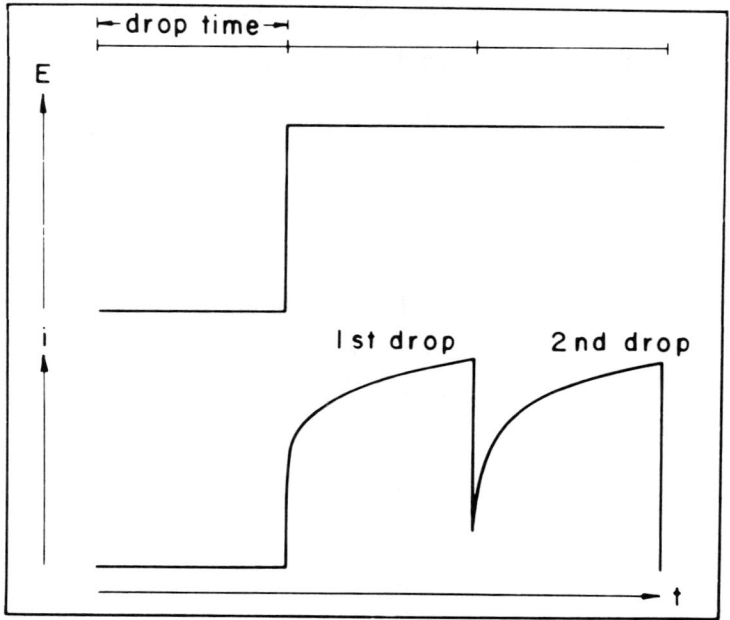

FIG. 4.4   CURRENT/TIME MEASUREMENTS FOR FIRST AND SECOND DROP

4.3 REDUCTION OF NICKEL ION

Prepare the following solutions: (1) 1 m$\underline{M}$ Ni^{2+}, 0.050 $\underline{M}$ NaClO$_4$; (2) 1 m$\underline{M}$ Ni^{2+}, 1.00 $\underline{M}$ NaClO$_4$; (3) 0.050 $\underline{M}$ NaClO$_4$; (4) 1.00 $\underline{M}$ NaClO$_4$. (The latter two solutions are for the measurements of the background current.)

Deaerate the solutions as in the previous section. Set the recorder at 5mm/sec and at the required sensitivity, as seen from the polarograms in the previous section. Set the potential-step generator on manual operation and apply the required step just as a drop falls. Allow the recorder to trace the current for two consecutive drops. The drop-life timer should be disconnected during these measurements, and the drops allowed to fall freely.

4.3.5 Treatment of Results

Collect your current/time data in the following table. (All currents are corrected for background, measured in the same concentration of supporting electrolyte. i_L and i_d are the limiting current and the diffusion current, respectively. The Roman numerals I, II signify the first and second drops. The parameter x' is calculated from Table 4.3.)

1 m$\underline{M}$ Ni^{2+}, 0.05 $\underline{M}$ NaClO$_4$			1 m$\underline{M}$ Ni^{2+}, 1 $\underline{M}$ NaClO$_4$		
t	$i_d^{\,I}$	$i_d^{\,II}$	t	$i_L^{\,I}$	$\dfrac{i_L^{\,I}}{i_d^{\,I}}$

Plot $\log i_d^I$, $\log i_d^{II}$ and $\log i_L^I$ as functions of $\log t$. Calculate the slopes. Show that in the dilute electrolyte solution the current is purely diffusion-controlled. Remark on the difference between the i/t plots of the first and second drop. What is the meaning of the value of the slope $\partial \log i_L^I / \partial \log t$?

Plot $\log \chi'$ against $\log t$ and calculate k_f' using the diffusion coefficient obtained above. Use the Ilkovic equation [II.85] to calculate the diffusion coefficient and comment on the accuracy of the method.

Take the data from each polarogram in the region where it is under mixed control of the chemical step and of diffusion and arrange in the following table:

1 mM $\underline{\text{Ni}}^{2+}$, _____ $\underline{\text{M}}$ $NaClO_4$

Potential vs SCE	Rational Potential	i_L	i_L/i_d	χ'	ϕ_2

The value of the potential of zero charge in these solutions should be obtained from literature data. The potential ϕ_2 may be taken from the data in Table 4.1.

Repeat this calculation for four solutions having the following concentrations of supporting electrolyte: 0.3; 0.5; 0.75; 1.0 $\underline{\text{M}}$.

Plot $\log \chi'$ against ϕ_2 and calculate the charge z_Y according to equation [4.22].

4.3 REDUCTION OF NICKEL ION

Choose four values of the potential in the region of interest and tabulate the results of the polarograms as follows:

E = _____

Concentration of $NaClO_4$	i_L	i_L/i_d	χ'	ϕ_2

Plot log χ' <u>vs</u> ϕ_2 and calculate the value of the charge z_Y as above.

Why is it permissible to use a two-electrode system in these measurements? What is the maximum permissible resistance of the reference electrode in this case?

Why is an electromechanical knocker used in these experiments? Should one use such a device to knock off the mercury drops in classical polarography? Discuss.

Show that $\lambda = k_f t^{1/2} D^{-1/2}$ for a heterogeneous (surface) chemical reaction. (Hint: This is analogous to the case of mixed diffusion and charge-transfer control.)

REFERENCES

1. A. N. Frumkin, Z. physik. Chem. A164, 121 (1933).
2. A. N. Frumkin, Disc. Faraday Soc. 1, 57 (1947).
3. G. M. Florianovich and A. N. Frumkin, Zhur. Fiz. Khim. 28, 473 (1954).
4. A. N. Frumkin, Z. Electrochem. 59, 807 (1955).
5. M. Breiter, M. Kleinerman and P. Delahay, J. Amer. Chem. Soc. 80, 5111 (1958).
6. P. Delahay and M. Kleinerman, J. Amer. Chem. Soc. 82, 4509 (1960).
7. L. Gierst, "Trans. Symp. Electrode Proc.", Philadelphia (1959), p. 109, J. Wiley, New York.
8. A. N. Frumkin, O. A. Petry and N. V. Nikolaeva-Fedorovich, Electrochim. Acta 8, 177 (1963).
9. R. Parsons, "Advances in Electrochemistry and Electrochemical Engineering", Vol. I, Chapter 1, P. Delahay, Ed., Interscience (New York), 1961.
10. C. D. Russell, J. Electroanal. Chem. 6, 486 (1963).
11. P. Delahay, "Double-Layer and Electrode Kinetics", Interscience (New York), 1965.
12. P. Delahay, "New Instrumental Methods in Electrochemistry", Chapter 4, Interscience (New York), 1954.
13. J. Koutecky, Chem. listy 47, 323 (1953); Coll. Czech. Chem. Comm. 18, 597 (1953).
14. J. Heyrovsky and J. Kuta, "Principles of Polarography", Academic Press, 1966.
15. E. Budevski, Compt. Rend. Acad. Bulg. Sci. 8, 25 (1955).
16. Ch. Yarnitzky, private communication.
17. L. Meites, "Polarographic Techniques", p. 240, Interscience (New York), 1965.
18. K. Asada, P. Delahay and A. Sundaram, J. Amer. Chem. Soc. 83, 3396 (1961).
19. H. Matsuda, J. Phys. Chem. 64, 336 (1960).

5. KINETICS OF THE HYDROGEN-EVOLUTION REACTION ON SOLID ELECTRODES

Introduction

5.1 <u>The Hydrogen-Evolution Reaction on Bright Platinum</u>

5.1.1 General

5.1.2 The Cell, Solutions, Electrodes and Electrical Setup

5.1.3 Measurements at Low Overpotential

5.1.4 Measurements at High Overpotential

5.1.5 Treatment of Results

5.2 <u>The Hydrogen-Evolution Reaction on Copper and Lead</u>

5.2.1 General

5.2.2 The Cell, Solutions, Electrodes and Electrical Setup

5.2.3 Determination of the Tafel Plots

5.2.4 Treatment of Results

5. KINETICS OF THE HYDROGEN-EVOLUTION REACTION ON SOLID ELECTRODES

Introduction

The hydrogen-evolution reaction has been studied more thoroughly than any other electrode reaction and a major part of our present understanding of electrode kinetics may be attributed to these studies. The early work of Tafel[1] on hydrogen evolution on mercury cathodes led to the experimental establishment of the Tafel equation (cf. equation [II.39]) relating the overpotential to the current density. Absolute reaction-rate theory was first applied in electrochemistry by Horiuti and Polanyi[2] who considered the various possible mechanisms of the hydrogen-evolution reaction. Early investigators believed that this was a simple reaction, the mechanism of which could be easily and quickly resolved and this would lead to a model for understanding the mechanisms of more complex electrode reactions. At present, several decades later and after several thousand papers have been published on the subject, many questions remain unanswered or are the subject of controversy. The hydrogen-evolution reaction has been discussed in a number of reviews[2-6] where numerous further references may be found.

The reaction is believed to proceed in two steps. Initial discharge of a water molecule or H_3O^+ ion leads to the formation of an adsorbed hydrogen atom

$$H_3O^+ + e_M \longrightarrow H_{ads} + H_2O \qquad [5.1]$$

This is removed from the surface either by a chemical surface-

INTRODUCTION

recombination step $\quad 2H_{ads} \longrightarrow H_2 \quad$ [5.2]

or by an ion-atom recombination step, namely

$$H_{ads} + H_3O^+ + e_M \longrightarrow H_2 + H_2O \quad [5.3]$$

These steps give rise to two possible reaction pathways and two possible rate-determining steps for each pathway. The kinetic parameters shown in Table 5.1 have been calculated for each of the four combinations, with the use of the quasi-equilibrium assumption (cf. section II.7.1). For limitingly low or high coverage the Langmuir isotherm can be employed[3], while for intermediate values of θ the Temkin isotherm applies[7] (cf. section II.7.3).

The free energy of chemisorption of hydrogen atoms or the dissociation energy of the M-H bond formed at the surface clearly affect the rate and mechanism of the hydrogen-evolution reaction, and correlations have been found between the exchange-current density observed and the M-H bond strength. Complications in the interpretation of such data arise because hydrogen is not only adsorbed on the surface of metals but it is also absorbed in the bulk phase in a number of important instances (Pd, Pt, Ni, Fe). Moreover, the energy of a hydrogen atom at the surface of a metal in solution depends in a complex manner on the structure and properties of the metallic crystal and may be poorly correlated with the chemical bond strength of the hydride or with the heat of adsorption in the gas phase.

The isotope separation factor (cf. section II.6.7) has been used for many years for the elucidation of the mechanism of the hydrogen-evolution reaction. Separation factors have been

TABLE 5.1

Kinetic Parameters in the hydrogen-evolution reaction

Kinetic parameter	Degree of coverage	[5.1] rds [5.2] -	[5.1] - [5.2] rds	[5.1] rds [5.3] -	[5.1] - [5.3] rds
b in units of (2.3 RT/F)	low	$1/\beta$	$1/2$	$1/\beta$	$1/(1+\beta)$
	interm.[1]	---	2β [2]	---	2β [2]
	high[1]	---	∞	---	2
$p_1(H_3O^+)$ [3]	low	$1-\beta$	0	$1-\beta$	$2(1-\beta)$
	interm.	---	0	---	$1-\beta$
	high	---	---	---	$1-\beta$
$p_2(H_3O^+)$ [3]	low	1	2	1	2
	interm.	---	2β	---	$1+\beta$
	high	---	---	---	1
γ	---	2	1	1	1

[1] Intermediate or high values of θ are unlikely when first discharge is rate-determining.

[2] Activated adsorption of molecular hydrogen is assumed (cf. p. 385 of ref. 7).

[3] The H_3O^+ ion is assumed to be the source of protons. The ionic strength in solution is maintained constant. (cf. Experiment 7).

measured for different electrodes and under a variety of experimental conditions. In most cases these were found to be independent of potential except when a clear change in mechanism occurred (which was evidenced by a change in other kinetic

INTRODUCTION

parameters). In certain cases, notably when mercury electrodes were used, a gradual decrease in the separation factor with increasing overpotential was observed. The interpretation of the experimental results observed is still a controversial subject. Recently, the role of the solvent in the elementary act of charge transfer at the electrode has been emphasized by Levich[8] and Dogonadze[9] and the importance of "barrierless" transition (which leads to a symmetry factor of unity) was pointed out by Krishtalik[5].

The rate of the hydrogen-evolution reaction depends primarily on the metal used as the working electrode, and to a lesser degree on the composition of the solution. For mercury and other soft metals (Pb, Tl), values of the exchange-current density in the range of $10^{-12} - 10^{-14}$ Amp/cm^2 have been reported. On noble metals such as Pt, Pd and Ir the exchange-current density is about ten orders of magnitude higher, in the range of $10^{-2} - 10^{-4}$ Amp/cm^2. (These metals are therefore suitable as reversible hydrogen electrodes, cf. Experiment 2.1). On metals such as Fe, Cu, Ag and Au, the rate of hydrogen evolution is intermediate, with exchange-current densities in the range of $10^{-6} - 10^{-8}$ Amp/cm^2.

The hydrogen-evolution reaction is highly sensitive to traces of impurities, particularly when the system being studied has a low exchange-current density. Compounds of arsenic, in quantities not sufficient to form one tenth of a monolayer can reduce the reaction rate by one or two orders of magnitude. Traces of organic impurities and molecular oxygen can also alter the reaction rate and render the results unreliable. When electrodes of the platinum group are used, mass transport of molecular hydrogen away from the electrode becomes important and it is best to conduct measurements under well controlled hydrodynamic conditions, *viz.* with a rotating-disc electrode. The effect of impurities is less important when the exchange-current density is high, since the rate of diffusion of an impurity to the surface becomes small compared to the rate of the reaction taking place at the electrode.

5.1 The Hydrogen-Evolution Reaction on Bright Platinum

5.1.1 General

Platinum is one of the best electrocatalysts for the hydrogen-evolution reaction and its reverse, namely, the electro-oxidation of molecular hydrogen. The activity of the metal depends on the method of pretreatment and it tends to decrease with time after pretreatment. Exchange-current densities as high as $0.03\ \text{Amp}/\text{cm}^2$ have been reported immediately after pretreatment and the exchange rate of the first charge-transfer step (equation [5.1]) is in the range of $0.1 - 0.2\ \text{Amp}/\text{cm}^2$. It is important to note that reaction [5.1] occurs on platinum at potentials _anodic_ to the potential of the reversible hydrogen electrode (cf. Experiment [11.1]), which shows that the heat of adsorption of atomic hydrogen is substantial. Instead of H_{ads} it might be better to write MH, indicating explicitly that a chemical bond has been formed with the surface. Thus, adsorbed hydrogen atoms should not be expected to behave as free radicals. The same argument applies, incidentally, to adsorbed intermediates formed in organic electrode

5.1 BRIGHT PLATINUM

reactions.

In this experiment the exchange-current density for hydrogen evolution on platinum will be measured in acid and in alkaline solutions. In the latter case equation [5.1] should be replaced by the equation

$$H_2O + e_M \longrightarrow H_{ads} + OH^- \qquad [5.4]$$

followed either by equation [5.2] or by the equivalent of equation [5.3], namely

$$H_{ads} + H_2O + e_M \longrightarrow H_2 + OH^- \qquad [5.5]$$

Replacing equations [5.1] and [5.3] by [5.4] and [5.5] affects the reaction orders calculated in Table 5.1. If the step shown in equation [5.4] is the rate-determining step, then $\rho_2(H_3O^+) = -\rho_2(OH^-) = 0$. On the other hand, the reversible potential is still a function of pH, hence $\rho_1(H_3O^+) = -\rho_1(OH^-) = -\beta.$ *

In view of the high exchange-current density in this system, the experimental conditions should be chosen so that a high rate of mass transport to and from the electrode is reached (cf. the "Introduction" section to Experiment 8). This can usually be achieved by employing a rotating-disc electrode. In the experiment below a <u>rotating-cone electrode</u> will be used. This behaves very much like a rotating disc, but has the added advantage (particularly useful in reactions where the product is a sparingly

---

* This is true for limitingly low values of θ, which is the only case of interest when the first charge-transfer step is rate-determining, as pointed out in Table 5.1.

soluble gas) that a gas bubble cannot be stabilized at the center of the rotating electrode where the radial force is zero.

5.1.2 The Cell, Solutions, Electrodes and Electrical Setup

The cell shown in Fig. 5.1 will be used in this experiment. The working-electrode compartment is covered with a thick teflon plate (2) with a hole in it. Through this hole the rotating electrode is inserted. A teflon disc (1) is attached to the shaft of the rotating electrode and is situated about 1-2 mm above the teflon cover. During operation this acts as a pump pulling the gas above the solution out of the cell. In this way diffusion of air into the cell is minimized without the need for an air-tight seal between the teflon cover and the rotating shaft.

The working electrode is a <u>rotating-cone electrode</u> (3) which can be fabricated by machining a rotating-disc electrode to the form of a cone with any desired angle. Mass transport to such an electrode is similar to that observed at a rotating disc and the linear relationship between the limiting current density i_L and the square root of the angular velocity (ω) is maintained.[10] (cf. Experiment 6.1). Here this electrode will be used only to maintain a high and reproducible rate of mass transport.

A bright platinum wire or gauze, flushed continuously with hydrogen, will serve as the reference electrode (4) and a platinum gauze will serve as the counter electrode (5). Alkaline solutions will be 0.01 $\underline{M}$, 0.1 $\underline{M}$ and 0.9 $\underline{M}$ NaOH, with $NaClO_4$ added to reach a total ionic strength of unity. Acid solutions will be 0.01 $\underline{M}$, 0.1 $\underline{M}$ and 0.9 $\underline{M}$ $HClO_4$ with $NaClO_4$ added to reach the same total ionic strength. All solutions will be made up of analytical-

5.1 BRIGHT PLATINUM

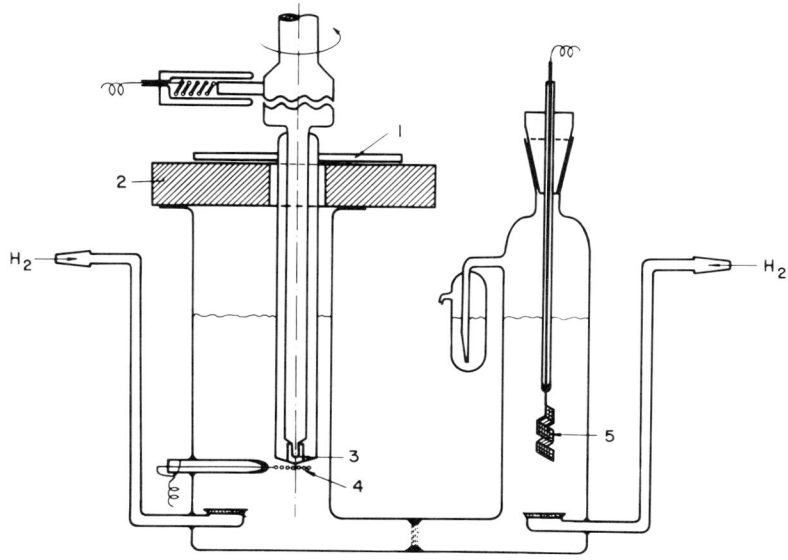

FIG. 5.1 TWO-COMPARTMENT CELL FOR WORK WITH ROTATING-DISC OR -CONE ELECTRODE.
1 - TEFLON DISC ATTACHED TO THE ROTATING ELECTRODE;
2 - TEFLON COVER;
3 - CONICAL WORKING ELECTRODE;
4 - PLATINUM-GAUZE REFERENCE ELECTRODE;
5 - PLATINUM COUNTER ELECTRODE.

grade reagents and triply distilled water.

Current/potential measurements will be made potentiostatically with the circuit shown in Fig. 6.1 (a strip-chart recorder can be used to replace the X-Y recorder). A sequence of potential steps such as shown in Fig. 8.5 will be applied, but the potential will be held constant at each measured value for about 40 seconds, until the current no longer changes with time.

5.1.3 Measurements at Low Overpotential

Introduce the electrodes into the cell and add a solution of 0.9 $\underline{M}$ $HClO_4$ and 0.1 $\underline{M}$ $NaClO_4$. Bubble hydrogen for 10 minutes before the experiment and continue passing the gas through the solution during the experiment. Pretreat the working and the reference electrodes electrochemically as outlined in Experiment 6.1. The potential difference between these electrodes should not exceed 1 mV at the end of this pretreatment. Rotate the electrode at about 4000–6000 rpm. Apply the potential sequence shown in Fig. 8.5 to measure the steady-state current density in the overpotential range of −10 to +10 mV at 2 mV intervals. Do this for all six solutions.

5.1.4 Measurements at High Overpotential

Use a solution of 0.9 $\underline{M}$ NaOH and 0.1 $\underline{M}$ $NaClO_4$. Pretreat the electrodes electrochemically as in the previous section. Bubble hydrogen for 10 minutes before the experiment and pass the gas above the solution during the experiment. Rotate the electrode at 4000–6000 rpm. Make steady-state polarization measurements in the cathodic direction at intervals of 10 mV up to an overpotential of −250 mV. Apply the sequence of potential steps shown in Fig. 8.5 but hold the potential constant at the final desired value for each step until the current has reached a constant value.

5.1.5 Treatment of Results

Calculate the values of the kinetic parameters in Table 5.1 for the different mechanisms for the hydrogen-evolution reaction

5.2   COPPER AND LEAD

in alkaline solutions.

Determine the Tafel slope and the exchange-current density from the measurements at high overpotentials.

Determine the ratio i_o/ν in all six solutions from the measurements at low overpotentials, assuming that the same mechanism applies in the three alkaline solutions (this permits you to use the same value of ν in each case). Calculate the stoichiometric number and the reaction-order parameter $\partial \log i_o / \partial \log C_{OH^-}$ (cf. Experiment 7).

Can you propose a mechanism for the hydrogen-evolution reaction on bright platinum in alkaline solutions on the basis of the data collected in this experiment?

Calculate the reaction-order parameter $\partial \log i_o / \partial \log C_{H_3O^+}$ in acid solution, assuming that the same mechanism applies in all three acid solutions employed here.

5.2   The Hydrogen-Evolution Reaction on Copper and Lead

5.2.1   General

Copper and lead represent the groups of metals upon which the hydrogen-evolution reaction takes place at intermediate and low rates, respectively. The mechanism reported in the literature for both metals is that of a slow-discharge rate-determining step (equation [5.1]) followed by fast electrochemical desorption (equation [5.3]). The exchange-current density for copper is about 10^{-7} Amp/cm^2 and for lead about 10^{-14} Amp/cm^2, the exact value depending on the constitution of the solution. Mass transport does not pose a major problem when kinetic measurements are made with these metals in the linear Tafel region and therefore stationary electrodes may be used. On the other hand, the results are very sensitive to traces of oxygen or other impurities and reliable results can only be obtained in systems which have been rigorously purified.

In the experiment below only the usual purification techniques will be employed. As a result, the data may be expected to deviate from the best values reported in the literature.

5.2.2 The Cell, Solutions, Electrodes and Electrical Setup

A cell such as shown in Fig. 9.3 or 9.4 will be used in this experiment. Solutions will be 0.01 $\underline{M}$, 0.1 $\underline{M}$ and 0.4 $\underline{M}$ H_2SO_4 with Na_2SO_4 added to make a total concentration of 0.5 $\underline{M}$ $SO_4^=$ ions. All solutions will be prepared from analytical-grade reagents and triply distilled water.

The copper working electrode is in the form of a rod 4-8 mm in diameter, press-fitted into a teflon rod. Its surface, which is flush with the teflon, is abraded and polished with powdered alumina and washed thoroughly with triply distilled water just before use. The lead electrode is prepared by tightly wrapping a rod of similar dimensions with teflon tape. A thin slice of this electrode is cut off with a sharp surgical knife just before use. There will be no electrochemical pretreatment of the copper and lead electrodes. Platinum counter and reference electrodes will be used as in Experiment 5.1.

Measurements will be taken potentiostatically with the use of the circuit shown in Fig. 6.1 with a strip-chart recorder to follow the variation of the current with time. The solutions will

5.2 COPPER AND LEAD

be deaerated by bubbling hydrogen for ten minutes through the solution before the experiment and continuing to bubble the gas through the solution during the experiment.

5.2.3 Determination of the Tafel Plots

Introduce the copper electrode into a solution of 0.4 $\underline{M}$ H_2SO_4 and 0.1 $\underline{M}$ Na_2SO_4 and determine the overpotential at which the current density is about 10 $\mu A/cm^2$. Starting from this overpotential, determine the steady-state current density at 25-50 mV intervals over a range of about 400 mV. Between measurements the potential should be set back to its initial value. Stir the solution throughout the experiment with a magnetic stirrer.

Repeat this measurement for both copper and lead electrodes in each of the three solutions. Make a rough Tafel plot while the experiment is in progress and continue the measurements until the deviation from linearity due to mass transport becomes marked. Pretreat each electrode mechanically whenever a new solution is used.

5.2.4 Treatment of Results

Plot the η/i results in semilogarithmic form (Tafel plot) for the two electrodes separately. Determine the Tafel slope and the exchange-current density for both metals and compare with literature data.

What can be said about the fractional surface coverage by adsorbed hydrogen atoms in the Tafel region on the basis of these results?

Determine the reaction-order parameters $\rho_1(H_3O^+)$ and $\rho_2(H_3O^+)$. Are they consistent with the mechanism proposed in the literature for the hydrogen-evolution reaction on these metals?

If a copper electrode were to be used as a reversible hydrogen electrode, what would be the maximum permissible level of electroactive impurities?

How will a small amount of dissolved oxygen affect the Tafel plot?

REFERENCES

1. J. Tafel, Z. phys. Chem. 50, 641 (1905).
2. J. Horiuti and M. Polanyi, Acta Phisicochim. U.S.S.R. 2, 505 (1935)
3. J. O'M. Bockris, Modern Aspects of Electrochemistry, Vol. I, Chap. 4. J. O'M. Bockris and B. E. Conway, Editors, Butterworths, London 1954.
4. A. N. Frumkin, Advances in Electrochemistry and Electrochemical Engineering, Vol. I, Chap. 2, 1961; Vol. III, Chap. 5, 1963, P. Delahay and C. W. Tobias, Editors, Interscience, N. Y.
5. L. I. Krishtalik, Advances in Electrochemistry and Electrochemical Engineering, P. Delahay, Editor, Vol. VII, Chap. 5, 1970.
6. D. B. Matthews and J. O'M. Bockris, Modern Aspects of Electrochemistry, Vol. VI, Chap. 4, J. O'M. Bockris and B. E. Conway, Editors, Plenum Press 1971.
7. E. Gileadi and B. E. Conway, Modern Aspects of Electrochemistry, Vol. III, Chap. 5, J. O'M. Bockris and B. E. Conway, Editors, Butterworths, London 1964.
8. V. G. Levich, Advances in Electrochemistry and Electrochemical Engineering, Vol. IV, Chap. 5, 1966, P. Delahay and C. W. Tobias, Editors, Interscience.
9. R. Dogonadze, "Reactions of Molecules at Electrodes", Chap. III, N. S. Hush, Editor, Wiley-Interscience 1971.
10. V. G. Levich, private communication.

6. THE ROTATING-DISC ELECTRODE

Introduction

6.1 The Diffusion Coefficient of Hydroquinone

6.1.1 General
6.1.2 Electrode Pretreatment
6.1.3 The Cell and Solutions
6.1.4 Determination of the Working Potential
6.1.5 Determination of D
6.1.6 Treatment of Results

6.2 Determination of the Rate Constant and Transfer Coefficient for the Fe^{2+}/Fe^{3+} System

6.2.1 General
6.2.2 Electrode Pretreatment
6.2.3 The Cell, Solutions and Electrical Setup
6.2.4 Range of Working Potentials
6.2.5 Measurement of Current/Potential Curves
6.2.6 Measurement of $D_{Fe^{3+}}$ and $D_{Fe^{2+}}$
6.2.7 Treatment of Results

6.3 Determination of the Rate Constant and Transfer Coefficient for the $[Fe(CN)_6]^{4-}/[Fe(CN)_6]^{3-}$ System

6.3.1 General
6.3.2 Electrode Pretreatment
6.3.3 The Cell, Solutions and Electrical Setup
6.3.4 Range of Working Potential
6.3.5 Measurement of Current as a Function of Rotation Speed
6.3.6 Treatment of Results

6. THE ROTATING-DISC ELECTRODE

Introduction

The rotating-disc electrode was developed following the mathematical solution given by Levich of the hydrodynamic equations describing the rate of transfer of a substance in solution to a rotating disc surface, in terms of the angular velocity of rotation ω, the diffusion coefficient D, the concentration C^o of the substance and the kinematic viscosity ν of the solution*. The detailed mathematical treatment of the rotating-disc electrode is given in Levich's book[1] and in several review articles.[2-5] There are two cases generally considered in the treatment of reactions at the rotating-disc electrode. In the first case, the reaction is relatively fast and the current is determined by mass transport. In such a case, the current (termed "the limiting current") is independent of potential over a wide range**. The limiting current density i_L is given by

$$i_L = 0.62 \, nFD^{2/3} \nu^{-1/6} \omega^{1/2} C^o \qquad [6.1]$$

---

* The kinematic viscosity is defined as the ratio between the viscosity and the specific gravity of the medium.

** The range of potential is limited at one end by the reversible potential and a small overpotential needed to drive even a very fast reaction to mass-transport limitation and at the other end by another reaction which may take place, usually the evolution of oxygen or hydrogen in aqueous solutions.

In the second case, the reaction at the electrode has a higher activation energy, its rate is correspondingly lower and the current is thus controlled both by mass transport and by activation. The measured current density i can be given in this case by

$$1/i = 1/i_{ac} + 1/B\omega^{1/2} \qquad [6.2]$$

where i_{ac} is the activation-controlled current density which can be evaluated by plotting $(1/i)$ vs $(1/\omega^{1/2})$ and extrapolating to zero, i.e. to infinite angular velocity. The constant B is related to the limiting current density i_L by the equation

$$B = i_L/\omega^{1/2} \qquad [6.3]$$

Design and construction factors of a rotating-disc electrode are critical and home-made electrodes should be tested by applying equation [6.1] to a known system. Several types of rotating-disc electrode assemblies are now commercially available.

In the experiments described below, the rotating-disc electrode is used to measure the diffusion coefficient of a compound and the specific rate constants of two reactions.

It may be pertinent to mention here that in recent years the same basic idea has been developed by the construction of rotating ring-disc electrodes[3,4] and more recently also rotating ring-ring electrodes[6,7]. These devices allow detection by the ring of intermediates formed at the disc. A determination of the half-life of such intermediates is also possible. The experiments using these devices are rather involved and hence were not

6.1 DIFFUSION COEFFICIENT OF HYDROQUINONE

included here.

6.1 The Diffusion Coefficient of Hydroquinone

6.1.1 General

In this experiment the diffusion coefficient of hydroquinone (QH_2) will be determined with the use of equation [6.1] which relates the mass-transport-limited current density i_L to the angular velocity. The numerical factor in this equation is correct when the following units are used: i_L[Amp cm^{-2}]; D[cm^2 sec^{-1}]; ν[cm^2 sec^{-1}]; ω[radians sec^{-1}]; C^o[mol cm^{-3}].

The two-electron oxidation of hydroquinone to quinone was chosen since it is known to be a fast reaction with no complications (e.g. kinetic currents or adsorption) at low concentrations.*

6.1.2 Electrode Pretreatment

The electrode should be treated before each experiment in an identical manner.

Abrade with powdered alumina (0.3) for about one minute with a polishing cloth held against the rotating electrode, or on a polishing wheel. Rinse thoroughly (if possible, in an ultrasonic bath) with distilled water. (Mechanical treatment may be by-passed in the case of other electrode geometries).

For electrochemical pretreatment of a platinum electrode,

---

* At concentrations higher than 0.1 M, adsorption and self-inhibition effects have been reported in the oxidation of hydroquinone on Pt in both acid and alkaline solutions. [8]

insert it, together with a mercurous sulfate reference electrode and a platinum counter electrode in 0.5 M H_2SO_4. Set the potential to +1.8 V vs RHE for 10 seconds to oxidize impurities. Change to +1.2 V for 30 seconds to remove oxygen formed at the higher potential. Switch to +0.05 V for 30 seconds to reduce surface oxides formed in the previous steps. Repeat this procedure at least five times.

To check the effectiveness of the pretreatment, obtain a cyclic voltammogram between zero and +1.5 V vs RHE. Compare the curve obtained with that in Fig. 12.1. If there is marked difference between the two, the electrode must be immersed for about three minutes in chromic-sulfuric acid cleaning solution, then rinsed thoroughly (if possible, in an ultrasonic bath) with distilled water and treated electrochemically, as above.

6.1.3 The Cell and Solutions

A cell of the type shown in Fig. 5.1 or 9.3 will be used for this experiment. Care must be taken to avoid turbulence at high rotation speeds. In this respect better results are obtained if the walls of the cell are conical or spherical rather than cylindrical. Bubbles may form during the experiment and cling to the center of the disc, causing large errors in measurement. These bubbles may be removed with the aid of a bent platinum wire.*

A solution which is 2.00 m$\underline{M}$ in hydroquinone and 2$\underline{M}$ in KCl

---

* Recently the use of a rotating ring instead of a rotating disc has been proposed to circumvent this difficulty. The radial forces acting at any point on the ring will not allow the formation of a stationary bubble during operation[7].

6.1 DIFFUSION COEFFICIENT OF HYDROQUINONE

is used in this experiment. Purified nitrogen is passed through the solution before the experiment and above the solution during the experiment.

The working electrode will be a rotating-disc platinum electrode. A commercial calomel electrode and a platinum wire will serve as reference and counter electrodes, respectively. The latter is separated from the bulk of the solution in a thirsty-glass tube or with a fritted disc.

6.1.4 Determination of the Working Potential

The equation to be used here for the determination of the diffusion coefficient of hydroquinone is valid only in the potential region where the current is controlled by mass transport.* To determine this potential region, set the rotating-disc electrode at 3600 rpm and bubble purified nitrogen through the solution for 10 minutes and over the solution during the experiment. Follow the electrochemical pretreatment described above and then obtain the current/potential curve for the oxidation of hydroquinone. This may be done automatically by carrying out a linear potential sweep at a low sweep rate (1-3 mV sec^{-1}) or by taking point-by-point potentiostatic measurements at intervals of 10 mV starting from the open-circuit potential. Block diagrams for automatic and manual measurements are given in Figs. 9.5 and 6.1, respectively. For the determination of the diffusion coefficient,

---

* The term "diffusion-controlled current" or "diffusion-limited current" is often loosely applied to stirred solutions where the correct term should be "mass-transport controlled".

choose a potential in the limiting-current plateau region, before the appearance of a second current wave due to oxygen evolution.

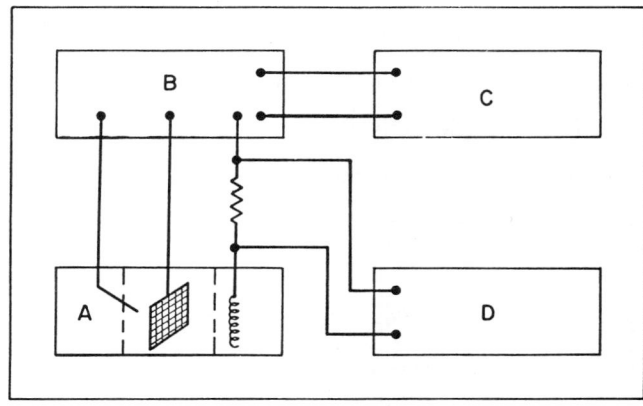

FIG. 6.1   BLOCK DIAGRAM FOR SIMPLE POTENTIOSTATIC CIRCUIT
A - CELL;   B - POTENTIOSTAT;
C - PROGRAMMER   D - OSCILLOSCOPE OR STRIP-CHART RECORDER

6.1.5 Determination of D

Set the potential of the rotating-disc electrode at the value chosen above, and measure the current density as a function of rotation speed in the range 400-8000 rpm. If the rotation speed is continuously variable, use at least 10 speeds over this range. If not, use all the available speeds between these limits.

Determine the geometric area of the electrode.

6.2 Fe^{2+}/Fe^{3+} SYSTEM

6.1.6 Treatment of Results

Plot the current density i vs $\omega^{1/2}$. There may be deviation from the straight line at high rotation speed. Explain.

Calculate D from the slope of the plot, assuming that $\nu = 0.01$ cm^{2}sec^{-1}. Compare your result with those found in the literature.

Discuss the errors which may arise in the determination of D by the rotating-disc method. Compare with other electrochemical methods for the determination of D. How is the rotating-disc electrode limited with respect to electrode area and speed of rotation?

Could the rotating-disc electrode be used as an analytical tool for the determination of the concentration of an electroactive reactant? What would be the accuracy and the lower limit of detection in the case of hydroquinone?

6.2 Determination of the Rate Constant and Transfer Coefficient for the Fe^{2+}/Fe^{3+} System

6.2.1 General

In this experiment the kinetic parameters for the Fe^{2+}/Fe^{3+} redox system will be determined by a method proposed by Randles[9]. This method is best suited to systems in which the reduced and the oxidized forms are present in comparable concentrations. The equations derived below are strictly correct only if the reaction occurs in a single step as must be the case for the system Fe^{2+}/Fe^{3+}. The rotating-disc electrode was used by Galus and Adams[10] in the study of this system. For the simple reaction

$$Fe^{3+} + e_M \rightleftharpoons Fe^{2+} \qquad [6.4]$$

the rate constants k_c and k_a for the cathodic and anodic reactions*
may be written as

$$k_c = k_c^r \exp(-\beta \eta F/RT) \qquad [6.5]$$

$$k_a = k_a^r \exp[(1-\beta)\eta F/RT] \qquad [6.6]$$

where k_c^r and k_a^r are the values of the specific rate constants for the reduction and oxidation reactions, respectively, at the reversible potential. Also, at $\eta = 0$ the rates of the anodic and cathodic reactions are equal, hence,

$$k_c^r C_{Ox}^o = k_a^r C_{Red}^o \qquad [6.7]$$

where C_{Ox}^o and C_{Red}^o are the bulk concentrations of Fe^{3+} and Fe^{2+}, respectively. Combining equations [6.5] and [6.7] one obtains the ratio of the specific rate constants

$$k_a/k_c = (k_a^r/k_c^r)\exp(\eta F/RT) = (C_{Ox}^o/C_{Red}^o)\exp(\eta F/RT) \qquad [6.8]$$

The net current density i is given by

$$i = F(k_a C_{Red} - k_c C_{Ox}) \qquad [6.9]$$

---

* The anodic current and overpotential are taken as positive, while the cathodic current and overpotential are taken as negative.

6.2 Fe^{2+}/Fe^{3+} SYSTEM

where C_{Ox} and C_{Red} are the concentrations of Fe^{3+} and Fe^{2+} at the electrode surface when current is passing. These concentrations can be related to the corresponding bulk concentrations by equations of the type (cf. equation [II.80]).

$$C_{Ox}/C^o_{Ox} = 1 - i/i_{L,c} \quad ; \quad C_{Red}/C^o_{Red} = 1 - i/i_{L,a} \qquad [6.10]$$

where $i_{L,c}$ and $i_{L,a}$ are the limiting current densities for the cathodic and anodic processes, respectively. Substituting equations [6.8] and [6.10] into [6.9] one has

$$i = Fk_c C^o_{Ox}[(1 - i/i_{L,a})\exp(\eta F/RT) - (1 - i/i_{L,c})] =$$

$$= Fk_a C^o_{Red}[(1 - i/i_{L,a}) - (1 - i/i_{L,c})\exp(-\eta F/RT)] \qquad [6.11]$$

Rearranging equation [6.11] gives rise to the expressions

$$-\log k_c = \log FC^o_{Ox}\left[\left(\frac{1}{i} - \frac{1}{i_{L,a}}\right)\exp\left(\frac{\eta F}{RT}\right) - \left(\frac{1}{i} - \frac{1}{i_{L,c}}\right)\right] \qquad [6.12]$$

$$-\log k_a = \log FC^o_{Red}\left[\left(\frac{1}{i} - \frac{1}{i_{L,a}}\right) - \left(\frac{1}{i} - \frac{1}{i_{L,c}}\right)\exp\left(-\frac{\eta F}{RT}\right)\right] \qquad [6.13]$$

These equations will be used to evaluate k_c and k_a as functions of the overpotential η for the Fe^{3+}/Fe^{2+} redox system. It may be noted that equations [6.12] and [6.13] are not specific to the rotating-disc electrode. They are applicable whenever the i <u>vs</u> η curve and the two limiting currents can be determined under reproducible conditions of mass transport. Such conditions are, however, best achieved with the aid of a rotating-disc electrode.

6.2.2 Electrode Pretreatment

Clean the electrode mechanically as described in Experiment 6.1 and rinse throughly with triply distilled water. Introduce the electrodes into a cell which contains 0.5 M H_2SO_4. Follow the electrochemical pretreatment procedure outlined in Experiment 6.1. Transfer the electrodes into the cell which contains the freshly prepared test solution, bubble purified nitrogen for ten minutes and continue passing nitrogen over the solution during the experiment.

6.2.3 The Cell, Solutions and Electrical Setup

The cell is identical with that used in Experiment 6.1. Current/potential curves will be determined potentiostatically, also as in Experiment 6.1.

The working solution will be 3 m$\underline{M}$ in Fe^{3+} and in Fe^{2+} and 1$\underline{M}$ in H_2SO_4. Solutions should be freshly prepared with triply distilled water. All reagents should be of analytical grade.

An electrochemically pretreated platinum wire (see Experiment 6.1) can be used as an indicator-type reference electrode in this system. The advantage of this is that one measures overpotential directly and in the low-overpotential region better accuracy can be achieved.

6.2.4 Range of Working Potentials

The current/potential curves for the Fe^{3+}/Fe^{2+} redox system at a platinum electrode are bounded on the anodic side by oxygen evolution and on the cathodic side by hydrogen evolution. Determine the useful working range by setting the electrode at a

6.2 Fe^{2+}/Fe^{3+} SYSTEM

medium rotation speed (1000-3000 rpm) and sweeping the potential slowly (at 1-3 mV/sec). Oxygen and hydrogen evolution will be evidenced by the appearance of additional current waves on the anodic and cathodic limiting-current plateaus, respectively. Confine all further measurements to the useful potential range for the study of the Fe^{3+}/Fe^{2+} systems determined in this section.

6.2.5 Measurement of Current/Potential Curves

Measurements of the current/potential curves will be performed potentiostatically. Best results are obtained by manual, point-by-point determination of the current/potential relationship. Automatic recording of the current/potential plot will yield good results in this system provided the potential is changed slowly (at a rate of, say, 1-3 mV/sec).[*] If this method is employed, it is advisable to repeat the measurement at different sweep rates (i.e. at different rates of change of potential with time) to find a region where the results obtained are independent of sweep rate.

Set the electrode to rotate at 900 rpm and apply the electrochemical pretreatment described above. Allow one minute at open circuit and then measure the current/potential curve from limiting cathodic to limiting anodic currents. If point-by-point measurements are taken, it is best to pretreat the electrode before each measured point. If the current/potential

---

[*] This is so because the reaction studied is fast and complications due to adsorption of reactants, products or intermediates are largely absent.

curve is recorded automatically, only a single pretreatment, before commencement of the sweep, can be done. The latter procedure thus leads to somewhat lower accuracy of the results. Care must be taken that the two limiting-current densities $i_{L,c}$ and $i_{L,a}$ are measured in the true plateau region and with the highest accuracy, since their values affect the calculated values of k_c and k_a at each potential.

Repeat at 3600 rpm and at the highest speed at which no turbulence occurs. In the last case determine the i/η relationship in the low-overpotential region (±10 mV) by taking measurements every 2 mV, or by expanding the current scale of the recorder if measurement is made by the potential-sweep technique.

Determine the geometric area of the working electrode.

6.2.6 Measurement of $D_{Fe^{3+}}$ and $D_{Fe^{2+}}$

The diffusion coefficients of Fe^{3+} and Fe^{2+} may be easily determined, as in Experiment 6.1, by measuring the anodic and cathodic diffusion-limited currents $i_{L,a}$ and $i_{L,c}$ as a function of the speed of rotation. A large number of rotation speeds should be used to obtain accurate results.

6.2.7 Treatment of Results

Use equation [6.12] and [6.13] to evaluate k_c and k_a at different overpotentials.

Plot $\log k_c$ and $\log k_a$ <u>vs</u> η. The two lines should intersect at $\eta = 0$. At this point $k_c = k_a = k_s$ where k_s is the standard rate constant at this concentration (since equal concentrations of

6.2 Fe^{2+}/Fe^{3+} SYSTEM

Fe^{3+} and Fe^{2+} were used). Calculate the exchange-current density i_o and the Tafel slopes, b_a and b_c. Estimate the error introduced by the iR drop.

Use the results obtained at the highest rotation speed to plot current vs potential at low overpotentials on a linear scale and estimate the exchange-current density from the equation

$$\left(\frac{\partial \eta}{\partial i}\right)_{\eta \to 0} = \frac{RT}{F} \cdot \frac{1}{i_o} \qquad [6.14]$$

which is applicable under purely activation-controlled conditions. Compare this value of i_o with that calculated from k_s in the previous section. Note that if the exchange-current density is not negligible compared to the limiting currents, equation [6.14] must be replaced by the equation

$$\left(\frac{\partial \eta}{\partial i}\right)_{\eta \to 0} = \frac{RT}{F}\left[\frac{1}{i_{L,a}} - \frac{1}{i_{L,c}} + \frac{1}{i_o}\right] \qquad [6.15]$$

Derive this equation (cf. "Double Layer and Electrode Kinetics" by P. Delahay, Interscience 1967). Estimate the error in i_o introduced by employing equation [6.14] instead of equation [6.15].

Describe the methods available for the determination of i_o (or k_s) with emphasis on the range of values of i_o (or k_s) which can be measured by each method.

Show how the rotating-disc electrode can be used to distinguish between mass transfer and charge transfer as the rate-determining step.

6.3 Determination of the Rate Constant and Transfer Coefficient for the $[Fe(CN)_6]^{4-}/[Fe(CN)_6]^{3-}$ System

6.3.1 General

In this experiment the kinetic parameters of the $[Fe(CN)_6]^{4-}/[Fe(CN)_6]^{3-}$ redox system will be measured by the method proposed by Levich[1]. This is based on the use of equation [6.2] in which the activation-controlled current (or the potential-dependent specific rate constant) is evaluated from a plot of $1/i$ vs $1/\omega^{1/2}$. The derivation of equation [6.2] in a form applicable to a redox system near its equilibrium potential, where both forward and reverse reactions must be considered, will be outlined, following the treatment given by Jahn and Vielstich[11].

The net current density for reduction is given by

$$i = nF(k_a C_{Red} - k_c C_{Ox}) \qquad [6.9]$$

The same current density is given also by the equation

$$i = nFD_{Red}(C^o_{Red} - C_{Red})/\delta_{Red} = -nFD_{Ox}(C^o_{Ox} - C_{Ox})\delta_{Ox} \qquad [6.16]$$

where C^o_{Red} and C^o_{Ox} are the concentrations of the species in the bulk and δ_{Red}, δ_{Ox} are the respective diffusion-layer thicknesses given, in the case of the rotating-disc electrode, by

$$\delta = 1.61 \, D^{1/3} \, \nu^{1/6} \, \omega^{-1/2} \qquad [6.17]$$

From equation [6.16] one obtains

6.3 $[Fe(CN)_6]^{4-}/[Fe(CN)_6]^{3-}$ SYSTEM

$$C_{Red} = C^o_{Red}\left(1 - i\frac{\delta_{Red}}{nFD_{Red}C^o_{Red}}\right) \qquad [6.18]$$

and

$$C_{Ox} = C^o_{Ox}\left(1 + i\frac{\delta_{Ox}}{nFD_{Ox}C^o_{Ox}}\right) \qquad [6.19]$$

Substituting these values of the surface concentrations into equation [6.9] and rearranging, one has

$$\frac{1}{i} = \frac{1}{nF(k_a C^o_{Red} - k_c C^o_{Ox})}\left[1 + \frac{k_a \delta_{Red}}{D_{Red}} + \frac{k_c \delta_{Ox}}{D_{Ox}}\right] \qquad [6.20]$$

and upon substitution of the values of δ from equation [6.17] into [6.20] one has

$$\frac{1}{i} = \frac{1}{nF(k_a C^o_{Red} - k_c C^o_{Ox})}\left[1 + \left(k_a D_{Red}^{-2/3} + k_c D_{Ox}^{-2/3}\right)\frac{1.61\nu^{1/6}}{\omega^{1/2}}\right] \qquad [6.21]$$

The specific rate constants k_a and k_c are obtained from the intercept and the slope of a plot of $1/i$ vs $1/\omega^{1/2}$, provided that the diffusion coefficients of both species are known. It may be recalled that the activation-controlled current density i_{ac} is given by

$$i_{ac} = nF(k_a C^o_{Red} - k_c C^o_{Ox}) \qquad [6.22]$$

so that equation [6.21] has the form of equation [6.2]. For conditions of high overpotential, where $k_a \gg k_c$, equation [6.21] can be simplified to

$$\frac{1}{i} = \frac{1}{nFk_a C_{Red}^o} + \frac{1}{0.62nFD_{Red}^{2/3} \nu^{-1/6} C_{Red}^o} \left(\frac{1}{\omega^{1/2}}\right) \quad [6.23]$$

or

$$1/i = 1/i_{ac} + 1/_{L,a} \quad [6.24]$$

6.3.2 Electrode Pretreatment

The platinum rotating-disc electrode should be pretreated mechanically and electrochemically as discussed in Experiment 6.1.

6.3.3 The Cell, Solutions and Electrical Setup

The cell, reference and counter electrodes are identical to those used in Experiment 6.1. The solution is 3 mM in $[Fe(CN)_6]^{4-}$ and in $[Fe(CN)_6]^{3-}$ and 1 M in KCl. Electrical circuits are as in Experiment 6.1. The commercial reference electrode may be replaced by a platinum wire (pretreated electrochemically) which will serve as an indicator-type reference electrode, as in Experiment 6.2.

6.3.4 Range of Working Potential

Introduce the freshly pretreated working electrode into the purified nitrogen for 10 minutes. Continue passing nitrogen above cell and bubble the solution throughout the experiment. Set the electrode to rotate at an intermediate speed (about 2500 rpm). Obtain the current/potential curve from the limiting cathodic to

6.3 $[Fe(CN)_6]^{4-}/[Fe(CN)_6]^{3-}$ SYSTEM

the limiting anodic current-density region by applying a slow linear sweep of a few mV per second. Choose about twelve values of the potential between the two limiting current regions for your measurements.

6.3.5 Measurement of Current as a Function of Rotation Speed

Set the electrode in motion, allow one minute on open circuit and then apply one of the potentials chosen in the previous section. Measure the current density i as a function of rotation velocity ω over as large a range of ω as possible. Repeat at all the other potentials chosen above. Carry out duplicate experiments at two or three of the potentials, to check the reproducibility of the results.

Determine the geometric area of the working electrode.

6.3.6 Treatment of Results

Plot $1/i$ vs $1/\omega^{1/2}$ at all measured potentials and calculate the slope and the intercept by the method of least squares. Obtain k_a and k_c from these data according to equation [6.21].* Plot log k_c and log k_a vs η and determine k_s and b. Compare with results in the literature. Estimate the error introduced by the iR drop.

Plot the current density i vs $\omega^{1/2}$ at different anodic and

---

* The calculation may be simplified by assuming an average value of the diffusion coefficient $D_{Ox} = D_{Red} = 6.2 \times 10^{-6}$ cm^2/sec. Use $\nu = 0.01$ cm^2/sec.

cathodic overpotentials. Discuss the effect of overpotential on the shape of these curves.

REFERENCES

1. V. G. Levich, "Physico-Chemical Hydrodynamics", Prentice-Hall, Englewood Cliffs, New Jersey 1962.
2. A. C. Riddiford, "Advances in Electrochemistry and Electrochemical Engineering", Vol. 4, C. Tobias, Ed., Interscience 1966.
3. A. N. Frumkin and L. N. Nekrasov, Dokl. Akad. Nauk SSSR 126, 1029 (1959).
4. W. J. Albery and S. Bruckenstein, Trans. Faraday Soc. 62, 1920 (1966).
5. R. N. Adams, "Electrochemistry at Solid Electrodes". pp. 80-102. Marcel Dekker Inc., New York 1969.
6. K. E. Heusler and H. Schuring, Z. physik. Chem. (Frankfurt) 47, 117 (1963).
7. I. V. Kadija and V. M. Nakic, J. Electroanal. Chem. 35, 177 (1972).
8. E. Zeigerson and E. Gileadi, J. Electroanal Chem. 28, 421 (1970).
9. J. E. B. Randles, Canad. J. Chem. 37, 238 (1959).
10. Z. Galus and R. N. Adams, J. Phys. Chem. 67, 866 (1963).
11. D. Jahn and W. Vielstich, J. Electrochem. Soc. 109, 849 (1962).

7. REACTION-ORDER STUDIES

Introduction

7.1 Variation of i_o with E_r and with Concentration

7.1.1 General
7.1.2 The Cell, Solutions, Electrodes and Electrical Setup
7.1.3 Determination of the Exchange-Current Density
7.1.4 Variation of i_o with E_r and Concentration
7.1.5 Treatment of Results

7.2 Measurements at High Overpotentials

7.2.1 General
7.2.2 The Cell, Solutions, Electrodes and Electrical Setup
7.2.3 Measurement of the η/i Relationship
7.2.4 Measurement of the Current at Constant η and at Constant E
7.2.5 Treatment of Results

INTRODUCTION   329

7.   REACTION-ORDER STUDIES

Introduction

In section II.6.5 the electrochemical reaction order was defined in terms of the change in current density with change in concentration of a species in solution. It was noted that electrochemical kinetics differs from chemical kinetics in that two types of reaction orders must be defined, one for conditions of constant overpotential, η and one for which the absolute metal/solution potential difference, $\Delta\phi$, is kept constant. The latter condition is realized experimentally if the potential E with respect to a particular reference electrode in the solution is maintained constant. Thus we define

$$\rho_1 = (\partial \log i / \partial \log C_k)_{\eta, C_{j \neq k}} \qquad [7.1]$$

and

$$\rho_2 = (\partial \log i / \partial \log C_k)_{E, C_{j \neq k}} \qquad [7.2]$$

Several other parameters have been employed[1,2,3] to describe the rate of change of i, i_o, E, E_r and η with the change in concentration of one of the reacting species. Some of these are listed below. It will be seen that all these experimental parameters can be related to the two fundamental reaction-order parameters ρ_1 and ρ_2 defined above and to the Tafel slope, b, by the laws of partial derivatives.

(i) The rate of change of overpotential with concentration at constant current density is given by

$$\left(\frac{\partial \eta}{\partial \log C_k}\right)_{i, C_{j \neq k}} = -\left(\frac{\partial \eta}{\partial \log i}\right)_{C_j} \cdot \left(\frac{\partial \log i}{\partial \log C_k}\right)_{\eta, C_{j \neq k}} = -b\rho_1(k) \quad [7.3]$$

(ii) Similarly, the rate of change of potential (with respect to a particular reference) with concentration at constant current density is given by:

$$\left(\frac{\partial E}{\partial \log C_k}\right)_{i, C_{j \neq k}} = -\left(\frac{\partial E}{\partial \log i}\right)_{C_j} \cdot \left(\frac{\partial \log i}{\partial \log C_k}\right)_{E, C_{j \neq k}} = -b\rho_2(k) \quad [7.4]$$

(iii) The rate of change of i_o with concentration is not a new type of parameter, since

$$\left(\frac{\partial \log i_o}{\partial \log C_k}\right)_{C_{j \neq k}} = \left(\frac{\partial \log i}{\partial \log C_k}\right)_{C_{j \neq k}, \eta=0} = \rho_1(k) \quad [7.5]$$

(iv) Often it is convenient to determine experimentally the rate of change of i_o with the reversible potential, the latter being varied by changing the concentrations of the reactants and/or the products. Under these conditions one has

$$\left(\frac{\partial \log i_o}{\partial E_r}\right)_{C_{j \neq k}} = \left(\frac{\partial \log i_o}{\partial \log C_k}\right)_{C_{j \neq k}} \cdot \left(\frac{\partial \log C_k}{\partial E_r}\right)_{C_{j \neq k}} \quad [7.6]$$

The Nernst equation may be written in the form

$$E_r = E^o + \frac{RT}{nF}\left[\Sigma \ln C_{Ox}^{\nu_{Ox}} - \Sigma \ln C_{Red}^{\nu_{Red}}\right] \quad [7.7]$$

where the sums are taken over all oxidized and reduced forms of

INTRODUCTION

the reactants or products present in solution.* The exponential coefficients ν_{Ox} and ν_{Red} are the respective stoichiometric coefficients in the charge-transfer equilibrium determining the reversible potential E_r.** Combining equations [7.6] and [7.7], one has

$$\left(\frac{\partial \log i_o}{\partial E_r} \right)_{C_{j \neq k}} = \rho_1(k) \cdot \frac{nF}{RT} \cdot \frac{1}{\nu_k} \qquad [7.8]$$

(v) In certain cases it is possible to maintain a constant value of E_r by changing the concentration of oxidized and reduced species in the same ratio and comparing current densities at the same overpotential. The parameter $(\partial \log i / \partial \log C_k)_{\eta, E_r}$ measured under these conditions may appear at first to be equal to ρ_1, since a constant overpotential is maintained. Closer examination, however, reveals that the condition that both E_r and η be constant is equivalent to the condition that $E = \text{const}$. Hence

---

* In practice it is better to use the experimental value of the differential coefficient $(\partial E_r / \partial \log C_k)_{C_{j \neq k}}$ rather than its theoretical value calculated from the Nernst relationship.

** Reaction-order studies are usually conducted in the presence of a supporting electrolyte at high concentration so that the ratio of activity coefficients in the Nernst equation may be regarded as constant.

$$\left(\frac{\partial \log i}{\partial \log C_k}\right)_{\eta, E_r} = \rho_2(k) \qquad [7.9]$$

Note that at least two concentrations must be changed if the reversible potential is to be kept constant. This complicates matters, since the current may be affected by the concentrations of both reactants and products, in which case equation [7.9] no longer holds and the experimental parameter $(\partial \log i/\partial \log C_k)_{\eta, E_r}$ cannot be interpreted in a simple way. In most reactions this is not the case and equation [7.9] may be safely used.

The choice of the experimental parameter from which the reaction order is calculated depends on the kind of reaction studied. For fast reactions, the Tafel slope is hard to measure and the variation of i_o with concentration may be obtained from η/i curves measured at low overpotentials. This still requires a knowledge of the Tafel slope, which may be obtained at some relatively high concentration for which the Tafel region is sufficiently long. Alternatively, the change of i_o with concentration at constant reversible potential $(\partial \log i_o/\partial \log C_k)_{E_r, C_{j \neq k}}$ gives ρ_2.

For slow reactions, where measurement can only be made at high overpotentials, it is best to determine the parameter ρ_2 at constant potential. Measurement at constant overpotential is convenient only if the reference electrode used is reversible with respect to the reactants and/or products of the reaction.

We shall now proceed to calculate the values of the reaction-order parameters ρ_1 and ρ_2 for a charge-transfer reaction of

INTRODUCTION

the type*

$$\text{Red} \rightleftarrows \text{Ox} + n e \qquad [7.10]$$

The partial currents in the anodic and cathodic directions may be written in the form

$$i_a = k_a C_{\text{Red}}^{z(a,\text{Red})} C_{\text{Ox}}^{z(a,\text{Ox})} \exp(E/b'_a) \qquad [7.11]^{**}$$

$$|i_c| = k_c C_{\text{Red}}^{z(c,\text{Red})} C_{\text{Ox}}^{z(c,\text{Ox})} \exp(-E/b'_c) \qquad [7.12]$$

These equations define the electrochemical reaction orders $z(a,\text{Red})$, $z(a,\text{Ox})$ for the species Red and Ox in the oxidation reaction and $z(c,\text{Red})$, $z(c,\text{Ox})$ for the same species in the reduction reaction. From equations [7.11] and [7.12] we have directly

$$\rho_2(a,\text{Red}) = z(a,\text{Red}) \quad ; \quad \rho_2(a,\text{Ox}) = z(a,\text{Ox})$$

and $\qquad\qquad\qquad\qquad\qquad\qquad\qquad\qquad\qquad\qquad$ [7.13]

$$\rho_2(c,\text{Red}) = z(c,\text{Red}) \quad ; \quad \rho_2(c,\text{Ox}) = z(c,\text{Ox})$$

The somewhat heavy nomenclature introduced by Vetter[1] is used here. What is meant by the equation $\rho_2(a,\text{Red}) = z(a,\text{Red})$, for example, is that the measured reaction-order parameter ρ_2 for the species Red in the anodic reaction is equal to the kinetically significant electrochemical reaction order with respect to the

---

* Eq. [7.10] represents the overall reaction, but this does <u>not</u> mean that it occurs in a single step.

** $b' = b/2.3$

same species.

The corresponding ρ_1 parameters may be obtained by substituting

$$E = E_r + \eta \qquad [7.14]$$

in equations [7.11] and [7.12]. Thus

$$i_a = k_a C_{Red}^{z(a,Red)} C_{Ox}^{z(a,Ox)} \exp(E_r/b_a') \exp(\eta/b_a') \qquad [7.15]$$

This gives us, for example, for $\rho_1(a, Red)$

$$\rho_1(a, Red) = z(a, Red) + \frac{1}{b_a'} \left(\frac{\partial E_r}{\partial \ln C_{Red}} \right)_{C_{Ox}} \qquad [7.16]$$

Combining this with the Nernst equation [7.7] we have

$$\rho_1(a, Red) = z(a, Red) - \frac{1}{b_a'} \cdot \frac{\nu_{Red}}{n} \cdot \frac{RT}{F} \qquad [7.17]$$

For the reaction order (at η = const) of the same species in the cathodic direction we have similarly, by combining equations [7.12], [7.14] and [7.7]

$$\rho_1(c, Red) = z(c, Red) + \frac{1}{b_c'} \cdot \frac{\nu_{Red}}{n} \cdot \frac{RT}{F} \qquad [7.18]$$

Equations [7.17] and [7.18] should preferably be used in this form, since they express the correlation between the experimentally determined parameters, $\rho_1(a, Red)$, $\rho_1(c, Red)$, b_a, b_c and the kinetically significant electrochemical reaction orders $z(a, Red)$

7.1   VARIATION OF i_o

and $z(c, Red)$. The overall number of electrons transferred, n, and the stoichiometric coefficient ν_{Red}, are known from the overall stoichiometry of the reaction studied.

Equations [7.17] and [7.18] may be transformed to a form more commonly encountered in the literature[1,2] by substituting (cf. equation [II.72]

$$\frac{1}{b'_a} = \frac{\alpha F}{RT} \quad ; \quad \frac{1}{b'_c} = \frac{(1-\alpha)F}{RT} \qquad [7.19]$$

Introducing equation [7.19] into [7.17] and [7.18] one has

$$\rho_1(a, Red) = z(a, Red) - \frac{\alpha}{n}\nu_{Red} \qquad [7.20]$$

and

$$\rho_1(c, Red) = z(c, Red) + \frac{1-\alpha}{n}\nu_{Red} \qquad [7.21]$$

The equations given in this introduction are summarized in Table 7.1 below.

7.1   Variation of i_o with E_r and with Concentration

7.1.1   General

The Fe^{2+}/Fe^{3+} redox system will be used in this group of experiments for the purpose of measuring the reaction orders by various methods and comparing the results. In this first experiment the concentration of either Fe^{2+} or Fe^{3+} will be varied and the dependence of the exchange-current density on the reversible potential and on log C will be evaluated. For this system, the

Table 7.1
Relation of Experimental Parameters to Kinetically Significant Reaction Orders

Experimental parameter	In terms of ρ_1 or ρ_2	In terms of z and b	In terms of z and α *
$\left(\dfrac{\partial \log i_a}{\partial \log C_{Red}}\right)_{\eta, C_{Ox}}$	$\rho_1(a, Red)$	$z(a, Red) - \dfrac{2.3 \nu_{Red}}{nf} \cdot \dfrac{1}{b_a}$	$z(a, Red) - \dfrac{\alpha}{n} \cdot \nu_{Red}$ **
$\left(\dfrac{\partial \log |i_c|}{\partial \log C_{Ox}}\right)_{\eta, C_{Red}}$	$\rho_1(c, Ox)$	$z(c, Ox) - \dfrac{2.3 \nu_{Ox}}{nf} \cdot \dfrac{1}{b_c}$	$z(c, Ox) - \dfrac{1-\alpha}{n} \cdot \nu_{Ox}$
$\left(\dfrac{\partial \log i_a}{\partial \log C_{Red}}\right)_{E, C_{Ox}}$	$\rho_2(a, Red)$	$z(a, Red)$	$z(a, Red)$
$\left(\dfrac{\partial \log |i_c|}{\partial \log C_{Ox}}\right)_{E, C_{Red}}$	$\rho_2(c, Ox)$	$z(c, Ox)$	$z(c, Ox)$
$\left(\dfrac{\partial \eta_a}{\partial \log C_{Red}}\right)_{i, C_{Ox}}$	$-b_a \rho_1(a, Red)$	$-b_a z(a, Red) + \dfrac{2.3 \nu_{Red}}{nf}$	$-b_a \left[z(a, Red) - \dfrac{\alpha}{n} \cdot \nu_{Red} \right]$
$\left(\dfrac{\partial \eta_c}{\partial \log C_{Ox}}\right)_{i, C_{Red}}$	$-b_c \rho_1(c, Ox)$	$-b_c z(c, Ox) + \dfrac{2.3 \nu_{Ox}}{nf}$	$-b_c \left[z(c, Ox) - \dfrac{1-\alpha}{n} \cdot \nu_{Ox} \right]$
$\left(\dfrac{\partial E_a}{\partial \log C_{Red}}\right)_{i, C_{Ox}}$	$-b_a \rho_2(a, Red)$	$-b_a z(a, Red)$	$-b_a z(a, Red)$
$\left(\dfrac{\partial E_c}{\partial \log C_{Ox}}\right)_{i, C_{Red}}$	$-b_c \rho_2(c, Ox)$	$-b_c z(c, Ox)$	$-b_c z(c, Ox)$

7.1 VARIATION OF i_o

Table 7.1 (cont'd) -2-

Experimental parameter	In terms of ρ_1 or ρ_2	In terms of z and b	In terms of z and α
$\left(\dfrac{\partial \log i_o}{\partial E_r}\right)_{C_{Ox}}$	$\rho_1(a, Red) \cdot \dfrac{nf}{\nu_{Red}}$	$\dfrac{nf}{\nu_{Red}} \cdot z(a, Red) - \dfrac{2.3}{b_a}$	$\left(\dfrac{z(a, Red)}{\nu_{Red}} - \dfrac{\alpha}{n}\right) nf$
$\left(\dfrac{\partial \log i_o}{\partial E_r}\right)_{C_{Red}}$	$\rho_1(c, Ox) \cdot \dfrac{nf}{\nu_{Ox}}$	$\dfrac{nf}{\nu_{Ox}} \cdot z(c, Ox) - \dfrac{2.3}{b_c}$	$\left(\dfrac{z(c, Ox)}{\nu_{Ox}} - \dfrac{1-\alpha}{n}\right) nf$
$\left(\dfrac{\partial \log i_o}{\partial \log C_{Red}}\right)_{E_r}$ ***	$\rho_2(a, Red)$	$z(a, Red)$	$z(a, Red)$
$\left(\dfrac{\partial \log i_o}{\partial \log C_{Ox}}\right)_{E_r}$ ***	$\rho_2(c, Ox)$	$z(c, Ox)$	$z(c, Ox)$

* It is assumed here that the sum of the anodic and cathodic transfer coefficients equals unity. The parameter α is defined by the equation $i = i_o[\exp \alpha f\eta - \exp(-(1-\alpha)f\eta)]$ where $f = F/RT$.

** ν_k is the stoichiometric coefficient in the Nernst equation describing equilibrium between reactants and products.

*** Consult the text for limitations of the validity of these equations.

equations take the form (cf. Table 7.1).

$$\left(\frac{\partial \log i_o}{\partial \log C_{Fe^{2+}}}\right)_{C_{Fe^{3+}}} = \rho_1(a, Fe^{2+}) = z(a, Fe^{2+}) - \frac{2.3\, RT}{b_a F} \quad [7.22]$$

$$\left(\frac{\partial \log i_o}{\partial \log C_{Fe^{3+}}}\right)_{C_{Fe^{2+}}} = \rho_1(c, Fe^{3+}) = z(c, Fe^{3+}) - \frac{2.3\, RT}{b_c F} \quad [7.23]$$

and

$$\left(\frac{\partial \log i_o}{\partial E_r}\right)_{C_{Fe^{3+}}} = \frac{nF}{RT} \cdot \frac{1}{\nu_{Fe^{2+}}} \cdot z(a, Fe^{2+}) - \frac{2.3}{b_a} \quad [7.24]$$

$$\left(\frac{\partial \log i_o}{\partial E_r}\right)_{C_{Fe^{2+}}} = \frac{nF}{RT} \cdot \frac{1}{\nu_{Fe^{3+}}} \cdot z(c, Fe^{3+}) - \frac{2.3}{b_c} \quad [7.25]$$

To obtain the reaction order from these equations, it is necessary to determine the Tafel parameters b_a and b_c by independent measurements. This will be done in the next experiment (7.2).

7.1.2 The Cell, Solutions, Electrodes and Electrical Setup

The cell shown in Fig. 8.4, in which the auxiliary electrode has been replaced by a commercial reference electrode, will be used in this experiment. The supporting electrolyte is 1 $\underline{M}$ H_2SO_4 and the required solutions are made up from 1 $\underline{M}$ stock solutions of Fe^{2+} and Fe^{3+}. All solutions are prepared from analytical-grade reagents and triply distilled water. A platinum wire (area 0.1–0.3 cm^2) will serve as the working electrode.

7.1 VARIATION OF i_o

Two reference electrodes will be employed. A commercial electrode, (saturated calomel or other) will be used to measure the reversible potential in different solutions and its variation with concentration. A platinum wire placed in the solution will serve as a reversible Fe^{2+}/Fe^{3+} indicator-type reference electrode and all overpotentials will be measured directly against it. The counter electrode will be a helical platinum wire or cylindrical wire mesh surrounding the working electrode.

It will be necessary to pass very small currents (0.05 - 12μA) in this experiment. This is outside the range of most commercial galvanostats but can be easily attained by employing a fresh dry-cell and a suitable set of resistors (to give about six values of the current in the above range). The galvanostatic circuit (cf. Fig. 2.4) will be employed. The recorder or digital voltmeter should be accurate to 0.1 mV.

7.1.3 Determination of the Exchange-Current Density

Introduce exactly 50.0 ml of 1 $\underline{M}$ H_2SO_4 into the cell and de-aerate by bubbling purified nitrogen for 10 minutes. Introduce the four electrodes and pretreat the platinum working and platinum reference electrodes, following the procedure given in Experiment 6.1.

Add 0.1 ml of the Fe^{2+} stock solution and 0.01 ml of Fe^{3+} stock solution. Stir with a magnetic stirrer, stop the bubbling and pass the gas over the solution. Measure the rest potential of the working electrode with respect to the commercial reference electrode.

Apply low currents alternately in the anodic and cathodic

directions and measure, for each current, the steady-state overpotential with respect to the platinum reference electrode. Measurements should be made to an accuracy of 0.1 mV. Between measurements, allow the open-circuit potential to return to zero. The currents passed in this experiment are such that the overpotential does not exceed about 10 mV in either direction. Measure the geometrical area of the working electrode.

7.1.4 Variation of i_o with E_r and Concentration

Repeat the above experiment, after adding to the solution in turn, the following volumes of 1 M stock solution of Fe^{3+}:

0.02, 0.04, 0.1, 0.2, 0.4 ml.

In addition to the current/overvoltage measurements, determine accurately the reversible potential in each solution (vs the commercial reference electrode).

Repeat the measurements, this time keeping the amount of Fe^{3+} added constant (0.1 ml of the stock solution) and changing the amounts of Fe^{2+} stock solution from 0.01 to 0.4 ml in the same steps as above. Measure E_r carefully in each solution.

7.1.5 Treatment of Results

Plot i vs η for each of the solutions studied and determine the exchange-current density i_o. The i/η plot should be a straight line going through the origin. Discuss the causes of any possible deviations from linearity which you may have observed in this experiment.

Collect your results in the form of a table as shown below:

7.2 MEASUREMENTS AT HIGH OVERPOTENTIALS

$C_{Fe^{2+}}$ mM	$C_{Fe^{3+}}$ mM	$C_{Fe^{2+}}/C_{Fe^{3+}}$	E_r mV(SCE)	i_o mA/cm^2

Plot log i_o <u>vs</u> E_r at constant $C_{Fe^{2+}}$ and at constant $C_{Fe^{3+}}$ and determine the slopes. Calculate the reaction orders from equations [7.24] and [7.25], with the values of b_a and b_c as determined in the next experiment.

Plot log i_o <u>vs</u> log $C_{Fe^{3+}}$ at constant $C_{Fe^{2+}}$ and <u>vs</u> log $C_{Fe^{2+}}$ at constant $C_{Fe^{3+}}$ and determine the slopes. Calculate the reaction orders from equations [7.22] and [7.23], again with the values of b_a and b_c as measured in Experiment 7.2.

Make a Nernst plot and determine the slope and the intercept (at $C_{Fe^{3+}}/C_{Fe^{2+}} = 1$). Compare these with the theoretically expected value of the slope (2.3 RT/F) and the literature value of E^o for this redox system. How is the measured value of E^o affected by the liquid-junction potential?

7.2 Measurements at High Overpotentials

7.2.1 General

This experiment is divided into two parts. In the first part, the Tafel slope for the oxidation of Fe^{2+} and for the reduction of Fe^{3+} will be determined from measurements at relatively high overpotentials. The concentration of reactants will be increased so that complications due to mass-transport limitation can be avoided. The values of the two slopes, b_a and b_c, can then be used to evaluate the reaction order from measurements of the change of log i_o with log C and with E_r, that were performed in Experiment 7.1. Measurements will be made potentiostatically and the potential will be allowed to return to its equilibrium value after each measurement.

In the second part of this experiment, the two kinetic parameters ρ_1 and ρ_2, defined by equations 7.1 and 7.2, will be determined for both species in solution. Their relation to the electrochemical reaction orders z(a, Red) and z(c, Ox) is given in Table 7.1. For a complete set of measurements, it would be necessary to obtain the Tafel plot both in the anodic and cathodic branches at several concentrations. To shorten the procedure, the currents will be measured at fixed anodic and cathodic overpotentials and at fixed anodic and cathodic potentials, as a function of concentration of one reactant at a time, as indicated in Table 7.2 and 7.3 below. The values of the fixed potential were chosen so that with the change in reversible potential E_r, the overpotential η will be in the range of ±0.10 volt to ±0.16 volt, that is, in the linear Tafel region. Measurement at a fixed potential E with respect to a reference electrode amounts to measurement at a fixed metal/solution potential difference $\Delta\phi$, and hence directly yields the electrochemical reaction orders. The Tafel slope must also be determined when measurements of the reaction orders are made at constant overpotential. Values obtained in the first part of this experiment will be used under the assumption that b_a and b_c are both independent of the concentrations of Fe^{2+}

7.2   MEASUREMENTS AT HIGH OVERPOTENTIALS   343

and Fe^{3+}.

7.2.2   The Cell, Solutions, Electrodes and Electrical Setup

The cell, solutions and electrodes will be as in Experiment 7.1. A simple potentiostatic circuit such as shown in Fig. 6.1, will be used. The potential will be set manually for each step and currents will be recorded on a strip-chart recorder.

Table 7.2   Dependence of i_a and i_c on $C_{Fe^{2+}}$

$C_{Fe^{2+}}$ (mM)	$C_{Fe^{3+}}$ (mM)	E_r (mV vs SCE)	i_a		i_c	
			$\eta = 100$ mV	$E = 560$ mV	$\eta = -100$ mV	$E = 320$ mV
0.2	2.0					
0.4	2.0					
0.8	2.0					
2.0	2.0					
4.0	2.0					
8.0	2.0					
20	2.0					

Table 7.3   Dependence of i_a and i_c on $C_{Fe^{3+}}$

$C_{Fe^{3+}}$ (mM)	$C_{Fe^{2+}}$ (mM)	E_r (mV vs SCE)	i_a		i_c	
			$\eta = 100$ mV	$E = 560$ mV	$\eta = -100$ mV	$E = 320$ mV
0.2	2.0					
0.4	2.0					
0.8	2.0					
2.0	2.0					
4.0	2.0					
8.0	2.0					
20	2.0					

7.2.3 Measurement of the η/i Relationship

Introduce into the cell exactly 50.0 ml of 1 $\underline{M}$ H_2SO_4, deaerate by bubbling nitrogen for 10 minutes, then introduce the electrodes and pretreat the two platinum electrodes as in Experiment 7.1. Add 1 ml of Fe^{2+} stock solution and 0.5 ml of Fe^{3+} stock solution (thus providing a solution that is 0.02 $\underline{M}$ in Fe^{2+} and 0.01 $\underline{M}$ in Fe^{3+}). Stir the solution well throughout the experiment. Measure the reversible potential (potentiostat on "stand by") then apply overpotentials over the range -220 to +220 mV in the following order: +80, -80, +90, -90, ..., +210, -210, +220, -220. Allow sufficient time on open circuit between measurements for the system to reach the reversible potential. Measure the steady-state current at each potential.

7.2.4 Measurement of the Current at Constant η and at Constant E

The experimental procedure will be the same as in the preceding section, except that in each solution the current will be measured at only four potential settings, as indicated in Tables 7.2 and 7.3. The order of measurements is as follows: After deaeration of the solution and pretreatment of the electrode, amounts of stock solutions of both reactants needed to reach, say, the first composition given in Table 7.2 are added. The reversible potential is measured, then the four potentials given in the table are applied and the steady-state current measured at each potential. Between measurements, the electrodes are allowed to return to the reversible potential at open circuit. A further amount of stock solution of Fe^{2+} is added to reach the next concentration shown in the table and the measurement is repeated.

7.2 MEASUREMENTS AT HIGH OVERPOTENTIALS   345

The measurements shown in Table 7.3 should be performed in the same way but with successive amounts of Fe^{3+} stock solution added to reach the desired concentrations, as indicated.

7.2.5 Treatment of Results

a. Measurement of the η/i Relationship

Plot η vs log i for both the anodic and cathodic branches and determine b_a, b_c, i_o and α. Estimate the iR correction from the known geometry of the cell (cf. Introduction section to Experiment 2.) Estimate the correction required in this experiment as a result of the change of diffuse-double-layer potential with applied potential. Compare the steady-state method for determining Tafel parameters employed here with the transient (galvanostatic and potentiostatic) methods used in Experiments 8.1 and 8.2.

b. Measurement of the Current at Constant η and at Constant E

Use the data collected in Table 7.2 to evaluate the parameters $\rho_1(a, Fe^{2+})$, $\rho_1(c, Fe^{2+})$, $\rho_2(a, Fe^{2+})$, $\rho_2(c, Fe^{2+})$. Use the data in Table 7.3 to evaluate the same parameters for Fe^{3+}. Calculate from the above ρ_2 values, the four electrochemical reaction orders $z(a, Fe^{2+})$, $z(c, Fe^{2+})$, $z(a, Fe^{3+})$, $z(c, Fe^{3+})$. Calculate the same four reaction orders from the measured ρ_1 values and the Tafel slopes obtained in this experiment.

REFERENCES

1. K. Vetter, "Electrochemical Kinetics", p. 432, Academic Press, 1967.
2. P. Delahay, "Double Layer and Electrode Kinetics", p. 183, Interscience, 1966.
3. E. Gileadi and B. E. Conway in "Modern Aspects of Electrochemistry" Vol. 3 Chap. 5 p. 394. J. O'M. Bockris and B. E. Conway Ed. Butterworths, 1964.

8. DETERMINATION OF KINETIC PARAMETERS BY GALVANOSTATIC AND POTENTIOSTATIC TRANSIENTS

Introduction

8.1 Galvanostatic Transients
8.1.1 General
8.1.2 The Cell, Electrodes, Solutions and Electrical Setup
8.1.3 Preparation of Hanging Mercury Drop
8.1.4 Preparation of Amalgam
8.1.5 Determination of the Current/Potential Curve
8.1.6 Treatment of Results

8.2 Potentiostatic Transients
8.2.1 General
8.2.2 The Cell, Solutions, Electrodes and Electrical Setup
8.2.3 In-Situ Generation of Bromine
8.2.4 Measurements at Low Overpotentials
8.2.5 Measurements at High Overpotentials
8.2.6 Treatment of Results

8. DETERMINATION OF KINETIC PARAMETERS BY GALVANOSTATIC AND POTENTIOSTATIC TRANSIENTS

Introduction

In the study of electrode reactions one is often confronted by the problem of partial or complete mass-transport limitation of the process taking place at the interface under the experimental conditions chosen. Thus, as the current density increases, the concentrations of reactants and products deviate more and more from their values in the bulk of the solution and a suitable correction method must be employed to extract meaningful kinetic parameters from the directly measured quantities. The activation-controlled overpotential η_a is a function of the ratio (i/i_o) where i_o is the exchange-current density for the reaction studied. This means that for fast reactions (i.e., those having a large value of i_o) a large current density must be applied to reach a given overpotential. Thus, mass transport is of major importance in the study of fast electrode reactions, while its effect on slow reactions may be negligible under properly chosen experimental conditions. For measurement in the linear Tafel region, the current and overpotential should satisfy the conditions*

$$i/i_o \geq 5 \quad ; \quad \eta/b \geq 0.7 \qquad [8.1]$$

---

* This corresponds to a maximum deviation from linearity of about 4% with the simplifying assumption that $b_a = b_c$. The corresponding ratios for a 1% deviation from linearity are $i/i_o \geq 10; \eta/b \geq 1$.

INTRODUCTION

On the other hand, for the concentration at the electrode surface to be close enough (within 5%) to the concentration in the bulk, the current density must satisfy the condition

$$i/i_L \leqslant 0.05 \qquad [8.2]$$

where i_L is the mass-transport-limited current density. If, in addition, it is required that the linear Tafel region should extend over at least a tenfold increase in current, we can see from the above inequalities that we must have

$$i_L/i_o \geqslant 10^3 \qquad [8.3]$$

The ratio of i_L/i_o can be increased by efficient stirring. Increasing the concentration of reactants also helps, since in most reactions $(\partial \log i_o/\partial \log C) < 1$ while $(\partial \log i_L/\partial \log C) = 1$ (cf. equation [7.22]). When the above techniques are not sufficient, it is possible to measure at values of i exceeding $0.05\, i_L$ provided that a correction is made for partial control by mass transport. This can be achieved by one of the following techniques: (a) employing fast galvanostatic or potentiostatic transients and resorting to a suitable extrapolation technique to obtain the overpotential or current density, respectively, at time zero; (b) taking measurements with a rotating-disc electrode at various angular velocities and extrapolating to $\omega = \infty$ (c.f. Experiment 6); (c) solving the diffusion equation (cf. equation [8.8]) to obtain the potential-dependent rate constant from the value of i/i_d, where i_d is the diffusion-limited current density and i is the current density, (both measured in unstirred solutions) under conditions of semi-infinite linear diffusion.

A discussion of the scope and limitations of various methods for the study of electrode reactions has been given by Gerischer.[1] A.C. methods were discussed recently by Damaskin[2] and various transient methods have been given by Adams, with particular reference to application at solid electrodes.[3]

When a potential- or current-step function is applied to an electrochemical system, as in Experiments 8.1 and 8.2 below, three processes take place. The ionic double layer is charged, the charge-transfer reaction starts, causing the activation overpotential to reach its steady-state value and finally, concentration gradients are set up near the electrode surface, giving rise to concentration polarization and causing a further increase of overpotential. Charging of the double layer is a relatively fast process (which can be completed in less than 10 μsec) and, unless a very fast reaction is being studied, this process will not interfere with the measurement.

In the following experiments, several extrapolation methods will be used to separate the contributions due to activation- and mass-transport-controlled processes and to evaluate the corrected kinetic parameters for moderately fast electrode reactions.

8.1   Galvanostatic Transients

8.1.1   General

In this experiment, the electrodeposition of Zn^{++} from dilute solutions into a zinc amalgam and the reverse anodic

8.1 GALVANOSTATIC TRANSIENTS

dissolution of the amalgamated zinc will be studied. This is known to be a moderately fast electrode reaction (i_o of the order of 10^{-3} A/cm^2) and particularly suited to the galvanostatic-transient technique. When current steps are applied to the system initially at equilibrium, the potential/time curves have the shape shown in Fig. 8.1. The overpotential first rises rapidly while the double layer is being charged and at the same time the faradaic reaction takes place. This process will last at most 0.5 milliseconds under the experimental conditions chosen in this experiment. Next, the overpotential becomes constant at lower current densities or varies slowly with time at higher current densities. This lasts several times longer than the charging of the double layer, as seen in Fig. 8.1. Finally, the overpotential starts to rise rapidly as the concentrations at the surface of the electrode deviate markedly from the corresponding bulk values. The activation overpotential η_a is obtained from the measurements by extrapolation of the flat region back to zero time. By proper choice of the concentration in solution and in the amalgam and of the current densities employed, the flat region can be extended and made nearly horizontal, thus rendering the values of the extrapolated activation overpotentials more accurate.

8.1.2 The Cell, Electrodes, Solutions and Electrical Setup

The cell and electrode configuration shown in Fig. 3.2 will be used in this experiment. A mercury drop, attached to the end of a platinum wire, will serve as the working electrode. A symmetrical platinum counter electrode and a commercial reference electrode will be employed.

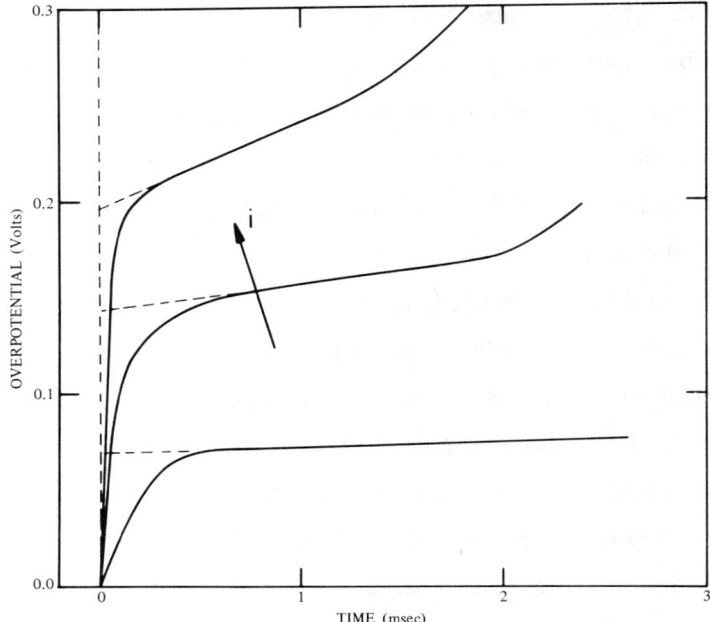

FIG. 8.1 VARIATION OF OVERPOTENTIAL WITH TIME DURING TYPICAL GALVANOSTATIC TRANSIENT FOR ELECTRODEPOSITION OF Zn^{2+} ON MERCURY.

The electrical setup consists of the galvanostatic circuit shown in Fig. 2.4 with an oscilloscope instead of a recorder connected at the output of the impedance-matching unit.*

The solution used in this experiment will be 1 $\underline{M}$ $NaClO_4$, 1 m$\underline{M}$ $HClO_4$ and 2×10^{-2} $\underline{M}$ Zn^{++}, made up of analytical-grade reagents and triply distilled water.

---

* Make sure that the response time of the impedance-matching unit is fast enough for the transients being measured. The 10^7 Ω probe supplied with the oscilloscope may be used to replace the impedance-matching unit if the latter is too slow.

8.1 GALVANOSTATIC TRANSIENTS

8.1.3 Preparation of Hanging Mercury Drop

The hanging mercury drop is formed at the flat end of a thin platinum wire of about 0.2 mm diameter sealed in glass following the design of Shain,[4] as shown in Fig. 8.2. The mercury drop attached to the platinum wire is a standard-size polarographic drop. The platinum is first cleaned and placed in a cell containing 1 $\underline{M}$ $HClO_4$ and a platinum counter electrode (a reference electrode is not required at this stage). The electrode is polarized, first anodically and then cathodically, for ten minutes at 0.50 A/cm^2. After this treatment, a few drops of freshly distilled mercury are introduced into the cell and the exposed end of the platinum is immersed in the mercury while it is still polarized cathodically.[5] The mercury-coated platinum electrode is now removed from the cell, washed in triply distilled water and tapped gently to remove any excess of mercury. When properly done, this coating should hold during the entire experiment.

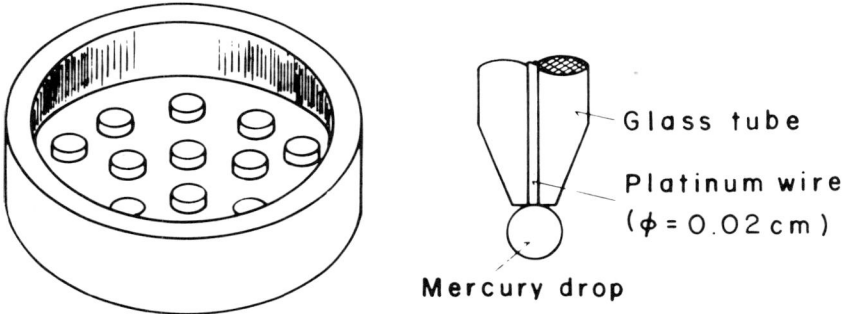

FIG. 8.2 HANGING MERCURY DROP (AFTER SHAIN[4]) AND CONTAINER FOR COLLECTION OF THE DROPS.

Mercury drops of uniform size can be obtained from a regular polarographic dropping-mercury electrode. A flat container with small depressions (shown in Fig. 8.2) is used to collect the drops. The container is filled with triply distilled water, the capillary introduced and allowed to drop for about one minute to reach a steady drop size. The container is then moved slowly so that a single drop falls into each of several depressions. A drop can be picked up by carefully touching the mercury-wetted platinum wire to the drop. The hanging-drop electrode produced in this way is quickly transferred to the cell and an experiment started as soon as possible. The mercury drops should be prepared just before the experiment. The area of the mercury drop should be determined by counting and weighing the drops collected for several minutes, as discussed in Experiment 3.1 (except that the drops are allowed to fall into triply distilled water rather than NaF solution as in Experiment 3.1)

8.1.4 Preparation of Amalgam

The zinc amalgam will be prepared *in situ* just before the experiment. It can be used for many pulses, but if the experiment lasts more than about ten minutes, it is advisable to replace the mercury drop. A particular drop can be used as long as its open-circuit potential (or rest potential which, in the present system, is equal to the reversible potential) is constant and a transient taken at the beginning of the series of measurements can be reproduced.

Place the solution and electrodes into the cell and bubble

8.1   GALVANOSTATIC TRANSIENTS 355

purified nitrogen for ten minutes.* Stir the solution and continue bubbling nitrogen through it all through the experiment. Adjust the oscilloscope trigger to respond to a galvanostatic pulse of 0.05 mA. (The oscilloscope is triggered by the rapid change in potential associated with charging of the double layer.) From the known weight of the drop (determined in the previous section) calculate the charge required to produce a concentration of 0.5 mole percent zinc in the amalgam, assuming 100% faradaic efficiency. Set the current density so that the charge required will be passed in about two minutes and electrolyze for the exact time to reach the above concentration in the amalgam. Begin measurements immediately on the freshly prepared amalgam.

8.1.5   Determination of the Current/Potential Curve

Measure the open-circuit potential of the freshly amalgamated drop in the solution containing Zn^{2+} and mark it on the screen of the storage oscilloscope. This is the reversible potential for the system studied and the potentials measured with respect to it during the transients may be read directly as the total overpotentials. Set the oscilloscope sensitivity to 5-50 mV/cm (depending on current density) and the time base at 0.5 msec/cm. Apply about ten galvanostatic steps in the range of 1.0-100 mA/cm^2 both in the anodic and the cathodic directions. About five transients can be recorded on the same picture. To

---

* It may be found preferable to introduce the working electrode into the cell only at this point, after the solution has been deaerated.

avoid large changes in the concentration of zinc in the amalgam, the transients should be made as short as possible by switching the current on and immediately switching it off manually. Stir the solution between transients. If the experiment lasts longer than ten minutes, or if the open-circuit potential is seen to change noticeably, a new mercury drop should be amalgamated and used.

Plot η vs log $(i)_{t=0}$ as suggested in the first paragraph of the next section. On the basis of this plot, decide which further experimental points should be taken to determine i_o and b from the linear and semi-logarithmic η/i relationships, respectively.

8.1.6 Treatment of Results

Measure the overpotential corresponding to each current density by extrapolating the plateau region of the transients to zero time. Calculate the iR correction as discussed in the introduction to Experiment 2. Plot the overpotential η vs log i for both anodic and cathodic polarization. Determine the Tafel slopes and the exchange-current density. Compare with results in the literature.[6]

Discuss the possible reasons for deviation of the points from the linear Tafel relationship.

On the basis of your results and the literature data, discuss the mechanism of the electrodeposition and dissolution of zinc amalgam from noncomplexing media.

Plot the overpotential vs the current density and evaluate the exchange-current density from the linear section of the curve near the equilibrium potential. Assume for the purpose of this calculation that the stoichiometric number is equal to unity.

8.2   POTENTIOSTATIC TRANSIENTS

Calculate the time which would be required at the lowest and highest current densities used to charge the double layer to the extrapolated value of the overpotential if no faradaic current passed during this time. (Assume an average value of the double-layer capacitance $C_{dl} = 16 \ \mu F/cm^2$).

8.2   Potentiostatic Transients

8.2.1   General

In this experiment, we will study the kinetics of the Br^-/Br_2 redox system in 0.1 $\underline{M}$ $HClO_4$, 0.9 $\underline{M}$ $NaClO_4$ supporting electrolyte on a platinum electrode. Potential steps of varying amplitude will be applied and the solution will not be stirred, so that conditions of semi-infinite linear diffusion during the transient are maintained.* The shape of the current/time transient is shown schematically in Fig. 8.3. After an initial sharp rise associated with double-layer charging, the current decays with time in a hyperbolic fashion, as a result of depletion of the reactant near the electrode surface. The dependence of current density on time has been derived for several specific cases.[7]

For a reaction of the type

$$Ox + ne_M \longrightarrow Red \qquad [8.4]$$

---

* For the relatively short transients employed here (less than 20 msec), semi-infinite linear diffusion in unstirred solution is maintained for almost all electrode configurations without taking special care to eliminate vibrations.

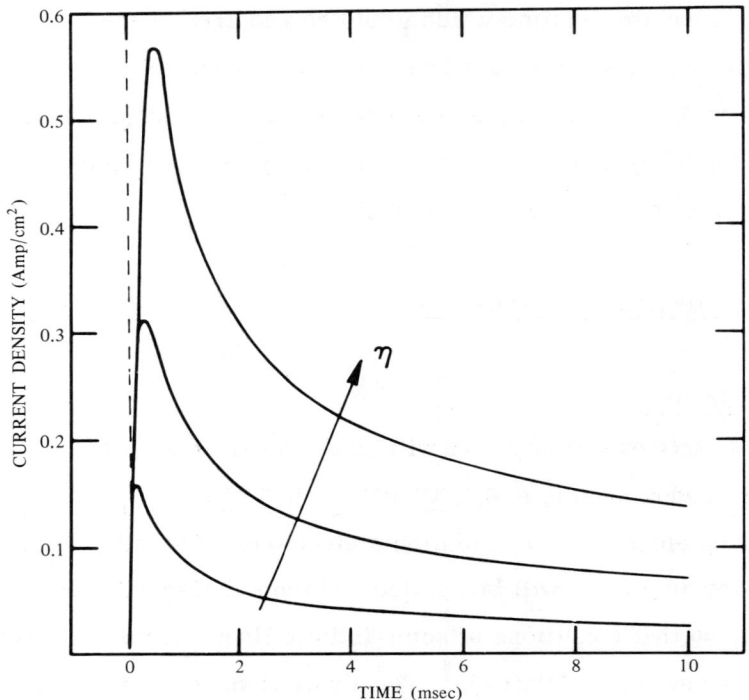

FIG. 8.3 CURRENT/TIME TRANSIENTS DURING POTENTIOSTATIC STEP FOR FAST CHARGE-TRANSFER REACTION WITH $n = 2$, $D = 6 \times 10^{-6}$ cm^2/sec AND $C^o = 5 \times 10^{-5}$ $mole/cm^3$.

under mixed activation and diffusion control (i.e., in the region where $0.05\ i_d < i < i_d$), the current density can be expressed by the equation[8]

$$|i| = nFC^o k_c \exp(\lambda^2)\ \text{erfc}(\lambda) \qquad [8.5]$$

where

8.2 POTENTIOSTATIC TRANSIENTS

$$\lambda = \frac{k_c t^{1/2}}{D_{Ox}^{1/2}} \quad [8.6]$$

in which k_c is the potential-dependent rate constant for reduction of Ox.

The diffusion-limited current density, i_d, is given by

$$|i_d| = nFD_{Ox}^{1/2} C^o (\pi t)^{-1/2} \quad [8.7]$$

Combining the last three equations, we have

$$\frac{i}{i_d} = \pi^{1/2} \lambda \exp(\lambda^2) \operatorname{erfc}(\lambda) \quad [8.8]$$

Tables of (i/i_d) as a function of λ according to equation [8.8] have been calculated and are given in Table 4.2. Thus, by measurement of i and i_d at the same time t after application of the transient, the parameter λ can be obtained. From this one can calculate the rate constant k_c corresponding to each value of the measured current density i, that is, to each value of the applied potential. The diffusion coefficient D_{Ox} may be found in the literature, or determined experimentally with the use of equation [8.7] above.

DETERMINATION OF KINETIC PARAMETERS

A different approach to the evaluation of the activation-controlled reaction rate from measurements taken under mixed activation and diffusion control is to consider the two processes as consecutive steps in a reaction sequence. The measured current density may then be written in the form

$$1/i = 1/i_{ac} + 1/i_{diff} \qquad [8.9]$$

where i_{diff} is the diffusion-controlled current density which would be observed if reaction [8.4] were ideally reversible. The quantity i_{ac} is the activation-controlled current density which would be observed if mass transport to the electrode would be very fast. The current density i_d is given [8] by

$$i_{diff} = nFD_{Ox}^{1/2} C^{o} (\pi t)^{-1/2} (1 + \xi\theta)^{-1} = i_d (1 + \xi\theta)^{-1} \qquad [8.10]$$

in which

$$\xi \equiv (D_{Ox}/D_{Red})^{1/2} \qquad [8.11]$$

8.2 POTENTIOSTATIC TRANSIENTS

and θ is the ratio of concentrations of reactants and products at the electrode surface

$$\theta = (C_{Ox}/C_{Red})_{x=0} \qquad [8.12]$$

θ varies with applied potential according to the Nernst equation.

Combining equations [8.9] and [8.10] we can write

$$1/i = 1/i_{ac} + t^{1/2}/K \qquad [8.13]$$

where K is a potential-dependent quantity given by

$$K = nFD^{1/2} C^o \pi^{-1/2} (1 + \xi\theta)^{-1} \qquad [8.14]$$

Equations [8.9] and [8.13] are of the same form as equation [6.2] for the rotating-disc electrode. The activation-controlled current density i_{ac} may be evaluated by plotting $(1/i)$ <u>vs</u> $t^{1/2}$ and extrapolating back to $t = 0$.

8.2.2 The Cell, Solutions, Electrodes and Electrical Setup

The cell shown in Fig. 8.4 will be used in this experiment. It should be kept tightly closed throughout the experiment to avoid loss of bromine. For this purpose the cell is fitted with a tapered glass joint and a matching teflon cover. All electrodes and tubes passing through the cover are also made to fit tightly.

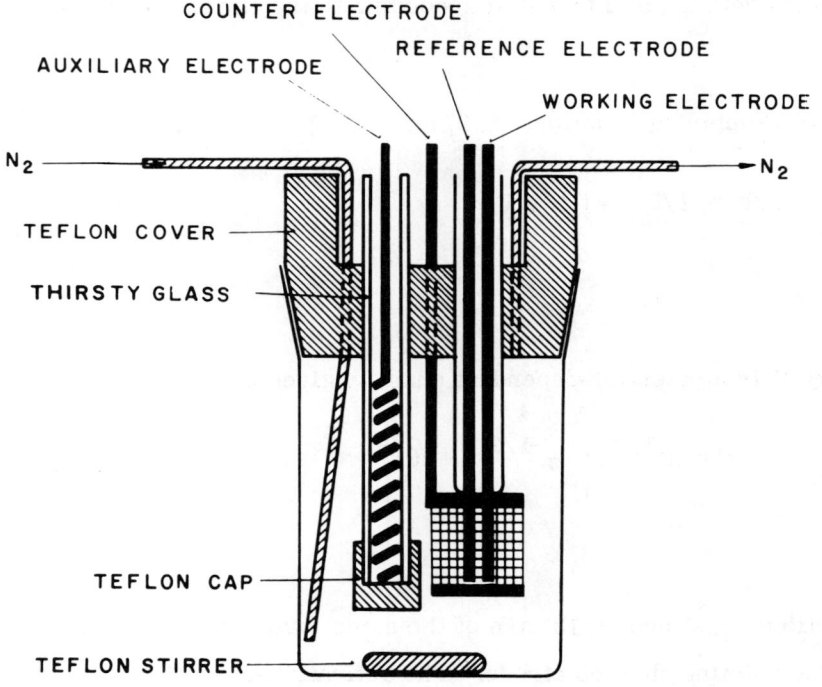

FIG. 8.4 CELL FOR MEASUREMENTS WITH THIN WIRE ELECTRODES MAINTAINING CYLINDRICAL SYMMETRY. DIMENSIONS: DIAMETER OF WORKING AND REFERENCE ELECTRODES 0.02 cm. DISTANCE BETWEEN THEM 0.2 cm.

8.2   POTENTIOSTATIC TRANSIENTS

The final composition of the solution will be 0.1 $\underline{M}$ $HClO_4$, 0.9 $\underline{M}$ $NaClO_4$, 100 m$\underline{M}$ NaBr and 25 m$\underline{M}$ Br_2, made up of reagent-grade chemicals and triply distilled water. Bromine will be generated <u>in situ</u> and so the initial concentration of NaBr must be 150 m$\underline{M}$.

Two platinum wires, about 1 mm apart, are the working and the reference electrodes. A cylinder of platinum gauze serves as the counter electrode. A fourth electrode, used only for the <u>in-situ</u> generation of bromine, will be a platinum wire in the shape of a small helix and separated from the rest of the solution in a thirsty-glass tube, to avoid reduction of the bromine which is formed at the other electrode.

Bromine is generated galvanostatically but all subsequent measurements will be taken potentiostatically. For best results, a low-noise, fast-rise-time potentiostat should be employed. The circuit shown in Fig. 6.1 will be used to apply a sequence of potential steps as shown in Fig. 8.5. The resulting current/time transients will be observed on an oscilloscope which is triggered by the initial rise of the current associated with the charging of the double layer (cf. Fig. 8.3). Between pulses, the overpotential is forced back to zero (the electrode is <u>not</u> left on open circuit). Since a reversible Br^-/Br_2 electrode is employed, this corresponds to zero setting of the reference potential on the potentiostat.

The working and reference electrodes are pretreated electrochemically in a 0.1 $\underline{M}$ $HClO_4$ solution, following the procedure outlined in Experiment 6.1.

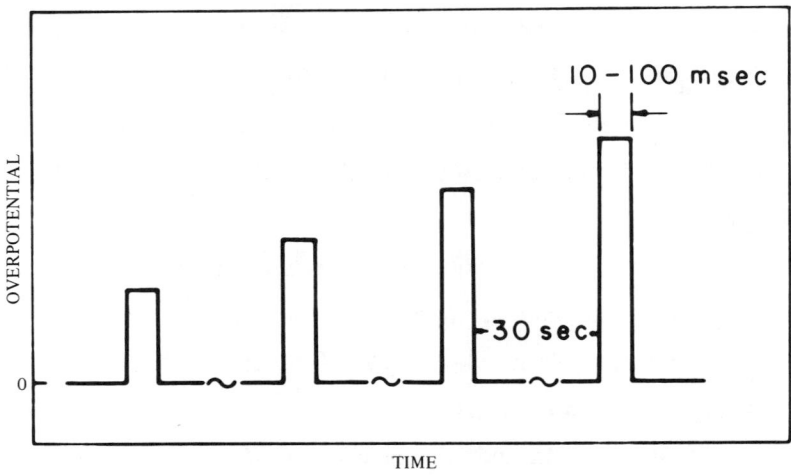

FIG. 8.5   SEQUENCE OF POTENTIAL STEPS APPLIED TO THE WORKING ELECTRODE.

8.2.3   **In-Situ Generation of Bromine**

Add to the cell 20.0 ml of the solution (which is 150 m$\underline{M}$ in NaBr and as yet contains no free Br_2), introduce all four electrodes and deaerate by bubbling purified nitrogen for 10 minutes. Stop the gas stream and do not pass nitrogen either through or above the solution during the rest of the experiment. Connect the counter electrode (which should have an area of at least 5 cm^2) and the auxiliary electrode (in the thirsty-glass tube) to the galvanostat and apply a current of 100 mA. Maintain the same constant current until exactly one third of the Br^- initially in solution has been oxidized to Br_2, in other words, until the final concentrations have become 25 m$\underline{M}$ Br_2 and 100 m$\underline{M}$ Br^-. Stir with a magnetic stirrer throughout this part of the experiment. The gas evolved at the auxiliary electrode may form

8.2   POTENTIOSTATIC TRANSIENTS   363

bubbles which increase the resistance to the point that the galvanostat can no longer pass the required current. This should be checked by measuring the current from time to time during the bromine generation or by recording its value continuously.

Experiments should be conducted as soon as possible after bromine has been generated, in order that changes of concentration due to loss of bromine may be minimized.

8.2.4 Measurements at Low Overpotentials

Disconnect the galvanostat and measure the potential difference between the working and reference electrodes. This should be less than 1 mV since both act as reversible electrodes for the Br^-/Br_2 redox couple. Stop stirring and conduct all further experiments in unstirred solutions. Apply a series of five potential pulses from 2.0 to 10.0 mV in the anodic direction and repeat in the cathodic direction. The pulses should be short (less than 0.1 sec) and the electrodes should be left for 30 seconds at zero overpotential between pulses, to allow sufficient time for the concentration gradients created at the surface of the electrode during the pulse to disappear.

Measure the i/t transients on the oscilloscope, setting the time base to 2 msec/cm and the sensitivity as required. The rise in current due to double-layer charging should trigger the oscilloscope. Take a single picture of the oscilloscope traces for all the anodic pulses and one for all the cathodic pulses. Mark the point corresponding to $i = 0$ on each picture.

8.2.5 Measurements at High Overpotentials

These measurements will be limited to the cathodic region. Continue the measurement as in the previous section and in the same solution. Apply increasingly cathodic potential pulses at 10 mV intervals up to 150 mV and record all the i/t transients on one or two pictures. (All transients recorded on a single picture should be taken at the same sensitivity.) Take four more measurements at 50 mV intervals (200, 250 . . .) and record each transient separately.

8.2.6 Treatment of Results

Low Overpotential

Measure the currents at t = 8 msec after application of the transient and determine i_o/ν from the linear i/η plot.

Extrapolate the initial parts of the transients back to t = 0 and plot η against the values of i(t = 0). Calculate i_o/ν as above and compare.

High Overpotential

Read the current density at t = 8 msec and plot η <u>vs</u> log i (t = 8 msec). Determine the potential at which the current has reached its limiting value. Plot log i <u>vs</u> log t for a transient at this potential and determine the slope. Is the reaction diffusion-limited? How would a preceding chemical step affect this plot?

Determine the diffusion coefficient for Br_2 from the above plot.

Read off the values of the current at different times for all the transients. Tabulate the data as shown below and plot (1/i)

8.2   POTENTIOSTATIC TRANSIENTS   365

$\underline{vs}\ t^{1/2}$ to determine i_{ac} (which is $i(t = 0)$).

η (mV) \ time (msec)	4	8	12	16	extrapolated to $t = 0$ from $(1/i)$ $\underline{vs}$ $t^{1/2}$
20					
30					
.					
.					
.					
.					
350					

Plot the Tafel line using the values of $i(t = 0)$ which were obtained from the plot of $(1/i)$ $\underline{vs}$ $t^{1/2}$. Determine the Tafel slope b and the exchange-current density i_o. For comparison, determine the same quantities from a plot of η $\underline{vs}$ log $i(t = 8$ msec).

Compare the exchange-current densities obtained from low- and high-overpotential measurements and comment on the value of the stoichiometric number which may be deduced.

Use the values of i measured at t = 8 msec to calculate

the parameter λ (cf. equation [8.8] and Table 4.2) and tabulate as shown below.

η(mV)	i (mA/cm^2)	i/i_d	λ
20			
30			
.			
.			
.			
.			
350			

Plot η vs log λ and determine the slope, which is equal to the Tafel slope, (why?) and the exchange-current density. The latter is calculated by extrapolating the plot to $\eta = 0$ and finding the value of i/i_d corresponding to the value of λ at $\eta = 0$.

Comment on the reliability of the Tafel parameters obtained by the three methods (from i(t = 8 msec), from i(t = 0) extrapolated by plotting 1/i vs $t^{1/2}$, and from λ(t = 8 msec)).

Comment on the applicability of the equation for semi-infinite linear diffusion to cylindrical electrodes employed in this experiment. (Hint: Consider the variation of the diffusion-layer thickness with time.)

Estimate the error in the values of b and i_o due to iR loss (cf. Introduction section to Experiment 2).

REFERENCES

1. H. Gerischer, Z. Elektrochem. $\underline{59}$, 604 (1955).
2. B. B. Damaskin, "The Principles of Current Methods for the Study of Electrochemical Reactions", McGraw-Hill, 1967.
3. R. N. Adams, "Electrochemistry at Solid Electrodes", Marcel Dekker Inc., New York, 1969.
4. I. Shain, Anal. Chem. $\underline{33}$, 1966 (1961).
5. L. Ramaley, R. L. Brubaker and C. G. Enke, Anal. Chem. $\underline{35}$, 1088 (1963).
6. H. Gerischer, Z. physik. Chem. $\underline{202}$, 302 (1953).
7. P. Delahay, "New Instrumental Methods in Electrochemistry", Chapter 3, Interscience Publishers, New York, 1954.
8. Ref. 7, Chapter 4.
9. E. Kirowa-Eisner, N. Tshernikovski and U. Eisner, J. Electrochem. Soc., $\underline{120}$, 361 (1973).

9. CYCLIC VOLTAMMETRY

Introduction

9.1 Effect of Medium and of Electrode Material
9.1.1 General
9.1.2 The Cell, Solutions and Electrical Setup
9.1.3 Background Current and Cycling History
9.1.4 The Effect of the Medium on Reversibility
9.1.5 The Effect of Electrode Material on Reversibility
9.1.6 Treatment of Results

9.2 Kinetic Parameters from Linear Potential Sweep
9.2.1 General
9.2.2 The Cell and Electrical Setup
9.2.3 The Effect of Sweep Rate on Current/Potential Transients
9.2.4 Treatment of Results

9.3 Mechanism of Oxidation of p-phenylenediamine (i) Determination of the Number of Electrons Transferred by Constant-Current Electrolysis
9.3.1 General
9.3.2 The Cell, Solutions and Electrical Setup
9.3.3 Following the Change in Concentration of PPD in Solution
9.3.4 Determination of the Electrolysis Current
9.3.5 Measurement of n

9.4 **Mechanism of Oxidation of p-phenylenediamine (ii)
Determination of the Number of Electrons Transferred by
Constant-Potential Electrolysis**

9.4.1 General

9.4.2 The Cell, Electrodes and Electrical Setup

9.4.3 Measurement of n

9.4.4 Treatment of Results

9.5 **Mechanism of Oxidation of p-phenylenediamine (iii)
Steps in the Overall Reaction**

9.5.1 General

9.5.2 The Cell, Electrodes and Electrical Setup

9.5.3 Identification of Species in Solution

9.5.4 Proof of Slow Hydrolysis in 1 $\underline{M}$ H_2SO_4

9.5.5 The Protonation of PPD and of PAP

9.5.6 The Catalytic Effect of the Electrode on the Rate of Hydrolysis

9.5.7 Treatment of Results

9. CYCLIC VOLTAMMETRY

Introduction

In cyclic voltammetry the potential applied to an electrode (by means of a potentiostat) is changed linearly with time in a repetitive manner. The current is measured as a function of potential or time. If only a single anodic or cathodic sweep is performed, the technique is usually referred to as the linear potential-sweep method. In both cases, the experiment is conducted in unstirred solutions, where convection is eliminated as far as possible.

The fundamental equations for linear potential sweep and cyclic voltammetry have been developed by Delahay[1], Shain[2,3] and others[4-7]. The basic feature of a voltammogram (i.e. a plot of current vs potential during cyclic voltammetry or linear potential sweep) is the appearance of a current peak at a potential characteristic of the electrode reaction taking place (Fig. 9.1). The position and shape of a given peak depends on such factors as sweep rate, electrode material, solution composition and the concentration of reactants. Of the two techniques, only linear potential sweep can provide accurate kinetic parameters, since the equations derived apply only if there are no concentration gradients in solution just before the sweep is started. Cycling the potential several times creates complex concentration gradients near the electrode surface and the boundary-layer problem has not been solved. Thus, cyclic voltammetry is better suited to the identification of steps in the overall reaction and of new species which appear in solution during electrolysis as a

INTRODUCTION

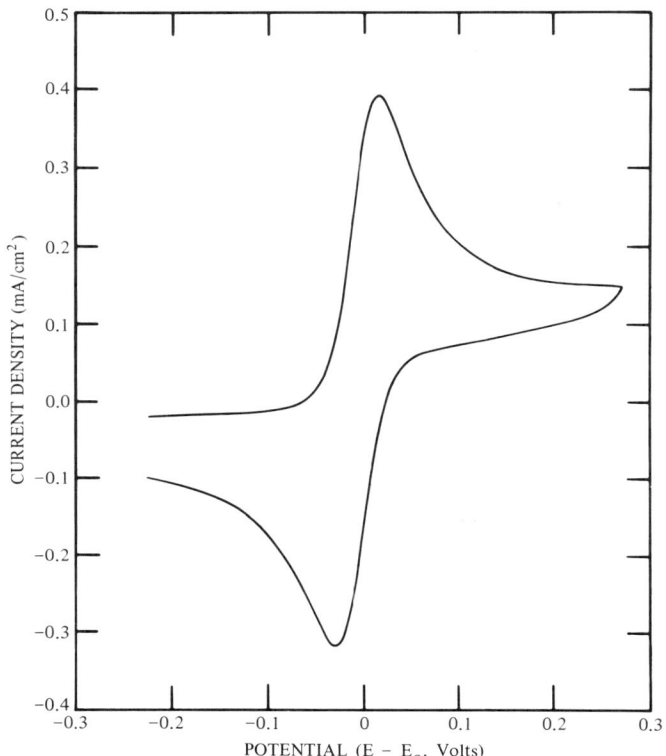

FIG. 9.1 TYPICAL CYCLIC VOLTAMMOGRAM
3 m$\underline{M}$ Fe^{2+} IN 1 $\underline{M}$ H_2SO_4 v = 50 mV/sec.

result of combined electrochemical and chemical steps. Nevertheless, an approximate value of the rate constant can be derived from the separation of the anodic and cathodic peak potentials at a given sweep rate for irreversible reactions.

The mechanisms derived from cyclic voltammetry alone are relevant only to the specific experimental conditions chosen, and may not be applicable to steady-state electrolysis of the same systems in well stirred solutions. Kinetic parameters can be evaluated correctly only if the mechanism of reaction is known and the equations relevant to this mechanism are employed. In spite of these limitations, cyclic voltammetry has been widely employed for the study of organic electrode reactions and it is particularly valuable for the qualitative detection of intermediates formed in complex reaction sequences. Numerous calculations of the shape of the voltammogram for various assumed mechanisms have been published and compared with the experimentally observed curves. A comprehensive list of references has been given recently (cf. p. 160 in ref. 9).

The basic equations of linear potential sweep relate the peak current density i_p and the corresponding potential E_p to k_s, the electrochemical specific rate constant at the standard potential E^o, the Tafel slope b, the concentration in solution C^o and the sweep rate $v = dE/dt$. For a simple cathodic charge-transfer process under reversible conditions, where both reactant and product are soluble, i_p in Amp cm^{-2} at 25°C is given by

$$|i_p| = 2.72 \times 10^5 \, n^{3/2} \, D^{1/2} \, C^o \, v^{1/2} \qquad [9.1]$$

where D is in cm^2 sec^{-1}, C^o in mole cm^{-3} and v in Volt sec^{-1}. E_p is related to the polarographic half-wave potential $E_{1/2}$ by:

$$E_p = E_{1/2} - 1.1\frac{RT}{nF} \qquad [9.2]$$

Note that E_p is quite close to $E_{1/2}$, which itself is related to the standard electrode potential by the equation

INTRODUCTION

$$E_{1/2} = E^o - \frac{RT}{nF} \ln\left(\frac{f_{Red}}{f_{Ox}}\right)\left(\frac{D_{Ox}}{D_{Red}}\right)^{1/2} \quad [9.3]$$

where the subscripts refer to the oxidized and reduced species and $\underline{f}$ stands for the activity coefficient.

Under totally irreversible conditions, i.e. when the rate of the reverse reaction is negligible throughout the potential region studied, the following equations apply[*]:

$$|i_p| = 3.01 \times 10^5 \, n \left(\frac{2.3RT}{bF}\right)^{1/2} D^{1/2} C^o v^{1/2} \quad [9.4]$$

where all units are as in equation [9.1] above and b is in Volts. The peak potential is given under these conditions by

$$E_p = E_{1/2} - b[0.52 - \tfrac{1}{2}\log(b/D) - \log k_s + \tfrac{1}{2}\log v] \quad [9.5]$$

The variation of E_p with sweep rate is an indication of the departure of the system from equilibrium. The specific rate constant k_s at the standard potential and the Tafel slope can be calculated from a plot of E_p <u>vs</u> log v according to equation [9.5], provided that the diffusion coefficient is known.

At sufficiently slow sweep rates a system will behave reversibly, while at high sweep rates it will behave irreversibly. The transformation from one type of behavior to the other occurs

---

[*] These equations are given in a form different from that commonly used, for reasons discussed elsewhere[8] (cf. section II 10.3). The transformation can easily be made by substituting $b = (2.3RT/\alpha n_a F)$ in the corresponding equations in, say, Ref. 1, p. 126.

at a characteristic sweep rate v_c which depends primarily on the value of k_s and, to a lesser degree, on D and b. At this sweep rate the values of E_p calculated from equations [9.2] and [9.5] are equal, that is

$$E_{1/2} - 1.1\frac{RT}{nF} = E_{1/2} - b[0.52 - \frac{1}{2}\log(b/D) - \log k_s + \frac{1}{2}\log v_c] \quad [9.6]$$

and k_s may be calculated from the point of intersection of the two straight portions of the curve in Fig. 9.2, i.e. from the value of v_c.

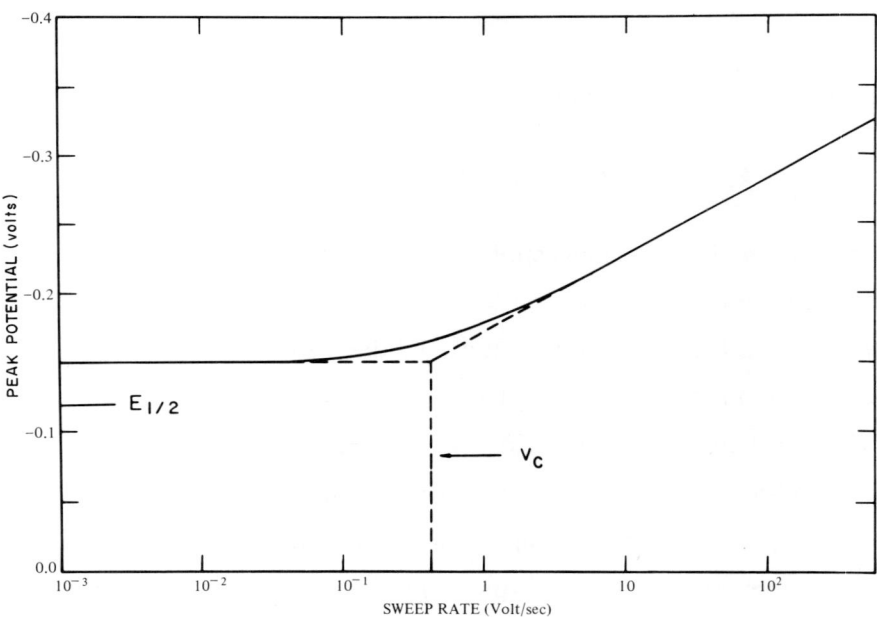

FIG. 9.2 VARIATION OF THE PEAK POTENTIAL E_p WITH SWEEP RATE IN THE REVERSIBLE AND IRREVERSIBLE REGIONS.

The equations given above apply quantitatively only to first-order charge-transfer reactions with no kinetic or catalytic complications. Equations for various more complex situations have been derived[2].

9.1 Effect of Medium and of Electrode Material

9.1.1 General

This experiment is intended to familiarize the student with the technique of cyclic voltammetry. The effects of electrode material and solution composition on the shape of the current/potential curve will be shown, and the reversibility of the reaction will be determined by measuring the distance between the anodic and cathodic peaks. This distance should be $2.2\, RT/nF$ for a reversible reaction, according to equation [9.2]. The "history" of the system, i.e. the difference between the first and subsequent cycles will also be examined.

9.1.2 The Cell, Solutions and Electrical Setup

The cell shown in Fig. 9.3 or in Fig. 9.4 will be used in this experiment. In view of the experimental conditions chosen (slow sweep rate, low concentration of reactant and high concentration of supporting electrolyte) the iR drop in solution is small and the position of the Luggin capillary is not critical.

Three types of working electrodes will be used: platinum, gold and pyrolytic graphite. Each is prepared by pressure-fitting into teflon and has a flat exposed surface which can be conveniently cleaned by abrading with fine alumina or diamond

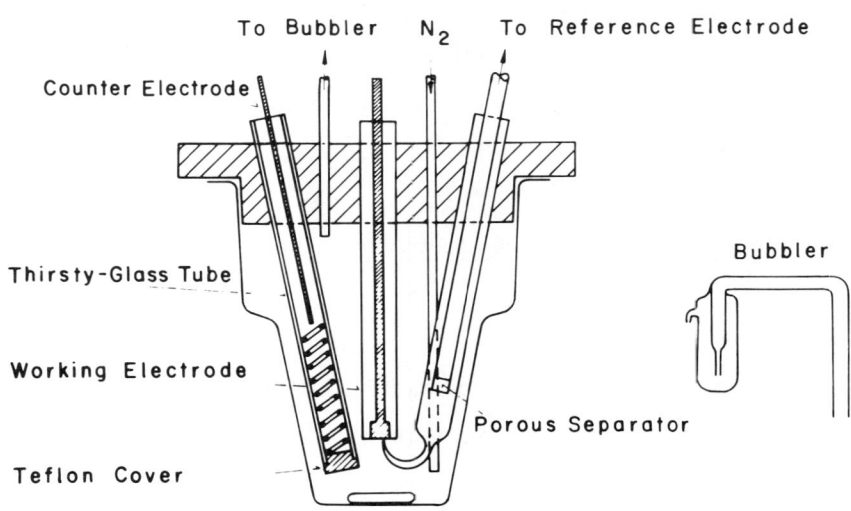

FIG. 9.3   CELL FOR CYCLIC VOLTAMMETRY.

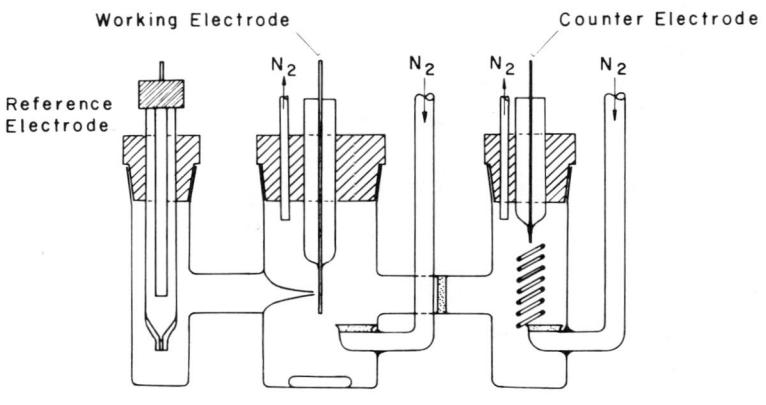

FIG. 9.4   THREE-COMPARTMENT CELL.

9.1 EFFECT OF MEDIUM AND ELECTRODE MATERIAL   377

powder. Contact with the pyrolytic-graphite electrode is made with a spring-loaded metal ribbon.

Measurements are made potentiostatically. The potential is varied linearly with time between fixed limits and the current is plotted as a function of potential on an X-Y recorder. The block diagram of the circuit is shown in Fig. 9.5.

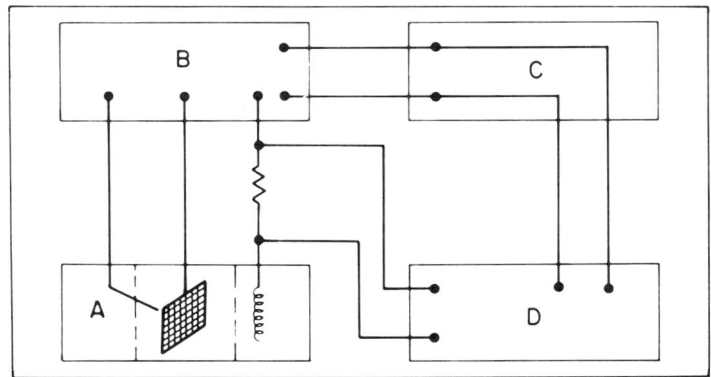

FIG. 9.5   BLOCK DIAGRAM OF POTENTIOSTATIC CIRCUIT FOR CYCLIC VOLTAMMETRY
A - CELL;   B - POTENTIOSTAT;
C - FUNCTION GENERATOR;
D - OSCILLOSCOPE OR X-Y RECORDER.

9.1.3 Background Current and Cycling History

The background current which flows under given experimental conditions, in the absence of the electroactive material, must be determined before meaningful experiments can be made. Measurements are usually confined to potential regions where the background current is negligible compared to the current due to the reaction being studied. If such a region cannot be found, a correction must be applied in the calculation of results. It must be remembered in such cases that the background current depends on the medium, the electrode potential and also on the sweep rate.

To determine the background current, use the platinum working electrode in 1 $\underline{M}$ H_2SO_4. Clean the electrode mechanically and pretreat electrochemically as outlined in Experiment 6.1. Deaerate the solution by bubbling purified nitrogen for 10 minutes before the experiment and continue to pass the gas above the solution during the experiment. Obtain the i/E plot in the range of 0.0-1.0 Volt <u>vs</u> SCE at a sweep rate of 40 mV/sec. Use a recorder sensitivity of 10 μA/cm per cm^2 of electrode area.

Repeat the same measurement with gold and pyrolytic-graphite working electrodes. Clean these electrodes mechanically as above but do not pretreat them electrochemically.

Prepare a solution of 3 m$\underline{M}$ Fe^{3+} in 1 $\underline{M}$ H_2SO_4. Deaerate as before and continue to pass the gas above the solution during the experiment. Use the platinum working electrode and obtain successive i/E curves at a sweep rate of 40 mV/sec., until the shape of the curves does not change with further cycling. This will be referred to as the steady-state voltammogram. Note particularly the difference between the first and subsequent transients.

9.1.4 The Effect of the Medium on Reversibility

Use a platinum working electrode to compare the voltammograms obtained in 3m$\underline{M}$ Fe^{3+} solution with 1$\underline{M}$ H_2SO_4 and with 1$\underline{M}$ HCl:5$\underline{M}$ $CaCl_2$ as supporting electrolyte. Use a sweep rate of 200 mV/sec in this experiment and plot both transients on the same graph paper. Measure the distance between the anodic and the cathodic peaks (ΔE_p) and compare with the value expected from equation

9.2 KINETIC PARAMETERS

[9.2] for a reversible charge transfer involving a single electron.

9.1.5 The Effect of Electrode Material on Reversibility

Record, on the same paper, the current/potential curves obtained in a solution 3m$\underline{M}$ in Fe^{3+} and 1$\underline{M}$ in H_2SO_4 for platinum, gold and pyrolytic-graphite electrodes at a sweep rate of 40 mV/sec. (Note that, because of the differences in electrode area, different sensitivities may have to be used on the Y axis of the recorder.) Repeat in a 3m$\underline{M}$ Fe^{3+} solution which is 1$\underline{M}$ in HCl and 5$\underline{M}$ in $CaCl_2$.

Measure the difference between the anodic and cathodic peaks in each case and discuss.

9.1.6 Treatment of Results

What are the partial currents which contribute to the background current? How will each depend on the sweep rate?

What are the reasons for the difference in the shape of the current/potential curves between the first and subsequent sweeps?

Discuss the effect of electrode material and of supporting electrolyte on the shape of the current/potential curves.

9.2   Kinetic Parameters from Linear Potential Sweep

9.2.1 General

The purpose of this experiment is to evaluate the behavior of a system during linear potential sweep over a wide range of sweep rates so that both reversible and irreversible behavior can be observed. The distance between cathodic and anodic peaks,

ΔE_p, or the dependence of the potential of the peak on sweep rate, may be used as tests for the reversibility of the reaction. Either the number of electrons taking part in the reaction, the diffusion coefficient, or the concentration of electroactive material may be evaluated from measurement of the peak current in the reversible region if the other two quantities are independently obtainable (cf. equation [9.1]). The Tafel slope may be obtained from the dependence of E_p on sweep rate in the irreversible region (cf. equation [9.5]) and from the same equation one may calculate k_s, the electrochemical rate constant at the standard reversible potential. The plot of E_p vs log v has two linear sections corresponding to equations [9.2] and [9.5]. The point of intersection gives the critical value of the sweep rate which marks the transition from reversible to irreversible behavior.

The experimental system studied will be 3m$\underline{M}$ Fe^{3+} in 1$\underline{M}$ H_2SO_4 at a platinum electrode. This system is chosen because the specific electrochemical rate constant has an intermediate value so that both reversible and irreversible behavior can be observed in the range of sweep rates available experimentally. Single-cycle voltammograms should be employed.

9.2.2 The Cell and Electrical Setup

The same cells as in Experiment 9.1 may be used in this experiment. Special care should be exercised in positioning the reference electrode or the Luggin capillary with respect to the working electrode, since a substantial iR drop may distort the shape of the current/potential curve, particularly in the vicinity

9.2 KINETIC PARAMETERS

of the peak, (cf. Fig. 1.26). At sweep rates greater than about 1 Volt/sec., the internal resistance of a commercial reference electrode may lead to considerable distortion in the shape of the transient. To avoid this problem, a palladium/hydrogen electrode prepared as in Experiment 2.2 or a platinized-platinum electrode* is introduced into the cell close to the working electrode. Either of these can serve as a low-resistance reference electrode during the short transient.

The electrical setup should be as in Experiment 9.1. The sweep generator employed should be capable of supplying pulses in the range 0.003 to 20 Volt/sec. A fast-response X-Y recorder (0.5 sec. for full-scale deflection) can be used for sweep rates of up to a few hundred mV/sec. At higher rates, a plug-in type oscilloscope (such as the Tektronix type 564) should be used in which the time base is replaced by an amplifier, so that the oscilloscope operates as a fast X-Y recorder.

9.2.3 The Effect of Sweep Rate on Current/Potential Transients

A platinum working electrode in a solution 3m$\underline{M}$ in Fe^{3+} and 1$\underline{M}$ in H_2SO_4 will be used in this experiment. Deaerate by bubbling purified nitrogen through the solution with stirring for about ten minutes. Continue to pass nitrogen over the solution

---

* The platinized-platinum electrode is not a suitable reference electrode for all measurements in this system (cf. Experiment 2). Its potential is sufficiently stable to permit its use as a reference electrode during a short transient, but should be checked against a commercial reference electrode just before the application of each transient.

during the experiment. Allow about 30 seconds for the solution to settle and then apply a single triangular sweep over the range 0.0 to 1.0 Volts vs SCE at the rate of 3 mV/sec., starting at 1.0 Volt.

Bubble nitrogen with stirring for 20 seconds, allow to settle for 30 seconds and apply a single triangular sweep over the same potential range, at the rate of 6 mV/sec.

Repeat the above procedure, each time doubling the sweep rate up to about 0.5 Volt/sec.

Replace the X-Y recorder by an oscilloscope and continue applying triangular sweeps at rates of from 0.1 Volt/sec. up to 10 Volt/sec. (the upper limit which can be achieved depends on the type of potentiostat used and on the cell configuration). For measurements at high sweep rates the palladium/hydrogen or platinized-platinum electrode should be used as the reference. The potential range is adjusted so that the sweep is between the same potential limits in all experiments.

Change the sensitivity of the Y axis of the recorder (or oscilloscope) as you increase the sweep rate. (Remember that i_p is proportional to $v^{1/2}$).

Measure the background current in the supporting electrolyte at several sweep rates. The relative value of this current decreases with decreasing sweep rate so that it may become negligible below a certain rate.

Measure the geometric area of the working electrode.

9.2.4 Treatment of Results

Plot the peak potential E_p (cathodic) as a function of log v.

9.3   OXIDATION OF P-PHENYLENEDIAMINE (i)   383

Evaluate the Tafel slope and the specific rate constant at E^o, with the equations given in the introduction. Note that k_s can be calculated from equation [9.6] even when measurements in the reversible region cannot be made, if E^o (and hence $E_{1/2}$, cf. equation [10.5]) is known.

Measure the distance between cathodic and anodic peaks, ΔE_p, and plot this against log v. Explain the deviations, if any, from the value given by equation [9.2] at low sweep rates.

Plot log i_p (cathodic) <u>vs</u> log v and determine the slope. Should there be a break in this plot? Explain.

Calculate the diffusion coefficient of Fe^{3+} in this system at low sweep rates (equation [9.1]) and at high sweep rates (equation [9.4]). In the second case use your experimental value of the Tafel slope, b.

Calculate the critical sweep rate v_c at 80°C from the value found in your experiment, assuming (arbitrarily) that the apparent energy of activation for this reaction is 10 kcal/mol and the energy of activation for the diffusion of Fe^{3+} ions is 5 kcal/mol.

9.3   <u>Mechanism of Oxidation of p-phenylenediamine (i)</u>
      <u>Determination of the Number of Electrons</u>
      <u>Transferred by Constant-Current Electrolysis</u>

9.3.1   <u>General</u>

This is the first of three experiments dealing with the mechanism of oxidation of p-phenylenediamine (PPD). In this experiment the number of electrons taking part in the overall reaction is determined. A solution of known concentration of PPD is electrolyzed at a constant current density and anodic linear potential-sweep measurements are made at fixed times during the electrolysis. The change in concentration of PPD is followed by measuring the peak current, i_p, in each voltammogram.

9.3.2 The Cell, Solutions and Electrical Setup

The cells shown in Fig. 9.3 will be used in this experiment. There are two working electrodes; a platinum-gauze electrode of large surface area is used for the constant-current electrolysis* and a small platinum electrode is employed for the linear potential-sweep measurements. A commercial reference electrode will be used. The platinized-platinum counter electrode should be effectively separated from the rest of the solution, in order to prevent reduction products formed on it from reaching the working electrode. A thirsty-glass tube (Corning type 7930) can serve as a low-resistance bridge. Platinizing the counter electrode serves to eliminate the formation of large gas bubbles at its surface during electrolysis, which might create a high resistance between the working and counter electrodes.

Commercially available, analytical-grade p-phenylenediamine dihydrochloride can be used without further purification. A 50 ml. stock solution, 0.10 $\underline{M}$ in PPD should be prepared just before use.

Two electrical circuits are used in this experiment: a simple galvanostatic circuit (Figure 2.4) for constant-current

---

* This should be large enough to enable performing the electrolysis in a reasonable time (about 20 minutes for 10% of the material to be oxidized).

9.3 OXIDATION OF P-PHENYLENEDIAMINE (i) 385

electrolysis and a potentiostatic circuit with a triangular-sweep generator (Figure 9.5) for linear potential sweep. The counter electrode is common to both circuits.

9.3.3 Following the Change in Concentration of PPD in Solution

Pretreat the small platinum working electrode electrochemically as outlined in Experiment 6.1 and determine the background current in the supporting electrolyte (0.05 $\underline{M}$ H_2SO_4, 1 $\underline{M}$ Na_2SO_4) as in Experiment 9.1.

Replace the solution with one containing 2 m$\underline{M}$ PPD in addition to the supporting electrolyte. Deaerate as before and carry out a triangular sweep between the same limits, beginning at 0.0 Volts. In order to test the reproducibility of measurement of i_p (and hence of concentration), stir the solution for 20 seconds, allow to settle for 30 seconds, then repeat the triangular sweep. Repeat until i_p is reproducible to within about 2%. Some difficulty may arise in the determination of i_p if the base line (i.e. the current at potentials before the beginning of the wave) is appreciable and depends on potential. A method for evaluating i_p graphically is shown in Figure 9.6 for those typical cases which may be encountered experimentally. Different methods of determining the base line and the peak current have been discussed by Adams[9].

9.3.4 Determination of the Electrolysis Current

Constant-current electrolysis is performed in a well stirred solution. To ensure 100% faradaic efficiency, the current should be well below the limiting current i_L throughout the

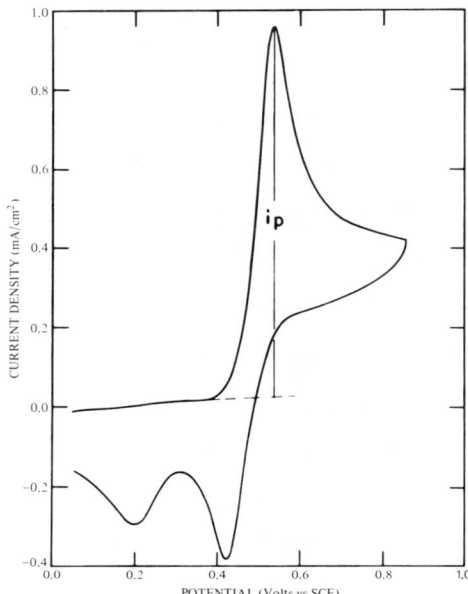

FIG. 9.6   CYCLIC VOLTAMMOGRAM OF 2 m$\underline{M}$ PPD IN 0.05 $\underline{M}$ H$_2$SO$_4$, v = 40 mV/sec. GRAPHICAL METHOD FOR THE EVALUATION OF i_p IS SHOWN.

electrolysis. To determine i_L use the large platinum-gauze working electrode and set its potential at the peak potential determined by linear potential sweep in the previous section. Determine i_L under given stirring conditions and electrolyze at a current of about 0.3 i_L. This value is chosen because electrolysis should be conducted until about half of the material present is oxidized (so that the limiting current at the end of electrolysis is about half of its value at the beginning) and an additional margin of safety is taken in case the conditions of stirring are slightly changed or the electrolysis is carried out

9.4 OXIDATION OF P-PHENYLENEDIAMINE (ii)

for a somewhat longer period.

9.3.5 Measurement of n

Measure exactly 50.0 ml of 2 m$\underline{M}$ PPD in 0.05$\underline{M}$ H_2SO_4, 1$\underline{M}$ Na_2SO_4 into a cell containing the two working electrodes that were used in the previous experiment. Bubble purified nitrogen for 10 minutes through the solution while stirring then continue passing the gas above the solution. Allow the solution to rest for 30 seconds then apply a linear potential sweep and determine i_p at the small platinum electrode. Carry out this measurement three times to obtain an average value. Stir the solution briefly between measurements.

Apply a constant current to the platinum-gauze electrode while stirring efficiently. The value of this current should be as determined in the previous section under similar conditions of stirring.

Continue the electrolysis until about 15% of the PPD has been oxidized (assume n = 2 to make this calculation). The duration of electrolysis should be measured accurately by a stop watch. Stop the electrolysis and stirring and measure i_p a few times as previously. Repeat twice more, oxidizing about the same amount of PPD each time.

Calculate n from the change in concentration (as determined by the decrease in i_p) and the number of coulombs passed.

Discuss the possible sources of error in this type of measurement.

9.4 Mechanism of Oxidation of p-phenylenediamine (ii)
Determination of the Number of Electrons
Transferred by Constant-Potential Electrolysis

9.4.1 General

In this experiment the potential is set at a high anodic value so that the concentration of reactant at the electrode surface is zero throughout the experiment. The current measured at any time during electrolysis under these conditions is the limiting current and its decrease with time reflects the decrease in bulk concentration due to electrolysis. The solution is stirred efficiently throughout the experiment.

The decrease in the number of moles, N, of reactant as a result of electrolysis is related to the current by the equation

$$iA = -nF\frac{dN}{dt} = -nF\frac{d(CV)}{dt} \qquad [9.7]$$

Where A is the area of the electrode, C is the concentration and V is the volume of solution. Rearranging equation [9.7] one has,

$$\frac{dC}{dt} = -\frac{iA}{nFV} \qquad [9.8]$$

Since the limiting current is proportional to the concentration, $i/C = i^o/C^o$ and

$$\frac{di}{dt} = \frac{i^o}{C^o} \cdot \frac{dC}{dt} = -\frac{i^o}{C^o} \cdot \frac{iA}{nFV} \qquad [9.9]$$

where i^o and C^o are the initial current and concentration, respectively. Upon integration, one obtains

9.4 OXIDATION OF P-PHENYLENEDIAMINE (ii)

$$\ln \frac{i}{i^o} = -\frac{i^o A}{nFC^o V} t \qquad [9.10]$$

The value of $i^o A$, which is the total current passed at $t = 0$, is determined by extrapolation of the plot of ln i <u>vs</u> t back to $t = 0$. The parameter n is determined from the slope of the same plot.

9.4.2 The Cell, Electrodes and Electrical Setup

The cell and electrodes are identical with those used in Experiment 9.3 except that only one working electrode, the large platinum gauze, will be used. The Luggin capillary should be placed close to the working electrode.

The potentiostatic circuit shown in Fig. 6.1 will be employed and the current will be recorded as a function of time.

9.4.3 Measurement of n

Measure exactly 50.0 ml of 2 m<u>M</u> PPD in 0.05<u>M</u> H_2SO_4, 1<u>M</u> Na_2SO_4 into the cell. Bubble nitrogen through the solution for ten minutes and then pass it over the solution during the course of the experiment.

Set the potential of the platinum-gauze working electrode to the peak value, E_p, obtained in the previous experiment. Set the recorder in motion and start the electrolysis as the pen passes a convenient reference point.

Electrolyze with efficient stirring until the current falls to about 70% of its initial value. Rinse the cell and electrodes thoroughly and determine the background current by repeating the experiment with a solution containing only the base electrolyte.

9.4.4 Treatment of Results

Plot log i against t, using values of the current corrected for background current. Calculate n from the slope and the intercept in each case.

Explain the small deviation from linearity of the points measured at short times (up to 10 seconds).

Assuming that $D = 7 \times 10^{-6}$ $cm^2 sec^{-1}$, calculate the diffusion-layer thickness δ under the present experimental conditions.

9.5 Mechanism of Oxidation of p-phenylenediamine (iii) Steps in the Overall Reaction

9.5.1 General

In this experiment, some of the possible uses of the technique of cyclic voltammetry in the study of complex electrode reactions will be illustrated. The reactant chosen, p-phenylenediamine (PPD) can be oxidized electrochemically to the corresponding diimine (PDI)[10,11]. This compound, however, is very unstable in aqueous solutions and immediately hydrolyzes to the quinoneimine (PQI). The quinoneimine produced in this way can either set up its own redox couple with p-aminophenol (PAP) or it can hydrolyze further to benzoquinone (Q) which will set up the quinone-hydroquinone redox couple. These reactions are shown below.

9.5 OXIDATION OF P-PHENYLENEDIAMINE (iii)

[Reaction scheme: PPD → PPDI (−2H⁺ −2e_M) → (+H₂O, −NH₃, very fast) PQI → (+2H⁺ +2e_M) PAP, and PQI → (+H₂O, −NH₃, slow) Q → (+2H⁺ +2e_M) QH₂]

The rate of the second hydrolysis step is sufficiently slow to be studied by the technique of cyclic voltammetry. This rate is affected by the medium in which the reaction is carried out, by the pH of the solution (the rate increases with decreasing hydrogen-ion activity), and by the type of electrode used, which, as will be seen, may catalyze the hydrolysis.

According to the reaction scheme shown above, the cyclic voltammogram in a solution of PPD should, after a sufficient number of cycles, have three anodic and two cathodic peaks. Figure 9.6 shows a typical voltammogram obtained in a solution of 2 m$\underline{M}$ PPD in 0.05 $\underline{M}$ H$_2$SO$_4$, which has only one anodic peak and two cathodic peaks. The reason for this is that the three oxidizable species, PPD, PAP and QH$_2$ are all oxidized at about the same potential, as verified by the anodic peaks observed in separate solutions of PPD, PAP and QH$_2$. Figure 9.7 shows the effect of increasing the acid concentration to 1$\underline{M}$. The cathodic peak due to the reduction of Q has disappeared (because of slow hydrolysis of PQI to Q) and the anodic peaks are shifted to higher anodic potentials as a result of protonation of PPD and PAP.

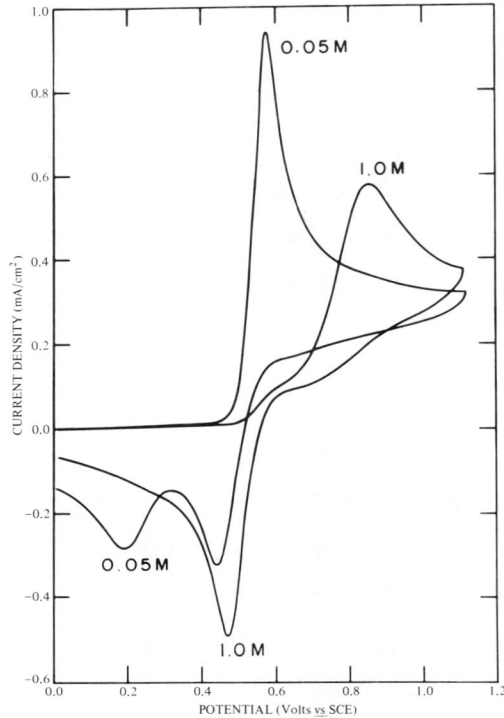

FIG. 9.7   CYCLIC VOLTAMMOGRAM OF 2 m$\underline{M}$ PPD IN TWO CONCENTRATIONS OF H_2SO_4.   v = 40 mV/sec.

Thus, although the peak potential observed in cyclic voltammetry is characteristic, under given experimental conditions, of the compounds being studied, caution must be exercised in using the observed peaks to identify a species in solution. When the peaks for two reactions overlap it is often possible, as will be shown, to separate them by changing the medium.

The experiment given below is rather long and may conveniently be divided into two parts, which, however, should be performed in the order presented.

9.5 OXIDATION OF P-PHENYLENEDIAMINE (iii)

9.5.2 The Cell, Electrodes and Electrical Setup

The cell, electrodes and electrical setup are identical to those described in Experiment 9.1. Solutions of PPD, PAP and QH_2 should be made up just before use. Platinum and pyrolytic-graphite working electrodes, constructed as in Experiment 9.1 will be employed. Both will be cleaned by abrading and the platinum electrode will also be pretreated electrochemically as discussed in Experiment 6.1. A large platinum-gauze electrode will also be employed in this experiment.

9.5.3 Identification of Species in Solution

Prepare 2 m$\underline{M}$ solutions of p-aminophenol (PAP), p-phenylenediamine (PPD) and hydroquinone (QH_2) in 0.05 $\underline{M}$ H_2SO_4, 1$\underline{M}$ Na_2SO_4. Obtain cyclic voltammograms for each solution in turn as follows. Bubble purified nitrogen through the solution for ten minutes and then pass it over the solution during the course of the experiment. Record the steady-state voltammogram (i.e. that observed after several cycles, when the shape of the curve no longer changes with time) obtained with the platinum working electrode, at a sweep rate of 40 mV/sec. Record the three voltammograms on the same paper.

Repeat these measurements in 0.2 $\underline{M}$ H_2SO_4 solutions with the same concentration of the reacting species. Note the changes in the relative position of the peaks with change in pH.

9.5.4 Proof of Slow Hydrolysis in 1 $\underline{M}$ H_2SO_4

Introduce 50.0 ml of 2 m$\underline{M}$ PPD in 1 $\underline{M}$ H_2SO_4 into the cell, deaerate as in the previous experiment and after allowing the solution to come to rest, perform a single triangular sweep at the small platinum electrode at 40 mV/sec. Now, with the large platinum-gauze working electrode, under conditions of stirring, determine the limiting current by setting the potential at E_p.

Electrolyze at about half the limiting current until 10% of the PPD has been oxidized to PQI. Stop stirring, wait for 30 seconds and perform a single triangular sweep at the small platinum electrode. Wait about 5 to 10 minutes and repeat the single sweep. Do this until you have about 3 or 4 voltammograms.

Note the appearance and growth of the peaks due to the Q/QH_2 couple, resulting from the hydrolysis of PQI to Q.

Continue the electrolysis, oxidizing an additional 10% of PPD.

Record the first two or three voltammograms in the electrolyzed solution. Explain the difference in the curves.

9.5.5 The Protonation of PPD and of PAP

Prepare a solution of 2 m$\underline{M}$ PPD in 0.05 $\underline{M}$ H_2SO_4, 1$\underline{M}$ Na_2SO_4, deaerate as before and record the voltammogram of a single triangular sweep at the small platinum electrode. Increase the concentration of sulfuric acid to 1.0 $\underline{M}$ in five to ten steps by adding the required amounts of 10 $\underline{M}$ acid to the solution. Measure the single-sweep voltammogram in each solution.

Repeat these experiments using PAP instead of PPD. Tabulate the values of the anodic and cathodic peak potentials at different acid concentrations.

9.5 OXIDATION OF P-PHENYLENEDIAMINE (iii)

9.5.6 The Catalytic Effect of the Electrode on the Rate of Hydrolysis

Prepare a solution of 2 m$\underline{M}$ PPD in 1$\underline{M}$ H$_2$SO$_4$, deaerate as before and record several consecutive voltammograms at a pyrolytic-graphite electrode. Repeat with PAP instead of PPD. Compare the peaks observed with those obtained with a platinum electrode in the same solution.

9.5.7 Treatment of Results

Tabulate the values of E_p for the anodic and the cathodic sweeps in each of the systems studied and assign the peaks to the various reactions given in the reaction scheme above. Compare with the corresponding half-wave potentials found in the literature.

If the peaks for oxidation or reduction of two different species overlap, under what conditions can they be separated by changing the sweep rate?

Discuss the mechanism of oxidation of PPD under different experimental conditions in view of the results obtained in this experiment.

REFERENCES

1. Delahay, "New Instrumental Methods in Electrochemistry", Chap. 6, Interscience Publishers Inc., New York, 1966.
2. R. S. Nicholson and I. Shain, Anal. Chem. 36, 706 (1964).
3. W. M. Schwarz and I. Shain, J. Phys. Chem. 69, 30 (1965).
4. A. Sevcik, Coll. Czech. Chem. Comm. 13, 349 (1948).
5. W. Kemula and Z. Kublik, Anal. Chim. Acta 18, 104 (1958).
6. H. Matsuda and Y. Ayabe, Z. Elektrochem. 59, 494 (1955).
7. J. E. B. Randles, Trans. Faraday Soc. 44, 322, 327 (1948).
8. U. Eisner and E. Gileadi, J. Electroanal. Chem. 28, 81 (1970).
9. R. N. Adams "Electrochemistry at Solid Electrodes", pp. 152-158, Marcel Dekker Inc., N. Y. 1969.
10. Z. Galus, H. Y. Lee and R. N. Adams, J. Electroanal. Chem. 5, 17 (1963).
11. D. Hawley and R. N. Adams, J. Electroanal. Chem. 10, 376 (1965).

10. VOLTAMMETRY AT CONTROLLED CURRENT AND AT CONTROLLED POTENTIAL

Introduction

10.1 Chronopotentiometry (Constant-Current Voltammetry)

10.1.1 General

10.1.2 The Cell, Solutions, Electrodes and Electrical Setup

10.1.3 Reversible Reactions:
- (i) The Tl^+/Tl reaction at the HMDE
- (ii) The $[Fe(CN)_6]^{4-}/[Fe(CN)_6]^{3-}$ reaction at a platinum electrode

10.1.4 Irreversible Reaction:

The Ni^{2+}/Ni reaction at the HMDE

10.1.5 Consecutive Reactions:

The reduction of oxygen in acetate buffer at the HMDE

10.1.6 Treatment of Results

10.2 Chronopotentiometry with Current Reversal

10.2.1 General

10.2.2 The Cell, Solutions, Electrodes and Electrical Setup

10.2.3 The Rate of a Chemical Reaction Following Charge Transfer:

The Oxidation of p-aminophenol and hydrolysis of p-quinoneimine

10.2.4 Treatment of Results

10.3 Voltammetry at Constant Potential

10.3.1 General

10.3.2 The Cell, Solutions, Electrodes and Electrical Setup

10.3.3 Determination of the Diffusion Coefficient of Hydroquinone

10.3.4 Treatment of Results

10.4 Two-Step Voltammetry at Controlled Potential

10.4.1 General

10.4.2 The Cell, Solutions, Electrodes and Electrical Setup

10.4.3 Determination of the Working Potentials

10.4.4 Determination of the Switching Time

10.4.5 Determination of the Rate Constant for Rearrangement

10.4.6 Measurement with No Following Chemical Reaction

10.4.7 Treatment of Results

10. VOLTAMMETRY AT CONTROLLED CURRENT AND AT CONTROLLED POTENTIAL

Introduction

Chronopotentiometry is a technique which was developed originally[1] for analytical purposes, but has since been extended to the study of the kinetics of charge-transfer and of following chemical reactions[2-4]. The subject has been described in great detail in several papers[2-7] and only the main points will be summarized below.

Consider a simple charge-transfer process

$$Ox + ne_M \rightleftharpoons Red \qquad [10.1]$$

Assume that equilibrium is maintained and the potential of an electrode immersed in a solution containing both species is given by the Nernst equation

$$E = E^o + \frac{RT}{nF} \ln \frac{f_{Ox} C_{Ox}}{f_{Red} C_{Red}} \qquad [10.2]$$

If a constant current is applied, the potential varies with time as shown in Fig. 10.1. Although the change in bulk concentrations is negligible during the transient, the concentrations of both reactant and product at the electrode surface vary markedly and the ratio $C_{Ox}(x=0)/C_{Red}(x=0)$ determines the observed potential. This ratio varies rapidly at the beginning of the transient, when $C_{Red}(x=0)$ grows from zero to a finite value; slowly when the two concentrations at the surface are comparable; and rapidly again towards the end of the transient, when $C_{Ox}(x=0)$ approaches zero.

Hence the rapid change in potential at the two ends in Fig. 10.1. If the current is maintained for a longer time, the potential will rise until it reaches another plateau, characteristic of a different reaction.

The diffusion equation was solved originally[1] for the simplest case discussed above, under the assumption of semi-infinite linear diffusion. Agreement with experiment was obtained for relatively short times (up to about 60 sec.) in solutions containing an excess of supporting electrolyte.

The important parameter in chronopotentiometry is the transition time τ which, for the present case is given by

$$\tau^{1/2} = \frac{\pi^{1/2} nFD^{1/2} C^o}{2i} \qquad [10.3]$$

where C^o is the bulk concentration of the reactant and i is the constant current density applied. The significance of the transition time τ can be seen by reference to Fig. 10.1. The application of chronopotentiometry for analytical purposes is based on equation [10.3]. A common way of testing the validity of the assumptions leading to this equation is to vary the applied current density and verify that the product ($i\tau^{1/2}$) is constant.

Introducing into the Nernst equation the time-dependent values of the surface concentrations of reactants and products (obtained by solving the diffusion equation) one has

$$E = E_{1/4} + \frac{RT}{nF} \ln \frac{\tau^{1/2} - t^{1/2}}{t^{1/2}} \qquad [10.4]$$

where

INTRODUCTION

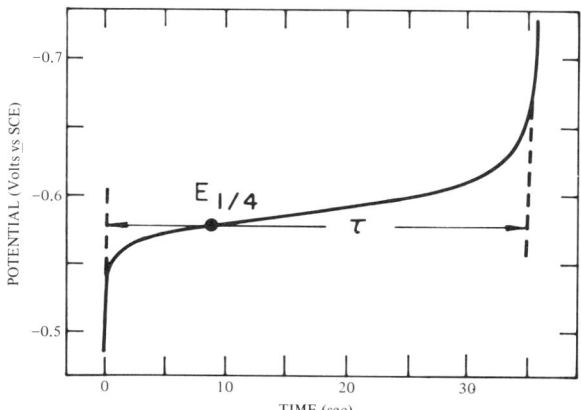

FIG. 10.1   CHRONOPOTENTIOGRAM AT A MERCURY ELECTRODE FOR 2 m$\underline{M}$ Cd^{2+} in 1$\underline{M}$ KNO$_3$. i = 163 μA/cm^2.

$$E_{1/4} = E^o + \frac{RT}{nF} \ln \frac{f_{Red} D_{Ox}^{1/2}}{f_{Ox} D_{Red}^{1/2}} \qquad [10.5]$$

The potential $E_{1/4}$ corresponding to the time $t = \tau/4$ is called the "quarter-wave potential"[11] and is equal numerically to the polarographic half-wave potential.

Equation [10.4] may be used to determine the number of electrons taking part in the overall reaction. It should be noted, though, that this equation only applies if equilibrium is maintained throughout the transient.* The validity of equation [10.4] can be

---

* Equation [10.3] for the transition time is valid for both reversible and irreversible reactions.

tested by repeating an experiment over a range of applied currents and determining the apparent value of n in each case. An apparent decrease in the value of n with increasing current density is indicative of irreversible behavior.

More complex cases involving two electroactive species reacting at different potentials or the same species undergoing two consecutive charge-transfer reactions will be discussed below.

The scope of chronopotentiometry has been widened by the introduction of the technique of chronopotentiometry with current reversal. This technique is particularly suited to the study of the rate of chemical reactions following charge transfer, such as

$$\text{Ox} + n e_M \rightleftharpoons \text{Red} \xrightarrow{k_f} Z \qquad [10.6]$$

It can be seen qualitatively that if the current is reversed at some time $t_f \leqslant \tau_f$, the transition time τ_r on the reverse pulse will depend on the rate constand k_f, since the species Red is consumed at the electrode surface both by its oxidation to Ox and by its transformation to Z.* Quantitative expressions have been derived[2] which allow the determination of the rate constant k_f by this technique (cf. Experiment 10.2 below). A similar technique,[12] in which the potential is controlled and is stepped from a value where Ox is reduced to Red, back to a value where the reverse reaction takes place, will also be discussed below (cf. Experiment 10.4).

---

* The species Z is assumed to be electrochemically inactive in the region of potential where the Ox/Red redox reaction occurs.

10.1   CHRONOPOTENTIOMETRY

It would appear that the above techniques are only applicable when the first-order rate constand k_f is of the order of the transition time τ (i.e. when the transition time can be adjusted to be of the same order of magnitude as the half-life of the reaction). This, however, is not a very serious limitation in most cases. The chemical reaction is often an acid/base catalyzed quasi-first-order reaction, the rate of which can be varied by adjusting the pH of the solution. Considering the range of more than three orders of magnitude of τ accessible experimentally, (10 msec $< \tau <$ 60 sec), the technique of chronopotentiometry with current reversal can be useful in the study of a great number of reactions.

10.1   Chronopotentiometry (Constant-Current Voltammetry)

10.1.1 General

The equations for the transition time and the variation of potential with time in a simple reversible charge-transfer process were given in the introduction. When an irreversible process is studied, the equation for the transition time (equation 10.3) remains unaltered, while the E/t plot is given by the equation*

$$E = E^o + b_c \log(nFk_s C^o/i) + b_c \log[1 - (t/\tau)^{1/2}] \quad [10.7]$$

hence

---

* These equations are given in a form different from that commonly used in the literature, for reasons discussed in section II.10.3.

$$E_{1/4}(\text{irrev.}) = E^o + b_c \log(nFk_s C_s^o/2i) \qquad [10.8]$$

where k_s is the rate constant for this process at the standard reversible potential E^o. This behavior can be understood qualitatively when the physics of the process is considered. The transition time is the time taken for the concentration of the reactant at the electrode surface to be reduced to zero. Since the rate of the reaction is controlled externally (by controlling the current), the transition time cannot depend on the specific rate constant of the reaction. On the other hand, the potential at any moment during the transient is determined by the Nernst equation in the reversible case and by the kinetic parameters of the reaction (k_s, b_c) in the irreversible case.

Equation 10.8 can be used to obtain the Tafel slope b_c by plotting $E_{1/4}$ (irrev) against log i. Once this is known, the specific rate constant k_s can also be evaluated.

Two further cases may be considered. When the solution contains two reducible substances, Ox_1 and Ox_2, reacting at different potentials, the transition time τ_2 for the species reacting at the more negative potential is given by the relationship

$$(\tau_1 + \tau_2)^{1/2} - \tau_1^{1/2} = \pi^{1/2} n_2 F D_2^{1/2} C_2^o/2i \qquad [10.9]$$

The interesting consequence of this equation is that τ_2 observed in a solution containing two electroactive species is greater than the value which would have been observed for the same concentration of the same species alone in solution. For example, for $C_1^o = C_2^o$ we find from equation [10.9] that $\tau_2 = 3\tau_1$. This

10.1 CHRONOPOTENTIOMETRY

happens because reduction of Ox_1 occurs throughout the transient, even at times longer than τ_1. Since the total current passed is fixed, the effective current density for the reduction of Ox_2 is diminished.

A special case occurs when the same species can undergo two consecutive charge-transfer reactions*

$$Ox + n_1 e_M \longrightarrow Red_1 \qquad [10.10]$$

$$Red_1 + n_2 e_M \longrightarrow Red_2 \qquad [10.11]$$

The first transition time τ_1 is given by equation [10.3] and τ_2 is related to it by the equation

$$\tau_2 = \tau_1 \left[\left(\frac{n_2}{n_1}\right)^2 + 2\frac{n_2}{n_1} \right] \qquad [10.12]$$

Equations [10.9] and [10.12], unlike equation [10.3], are valid for reversible as well as irreversible processes, although in the latter case the two waves are drawn out and may merge, making the experimental determination of τ_1 and τ_2 unreliable.

---

* The notion of consecutive steps employed here needs some clarification. It is widely accepted that only a unit charge can be transferred in an elementary step. In this sense, a sequence of consecutive steps must occur whenever n > 1. When two or more of the charges are transferred at the same potential, this will be considered a single step from the point of view of chrono-potentiometry, cyclic voltammetry or polarography. The implication of equations [10.10] and [10.11] is that Red_1 and Red_2 are formed at two different potentials, irrespective of the number of elementary steps involved in the formation of each.

Two reversible reactions will be studied here: The reduction of Tl^+ on a hanging-mercury-drop electrode (HMDE) and the oxidation of $[Fe(CN)_6]^{4-}$ to $[Fe(CN)_6]^{3-}$ on a platinum electrode. The reduction of Ni^{2+} on the HMDE will be used to exemplify irreversible behavior. In addition, an example of consecutive reactions will be studied: the reduction of molecular oxygen in acetate buffer at the HMDE.

10.1.2 The Cell, Solutions, Electrodes and Electrical Setup

The cell shown in Fig. 3.2 will be used in this experiment. The Luggin capillary should be placed close to the working electrode (cf. "Introduction" section to Experiment 2). Mechanical vibrations should be kept to a minimum to eliminate convective movement of the solution during the transient.[*]

Two types of working electrodes will be used in these experiments: a small platinum-disc electrode set in a teflon holder (like a rotating-disc electrode), having a total surface area of 0.01-0.3 cm^2, and a hanging-mercury-drop electrode. The latter is available commercially (the Kemula-type electrode) or can be constructed by attaching a small drop of known volume to the end of a platinum wire, according to the procedure described in Experiment 8.1. A platinum counter electrode and a commercial calomel reference electrode will be used in all these measurements.

---

[*] This is crucial if longer transition times, in the range of 10-100 sec, are required. No special precautions need be taken if the transition time is less than one second.

10.1 CHRONOPOTENTIOMETRY

Solutions should all be made from analytical-grade reagents, dissolved or diluted in triply distilled water.

The galvanostatic circuit shown in Fig. 2.4 will be used here. The potential will be recorded on a fast strip-chart recorder having a response time of less than 1 second for full-scale deflection.

10.1.3 Reversible Reactions:

(i) The Tl^+/Tl reaction at the HMDE

Prepare a hanging-mercury-drop electrode of known area (cf. Experiment 8.1) and introduce it into the cell, which contains an exactly known amount of 0.1 $\underline{M}$ KNO_3 solution which has been deaerated by bubbling purified nitrogen for 10 minutes. Introduce the counter and reference electrodes into the cell, stop nitrogen bubbling and pass the gas over the solution. Apply a current density of 350 $\mu A/cm^2$ and record the variation of potential with time. Add enough concentrated stock solution of $TlNO_3$ to reach a final concentration of 2.0 m$\underline{M}$ Tl^+ and record the E/t plot at 350 $\mu A/cm^2$. Increase the concentration of Tl^+ to 4, 6, 8 and 10 m$\underline{M}$ and repeat the measurement several times at each concentration. Prepare a fresh drop for each measurement. After each addition of $TlNO_3$ stock solution, bubble nitrogen for two minutes and allow the solution to rest for half a minute. Use the most concentrated solution to determine the transition time τ as a function of current density. Adjust the current density to give transition times in the range of 2-200 sec in about eight steps.

(ii) The $[Fe(CN)_6]^{4-}/[Fe(CN)_6]^{3-}$ reaction at a platinum electrode

Introduce an exactly known amount of 0.1 $\underline{M}$ KNO_3 solution into the cell and deaerate by bubbling purified nitrogen for 10 minutes. Continue passing the gas above the solution throughout the experiment. Introduce the platinum working electrode (area 0.01–0.3 cm^2), the counter and the reference electrodes. Pretreat the working electrode electrochemically as outlined in Experiment 6.1. Apply a current density of 350 $\mu A/cm^2$ and plot the E/t curve on the recorder. Add a small volume of concentrated stock solution of $K_4[Fe(CN)_6]$ to make a final solution 2 m$\underline{M}$ in $[Fe(CN)_6]^{4-}$. Bubble nitrogen for two minutes, allow half a minute for the solution to rest and apply a current density of 350 $\mu A/cm^2$. Increase the concentration of $[Fe(CN)_6]^{4-}$ to 4, 6, 8 and 10 m$\underline{M}$ and record several transients in each solution. Stir the solution for half a minute after each transient to destroy the concentration gradient built up near the electrode surface.

10.1.4 Irreversible Reaction:

The Ni^{2+}/Ni reaction at the HMDE

Set up the experiment as for the reduction of Tl^+ above, but use a 0.1 $\underline{M}$ KCl solution as supporting electrolyte and make the solution 5 m$\underline{M}$ with respect to Ni^{2+}. Record the first transient at a current density of 350 $\mu A/cm^2$, then repeat at 6–8 different current densities chosen to give transition times in the range of 3–60 sec. Use a fresh mercury drop for each experiment.

10.1.5 Consecutive Reactions:

The reduction of oxygen in acetate buffer at the HMDE

The reduction of oxygen on mercury in acetate buffer is known to take place in two steps:

$$O_2 + 2H_2O + 2e_M \longrightarrow H_2O_2 + 2OH^- \qquad [10.13]$$

$$H_2O_2 + 2e_M \longrightarrow 2OH^- \qquad [10.14]$$

This will be studied as an example of a reaction which takes place in two consecutive steps with an equal number of electrons transferred in each step.

Introduce into the cell a 1:1 acetic acid/sodium acetate buffer solution. Do *not* bubble nitrogen. Introduce the HMDE, counter and reference electrodes. Apply a current density of 350 $\mu A/cm^2$ and record the E/t plot. Repeat at several current densities giving total transition times $(\tau_1 + \tau_2)$ in the range of 10-60 sec. The same mercury drop may be used for all the experiments. Stir with a magnetic stirrer after each transient to ensure uniformity of the concentration. Choose a current density giving a total transition time of about 30 seconds and record the E/t curve. Bubble nitrogen for two minutes, allow to rest for half a minute, and repeat the transient. Bubble nitrogen for additional two-minute periods and record the transient after each such treatment until the waves due to oxygen reduction essentially disappear.

10.1.6 Treatment of Results

(i) Reversible reactions

Determine the transition time for the reduction of Tl^+ and the oxidation of $[Fe(CN)_6]^{4-}$ and plot $\tau^{1/2}$ vs concentration. Calculate the diffusion coefficients of the ions and compare with literature data.

Make a plot of E vs log $[(\tau^{1/2}-t^{1/2})/t^{1/2}]$ for one concentration of each of the ions. Find the value of n from the slope.

Determine the quarter-wave potential (average of all experiments) for each reaction and compare with the half-wave potential for Tl^+ and with the formal potential E^o for $[Fe(CN)_6]^{4-}$.

Calculate the product $i\tau^{1/2}$ and plot it against i. Discuss the reasons for departure, if any, from $i\tau^{1/2}$ = constant. What are the factors determining the measurable lower and upper limits of τ for which equation [10.3] still holds?

Discuss the advantages and disadvantages (in terms of accuracy, reproducibility, selectivity, ease of application, instrumentation required, etc.) of chronopotentiometry as an analytical tool compared to other electroanalytical techniques.

(ii) Irreversible reactions

Determine τ and $E_{1/4}$ in the reduction of Ni^{2+}. Plot $i\tau^{1/2}$ vs i. Calculate the value of $(i\tau^{1/2}/C^o)$ and compare with that obtained for the reduction of Tl^+.

Plot $E_{1/4}$ vs log i and calculate from this the Tafel slope b and the specific rate constant k_s.

Make a plot of E vs $\log[1 - (t/\tau)^{1/2}]$ and calculate the Tafel slope.

Calculate the current density below which this system would behave reversibly at this concentration. What is the corresponding transition time? (Assume for the purpose of this calculation that $E^{o'} \cong E_{1/4}$ (rev)).

(iii) Consecutive two-step reduction

Determine the two transition times and calculate the ratio of the numbers of electrons transferred in the two steps (n_2/n_1). Determine the quarter-wave potential of both waves and compare with the polarographic half-wave potentials available in the literature for these reactions. Plot the $E_{1/4}$ values against log i and determine the reversibility of the waves. Calculate the number of electrons transferred and the Tafel slope for the first wave.

10.2 Chronopotentiometry with Current Reversal

10.2.1 General

In this modification of chronopotentiometry, the current is reversed during the transient at a time t_f which is shorter than or equal to the forward transition time τ_f. A solution of the diffusion equation for the forward transient gives the concentrations of reactant and product at the electrode surface as a function of time, $C_{Ox}(0,t)$ and $C_{Red}(0,t)$. The values of these concentrations at $t = t_f$ serve as the initial conditions for the solution of the diffusion equation for the reverse transient.[8,13] For the simple charge-transfer step shown in equation [10.12], the relation between the time of current reversal t_f and the transition time τ_r

for the reverse transient is given by

$$\tau_r = t_f/3 \qquad [10.15]$$

if the current density in both directions is the same. This reverse transition time is independent of the diffusion coefficients of the reactant or the product since it depends on the rate of diffusion of the product away from the electrode during the forward transient and its rate of diffusion towards the electrode during the reverse transient. Like all other expressions for the transition time, equation [10.15] is independent of the reversibility of the reaction. A schematic E/t plot for chronopotentiometry with current reversal is shown in Fig. 10.2.

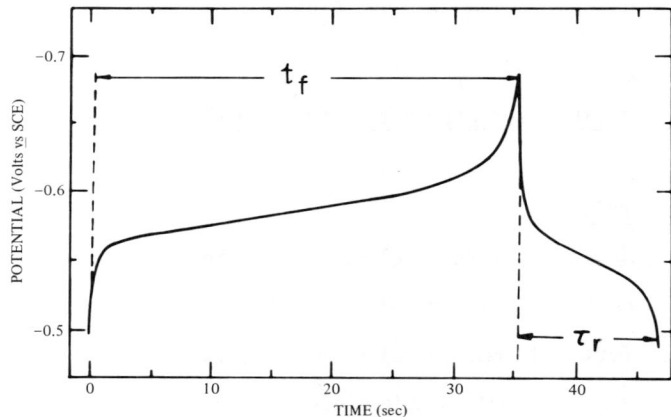

FIG. 10.2 REVERSE-CURRENT CHRONOPOTENTIOGRAM AT A MERCURY ELECTRODE FOR 2mM Cd^{2+} IN 1 M KNO_3. $i = 163\ \mu A/cm^2$.

10.2 CHRONOPOTENTIOMETRY WITH CURRENT REVERSAL

An interesting class of reaction which can be studied by this technique is that of charge transfer followed by an irreversible chemical reaction. A reaction of this type, which will be studied in this experiment, is the oxidation of p-aminophenol (PAP) followed by hydrolysis of the oxidation product (cf. equation [10.6]).

$$\underset{\substack{\text{(PAP)}}}{\text{HO-C}_6\text{H}_4\text{-NH}_2} \rightleftharpoons \underset{\substack{\text{(PQI)}}}{\text{O=C}_6\text{H}_4\text{=NH}} + 2\text{H}^+ + 2\text{e}_M \qquad [10.16]$$

$$\underset{\substack{\text{(PQI)}}}{\text{O=C}_6\text{H}_4\text{=NH}} + \text{H}_2\text{O} \xrightarrow{k_f} \underset{\substack{\text{(Q)}}}{\text{O=C}_6\text{H}_4\text{=O}} + \text{NH}_3 \qquad [10.17]$$

The fundamental equation relating the transition time τ_r on the reverse transient to the duration of the forward transient t_f and the specific rate constant k_f is[2,9,14,15]

$$2\ \text{erf}(k_f \tau_r)^{1/2} = \text{erf}[k_f(t_f + \tau_r)]^{1/2} \qquad [10.18]$$

The specific rate constant can be calculated from the experimentally determined values of t_f and τ_r by the method suggested by Testa and Reinmuth.[2] Equation [10.18] is first used to construct a working curve (Fig. 10.3) (or better, a table such as

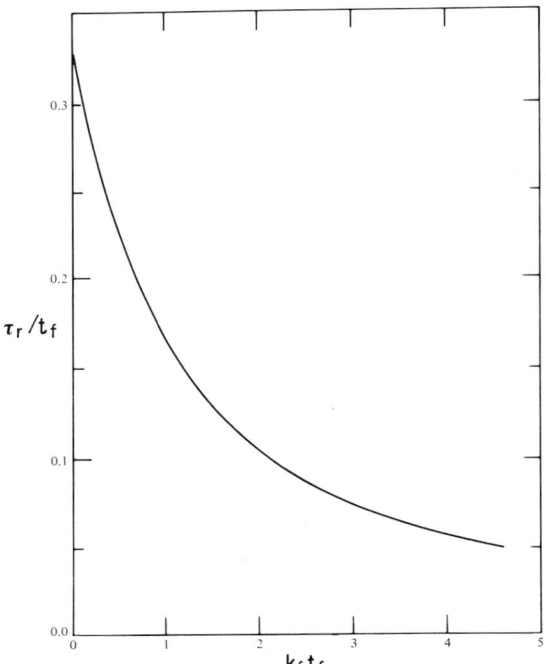

FIG. 10.3 WORKING CURVE FOR THE CALCULATION OF THE CHEMICAL RATE CONSTANT k_f FROM REVERSE-CURRENT CHRONOPOTENTIOMETRY.

Table 10.1) of τ_r/t_f vs $k_f t_f$.* The time of current reversal t_f is varied and the corresponding value of τ_r is measured. From this, $k_f t_f$ is obtained with the use of the working curve or table. This is plotted as a function of t_f in order to obtain k_f. It should be noted that equation [10.18] is independent of the applied current density and is applicable as long as the total time ($\tau_r + t_f$) does not exceed about one minute, after which mass transport by

---

* The method of constructing the working curve or table is explained in detail on page 182 of ref. 9.

Table 10.1

Data for Theoretical Working Curve for Current-Reversal Chronopotentiometry; Charge Transfer at a Plane Electrode Followed by an Irreversible First-Order Rate Process

$$2\,\mathrm{erf}(k_f \tau_r)^{1/2} = \mathrm{erf}[k_f(t_f + \tau_r)]^{1/2}$$

$k_f t_f$	τ_r/t_f	$k_f t_f$	τ_r/t_f	$k_f t_f$	τ_r/t_f	$k_f t_f$	τ_r/t_f
0.0152	0.329	0.336	0.254	0.865	0.180	1.970	0.106
0.0436	0.321	0.379	0.246	0.948	0.172	2.174	0.0977
0.0753	0.312	0.428	0.238	1.034	0.164	2.406	0.0898
0.107	0.304	0.481	0.230	1.131	0.156	2.673	0.0819
0.140	0.296	0.535	0.221	1.241	0.147	3.061	0.0725
0.176	0.288	0.590	0.213	1.339	0.140	3.470	0.0646
0.213	0.279	0.654	0.205	1.487	0.130	3.982	0.0566
0.251	0.271	0.719	0.197	1.623	0.123	4.631	0.0489
0.293	0.263	0.791	0.188	1.787	0.114		

convection becomes important.

A typical potential/time plot for the oxidation of PAP to PQI and the subsequent hydrolysis of PQI to Q (cf. equation [10.16] and [10.17]), is shown in Fig. 10.4. As the time of current reversal t_f is increased, the ratio τ_r/t_f decreases according to equation [10.18] (cf. Table 10.1). The second wave seen on the reverse transient is due to the reduction of benzoquinone to hydroquinone.

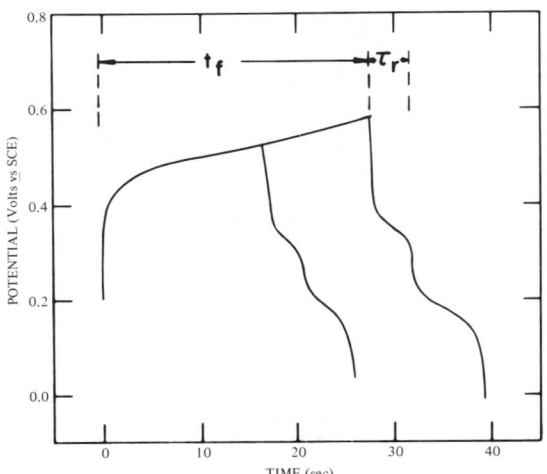

FIG. 10.4   REVERSE-CURRENT CHRONOPOTENTIOGRAM FOR THE OXIDATION OF 1 m$\underline{M}$ PAP IN 0.1 $\underline{M}$ H$_2$SO$_4$ AT A PLATINUM ELECTRODE. i = 100 μA/cm^2.

$$Q + 2H^+ + 2e_M \rightleftharpoons QH_2 \qquad [10.19]$$

This wave is well separated from the wave due to reduction of PQI and hence it does not interfere with the determination of the rate constant for the hydrolysis of PQI as outlined above.

10.2.2 <u>The Cell, Solutions, Electrodes and Electrical Setup</u>

The cell and electrical setup are the same as in Experiment 10.1. Current reversal can be effected manually since t_f

is several seconds, at least. The counter and reference electrodes will be as in Experiment 10.1 and the platinum working electrode will be employed. The solution will be 1 m$\underline{M}$ p-aminophenol (PAP) in 0.1 $\underline{M}$ H_2SO_4.

10.2.3 The Rate of a Chemical Reaction Following Charge Transfer: The Oxidation of p-Aminophenol and Hydrolysis of p-Quinoneimine

Introduce into the cell an exactly known volume of 0.1 $\underline{M}$ H_2SO_4. Deaerate by bubbling purified nitrogen for 10 minutes. Introduce the three electrodes and continue to bubble nitrogen for 2 minutes. Pretreat the platinum electrode electrochemically as outlined in Experiment 6.1. Add a small volume of concentrated solution of PAP in 0.1 $\underline{M}$ H_2SO_4 to make a final concentration of 1 m$\underline{M}$ in this substance. Continue to pass nitrogen above the solution throughout the experiment. Apply a current density of 100 $\mu A/cm^2$ and measure the transition time. Stir the solution for about half a minute to destroy concentration gradients near the electrode surface. Allow the solution to settle for about half a minute, then apply a current density calculated to give a transition time of about 30 seconds. Reverse the direction of the current after 3 seconds. Repeat 10-15 times at the same current density, increasing the time of current reversal t_f from 3 seconds to τ_f.

10.2.4 Treatment of Results

Use the data given in Table 10.1 to plot a working curve of τ_r/t_f $\underline{vs}$ $k_f t_f$ in the range needed for the data obtained in this measurement.

Obtain the value of $k_f t_f$ corresponding to each value of τ_r/t_f found experimentally and plot $k_f t_f$ vs t_f to calculate k_f.

Plot t_f against the sum of transition times of the two waves on the reverse transient and discuss the relationship found.

Discuss the advantages and limitations of the technique of chronopotentiometry with current reversal. What is the range of rate constants measurable by this technique and how could it be extended?

10.3 Voltammetry at Constant Potential

10.3.1 General

When a potential-step function is applied to an electrode/electrolyte interface so that the potential is changed from a value at which no faradaic process takes place, to one where the reaction is diffusion-limited, the current/time transient can be described by the equation (cf. equation [8.7])

$$i = nFC^o D^{1/2}/\pi^{1/2} t^{1/2} \qquad [10.20]$$

This equation is applicable only under conditions where equation [10.3] is applicable, namely, when mass transport takes place by semi-infinite linear diffusion, and when a simple charge transfer takes place, with no kinetic or catalytic currents. Furthermore, for both fundamental and practical reasons, equation [10.20] is not applicable at very short times. When the value of the potential is changed abruptly, the current at very short times is mainly due to charging of the ionic double layer and only after this process has been nearly completed will faradaic processes

10.3 VOLTAMMETRY AT CONSTANT POTENTIAL

become important.* On the practical side, specially designed high-speed potentiostats must be used and care must be taken in cell design in order that meaningful measurements at times of, say, 20 μsec or less may be obtained.

Voltammetry at controlled potential will be used here to evaluate the diffusion coefficient of hydroquinone and verify the fact that the oxidation of this substance on a platinum anode is a simple, diffusion-controlled process with no kinetic complications.

10.3.2 The Cell, Solutions, Electrodes and Electrical Setup

The cell and electrodes will be the same as in Experiment 10.1. The platinum working electrode will be used. It will be cleaned and pretreated electrochemically (in a solution containing 0.5M H_2SO_4 as outlined in Experiment 6.1.

The solution will be 2.00 m$\underline{M}$ hydroquinone (QH_2) in 2 $\underline{M}$ KCl.

The potentiostatic circuit shown in Fig. 9.5 or 6.1 will be used. Switching of the potential can be done manually, and the current/time plot can be recorded on a fast strip-chart recorder (response time less than 1 second).

10.3.3 Determination of the Diffusion Coefficient of Hydroquinone

Add to the cell a known volume of the solution containing only the base electrolyte, introduce the electrodes, (the working

---

* For example, if the potential is changed by 0.5 Volt in 5 μsec, the charging current during this interval of time will be 2 A/cm^2 for a double-layer capacity of 20 $\mu F/cm^2$.

electrode should be pretreated as in Experiment 6.1), deaerate and add a small volume of stock solution of hydroquinone to make a final concentration of 2.00 m$\underline{M}$. Continue to pass nitrogen over the solution.

Connect the electrodes for measurement of cyclic voltammetry (cf. Fig. 9.5). Set the initial potential to 0.05 Volt $\underline{vs}$ SCE and apply a sweep rate of about 50 mV/sec. Record the voltammogram and choose a potential $E \geqslant E_p$ at which the rate of electrooxidation of QH_2 is diffusion-controlled. With the potentiostat set at this potential but in the "standby" position, disconnect the potential-sweep function generator, stir the solution for a few seconds, then allow to stand for half a minute. Start the recorder at a speed of 1-2 cm/sec, turn the potentiostatic circuit to "on" and follow the variation of current with time. Repeat the experiment three more times, making sure to stir between transients. In one experiment decrease the chart speed to 0.1-0.2 cm/sec and follow the current decay for 50-100 sec. Change the sensitivity of the recorder at longer times as required.

10.3.4 Treatment of Results

Plot the current density $\underline{vs}$ $t^{1/2}$ for $t > 0.5$ sec.[*] and calculate the diffusion coefficient according to equation [10.20].

Plot $it^{1/2}$ $\underline{vs}$ t and calculate the diffusion coefficient from the value of $it^{1/2}$ at $t = 0$.

---

[*] This lower limit for t is chosen to allow for the finite response time of the recorder.

10.4 Two-Step Voltammetry at Controlled Potential

Plot $it^{1/2}$ vs t for the longest experiment and show for what length of time equation [10.20] holds.

10.4.1 General

The technique of two-step voltammetry at controlled potential, first suggested by Schwartz and Shain,[12] is analogous to chronopotentiometry with current reversal. It is best suited to cases in which charge transfer is followed by a homogeneous chemical reaction according to the general scheme

$$Ox + ne_M \rightleftharpoons Red \xrightarrow{k_f} Z \qquad [10.6]$$

In the first potential step the substance Ox is reduced at the electrode surface and Red is formed. The potential is then stepped to another value, where Red is oxidized back to Ox. Both potentials are chosen so that the respective reactions occur under conditions of diffusion control. If conditions are chosen so that $k_f \to 0$, then the current during the second potential step depends only on the relative amounts of the substance Red diffusing to the electrode and away from it. If k_f becomes substantial, this current is further diminished by the transformation of some Red to Z, which is assumed to be electrochemically inactive at the potential applied.

The mathematical treatment of this system is rather involved and has been given in the original paper.[12] Only the major results of the calculation are given here. Working curves have been calculated (cf. Fig. 10.5 and Table 10.2) and these permit easy application of the technique, without having to resort to complex mathematics. These curves do not depend on the reversibility of the reaction as long as the potentials for the two steps have been chosen properly, in the region of diffusion control for the respective reactions.

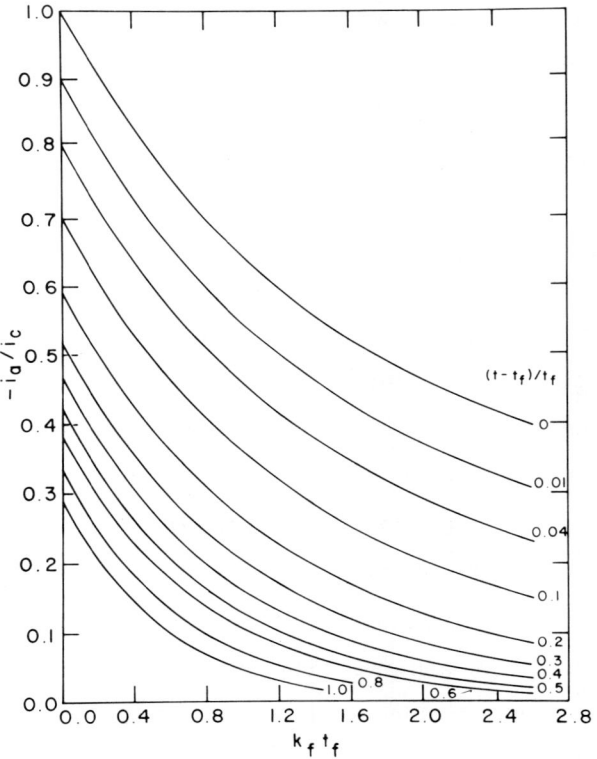

FIG. 10.5 WORKING CURVES FOR THE CALCULATION OF THE CHEMICAL RATE CONSTANT k_f FROM TWO-STEP VOLTAMMETRY AT CONTROLLED POTENTIAL[12]

10.4 TWO-STEP VOLTAMMETRY

Table 10.2: Data for Theoretical Working Curves for Two-Step Voltammetry at Controlled Potential; Charge Transfer at a Plane Electrode Followed by an Irreversible First-Order Rate Process. (by courtesy of Prof. I. Shain).

$$-\frac{i_a}{i_c} = \phi[k_f t_f, (t-t_f)/t_f] - \sqrt{\frac{(t-t_f)/t_f}{1+(t-t_f)/t_f}}$$

$\frac{t-t_f}{t_f}$	$\sqrt{\frac{t-t_f}{t_f}}$	$k_f t_f =$ 0.0	0.1	0.2	0.4	0.6	0.8	1.0	1.2	1.4	1.6	1.8	2.0	2.4	2.8	3.2	3.6	4.0
0.0000	0.000	1.000	0.952	0.907	0.827	0.758	0.697	0.645	0.599	0.559	0.524	0.493	0.466	0.420	0.383	0.353	0.329	0.309
0.0025	0.050	0.950	0.902	0.857	0.777	0.708	0.647	0.595	0.549	0.509	0.474	0.443	0.416	0.370	0.333	0.303	0.279	0.259
0.005	0.071	0.929	0.881	0.836	0.756	0.687	0.626	0.575	0.529	0.489	0.455	0.424	0.397	0.352	0.316	0.287	0.262	0.243
0.010	0.099	0.901	0.833	0.808	0.728	0.659	0.599	0.548	0.503	0.463	0.429	0.399	0.372	0.327	0.293	0.264	0.239	0.221
0.020	0.140	0.860	0.812	0.767	0.687	0.619	0.559	0.508	0.463	0.425	0.391	0.362	0.336	0.292	0.258	0.231	0.208	0.189
0.030	0.170	0.829	0.781	0.736	0.656	0.589	0.529	0.479	0.434	0.397	0.364	0.334	0.309	0.267	0.234	0.207	0.184	0.166
0.040	0.192	0.808	0.760	0.715	0.636	0.570	0.509	0.460	0.416	0.378	0.346	0.317	0.293	0.251	0.218	0.191	0.171	0.153
0.060	0.238	0.762	0.714	0.670	0.590	0.525	0.465	0.417	0.374	0.338	0.306	0.279	0.254	0.213	0.183	0.157	0.138	0.121
0.080	0.272	0.728	0.680	0.636	0.557	0.492	0.433	0.386	0.345	0.309	0.278	0.251	0.228	0.187	0.163	0.135	0.117	0.108
0.100	0.301	0.699	0.651	0.607	0.529	0.464	0.407	0.360	0.319	0.285	0.255	0.228	0.206	0.167	0.140	0.117	0.093	0.086
0.120	0.327	0.673	0.625	0.581	0.504	0.440	0.384	0.337	0.297	0.264	0.234	0.208	0.186	0.151	0.123	0.103	0.086	0.073
0.160	0.372	0.628	0.580	0.538	0.460	0.397	0.343	0.298	0.259	0.227	0.200	0.176	0.155	0.122	0.098	0.080	0.065	0.054
0.200	0.408	0.592	0.544	0.500	0.425	0.363	0.310	0.267	0.230	0.200	0.174	0.150	0.132	0.105	0.080	0.064	0.052	0.042
0.250	0.446	0.554	0.506	0.463	0.389	0.328	0.277	0.236	0.201	0.172	0.147	0.127	0.110	0.083	0.064	0.050	0.040	0.031
0.300	0.481	0.519	0.471	0.428	0.356	0.296	0.247	0.207	0.174	0.147	0.124	0.103	0.090	0.066	0.049	0.037	0.023	0.022
0.400	0.535	0.466	0.413	0.375	0.304	0.248	0.202	0.165	0.137	0.113	0.093	0.076	0.066	0.046	0.033	0.024	0.018	0.013
0.500	0.578	0.422	0.375	0.333	0.264	0.210	0.168	0.134	0.108	0.087	0.070	0.057	0.047	0.031	0.020	0.014	0.009	0.005
0.600	0.614	0.386	0.339	0.298	0.231	0.180	0.141	0.110	0.086	0.068	0.053	0.042	0.033	0.020	0.012	0.007		
0.800	0.667	0.333	0.287	0.247	0.183	0.137	0.103	0.077	0.059	0.045	0.033	0.026	0.019	0.011				
1.000	0.711	0.289	0.243	0.205	0.145	0.103	0.073	0.051	0.036	0.025	0.017	0.011	0.007					

The shape of the cathodic and anodic transients is shown schematically in Fig. 10.6. If the potential is switched after a time t_f, the anodic current at a time $t > t_f$ is given by the equation

$$i_a = nFC_{Ox}^o D_{Ox}^{1/2} \left[\frac{\phi(k_f, t, t_f)}{\sqrt{\pi(t-t_f)}} - \frac{1}{\sqrt{\pi t}} \right] \quad [10.21]$$

where $\phi(k_f, t, t_f)$ is a rather complicated function of the variables k_f, t, t_f (cf. equation [20] of reference 12). If the anodic and cathodic currents are measured at times t and $(t - t_f)$ respectively (i.e. at the same time after switching, cf. Fig. 10.6) one can write

$$\left| \frac{i_a}{i_c} \right| = \phi[k_f t_f, (t-t_f)/t_f] - \sqrt{\frac{(t-t_f)/t_f}{1 + (t-t_f)/t_f}} \quad [10.22]$$

This equation was used to calculate the working curves shown in Fig. 10.5 and the data given in Table 10.2. For the special case of $k_f = 0$, one has $\phi[k_f t_f, (t-t_f)/t_f] = 1$ and the ratio $|i_a/i_c|$ can be easily calculated from equation [10.22]. As k_f increases, the anodic current decreases, since some of the product Red is transformed to the species Z and cannot be reoxidized during the anodic transient.

In the present experiment, the reduction of azobenzene to hydrazobenzene and the subsequent rearrangement of the product will be studied.

10.4 TWO-STEP VOLTAMMETRY

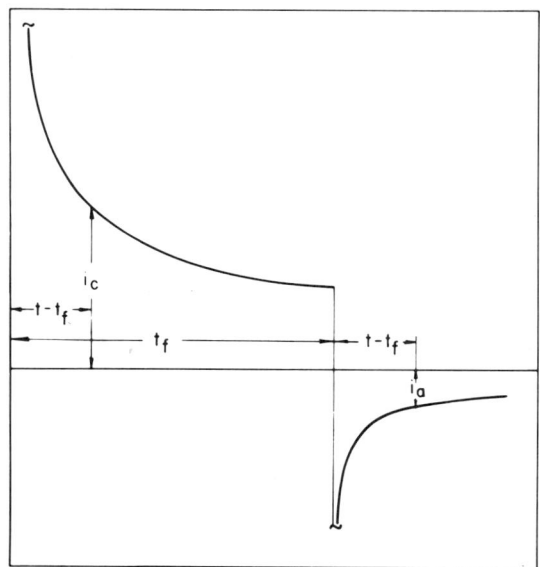

FIG. 10.6 CURRENT/TIME CURVE FOR TWO-STEP VOLTAMMETRY AT CONTROLLED POTENTIAL

azobenzene + $2H^+$ + $2e_M$ ⇌ hydrazobenzene [10.23]

$$\text{(PhNH-NHPh)} \xrightarrow[H^+]{k_f} \text{(H}_2\text{N-C}_6\text{H}_4\text{-C}_6\text{H}_4\text{-NH}_2\text{)} + \text{(2-NH}_2\text{-C}_6\text{H}_4\text{-C}_6\text{H}_4\text{-4-NH}_2\text{)} \quad [10.24]$$

The rearrangement reaction is acid-catalyzed and occurs at a significant rate only in strong acid solutions. Thus it is possible to choose a pH where k_f is small in comparison to the time scale of the experiment and only equation [10.23] need be considered. In very acidic solutions, equation [10.24] becomes important and the pseudo-first-order rate constant k_f can be evaluated with the aid of the working curves in Fig. 10.5.

10.4.2 The Cell, Solutions, Electrodes and Electrical Setup

The cell and electrodes will be identical to those used in Experiment 10.1 for the reduction of Tl^+ or Ni^{2+} (hanging mercury drop as the working electrode).

The solution will be 2 m$\underline{M}$ azobenzene freshly prepared and 2 $\underline{M}$ HClO$_4$ in a 50% (wt) ethanol-water mixture.

The potentiostatic circuit shown in Fig. 6.1 will be employed. The two-potential step will be produced by a square-wave function generator. This will be connected at the "external-modulation input" of the potentiostat. The current/time transients will be followed on an oscilloscope.

10.4 TWO-STEP VOLTAMMETRY

10.4.3 Determination of the Working Potentials

Introduce the working solution into the cell and deaerate by bubbling nitrogen for 10 minutes. Introduce the electrodes and connect them for measurement of cyclic voltammetry (cf. Fig. 9.5). Set the initial potential to +0.3 Volt (SCE) and apply a triangular wave of about 0.7 Volt amplitude at a sweep rate of about 50 mV/sec. Record the voltammogram and from it determine the potentials E_1 and E_2 at which reaction [10.23] proceeds to the right or to the left, respectively, under diffusion-controlled conditions.

10.4.4 Determination of the Switching Time

For the measurement of transients use a low-resistance reference electrode as in Experiment 9.2 (cf. Introduction to Experiment 2). Retain the SCE in the solution and check the potential of the Pd/H_2 electrode against it from time to time. Set the potential of the square-wave generator to the values E_1 and E_2 determined in the previous section (corrected for the different reference electrode used) and apply the square wave to the external-modulation input of the potentiostat. Set the switching time to 30 msec. Record the i/t curve on the oscilloscope and from the value of $|i_a/i_c|$ at times t and $(t-t_f)$, estimate the value of the rate constant k_f for the rearrangement reaction (equation [10.24]) with the aid of Table 10.2. Highest accuracy in the determination of k_f is achieved when the switching time t_f is of the order of the half-life of the quasi-first-order chemical reaction, that is, when $k_f t_f$ is in the range of 0.6-1.5.

10.4.5 Determination of the Rate Constant for Rearrangement

Apply the two-potential step with 3-4 switching times in the range determined in the previous section. Stir the solution by bubbling purified nitrogen after each transient, then allow to settle for half a minute. Pass the nitrogen above the solution throughout the experiment.

Prepare a solution differing from the previous one only in that the concentration of $HClO_4$ is 0.8 $\underline{M}$. Note that k_f depends very strongly on acid concentration. Literature data$^{(12)}$ indicate that in 0.8 $\underline{M}$ $HClO_4$ its value is smaller than in 2 $\underline{M}$ $HClO_4$ by a factor of about 30. Adjust the switching times accordingly and repeat the experiment.

10.4.6 Measurement with No Following Chemical Reaction

In view of the pronounced dependence of the specific rate constant on acidity, it is possible to conduct the two-step-voltammetry experiment in this system under conditions for which the chemical step following charge transfer may be completely neglected. Thus, if $k_f t_f \ll 1$, the fraction of hydrazobenzene molecules undergoing rearrangement will be approximately equal to $k_f t_f$. With the same solution as in the last experiment, make 4-6 measurements with t_f adjusted so that $k_f t_f$ is in the range 0.01-0.1. Stir the solution by bubbling purified nitrogen after each experiment and allow to settle for half a minute.

10.4 TWO-STEP VOLTAMMETRY

10.4.7 Treatment of Results

Calculate the value of k_f in 2 $\underline{M}$ and in 0.8 $\underline{M}$ $HClO_4$. For each switching time, calculate $k_f t_f$ for five different values of $(t-t_f)/t_f$. Obtain k_f from a plot of $k_f t_f$ $\underline{vs}$ t_f. Compare with results in the literature. Why is best accuracy achieved when $k_f t_f \simeq 1$? What would you expect to find experimentally in this system if $k_f t_f \gg 1$?

For measurements in 0.8 $\underline{M}$ $HClO_4$ plot i_c $\underline{vs}$ $t^{-1/2}$ and determine the diffusion coefficient of azobenzene.

For the same experiment, plot $|i_a/i_c|$ $\underline{vs}$ the second term on the r.h.s. of equation [10.22]. Discuss the reason for deviation from a straight line at higher values of t_f.

REFERENCES

1. H. J. S. Sand, Phil. Mag. 1, 45 (1901).
2. A. C. Testa and W. H. Reinmuth, Anal. Chem. 32, 1512 (1960); 32, 1518 (1960); 33, 1324 (1961).
3. W. H. Reinmuth, Anal. Chem. 32, 1514 (1960).
4. P. Delahay and T. Berzins, J. Amer. Chem. Soc. 75, 2486 (1953).
5. P. Delahay and G. Mamantov, Anal. Chem. 27, 478 (1955).
6. W. H. Reinmuth, Anal. Chem. 32, 1509 (1960).
7. M. Paunovic, J. Electroanal. Chem. 14, 447 (1967).
8. P. Delahay, "New Instrumental Methods in Electrochemistry", Chap. 8, Interscience, 1954.
9. R. N. Adams, "Electrochemistry at Solid Electrodes", Chap. 6, Marcel Dekker, Inc., 1969.
10. D. G. Davis, "Electroanalytical Chemistry", Vol. 1, ed. A. J. Bard, M. Dekker Inc., 1966.
11. C. N. Reilley, G. W. Everett and R. H. Johns, Anal. Chem. 27, 483 (1955).
12. W. M. Schwarz and I. Shain, J. Phys. Chem. 69, 30 (1965).
13. T. Berzins and P. Delahay, J. Amer. Chem. Soc. 75, 4205 (1953).
14. C. Furlani and G. Morpurgo, J. Electroanal. Chem. 1, 351 (1960).
15. D. Dracka, Coll. Czech. Chem. Comm. 25, 338 (1960).

11. ADSORPTION OF OXYGEN AND HYDROGEN ON NOBLE-METAL ELECTRODES

Introduction

11.1 Fast Linear Potential Sweep

11.1.1 General

11.1.2 The Cell, Solutions, Electrodes and Electrical Setup

11.1.3 Determination of the Voltammogram in Pure Solutions

11.1.4 The Effect of Impurities on the Voltammogram

11.1.5 Treatment of Results

11.2 Potential-Step Transients

11.2.1 General

11.2.2 The Cell, Solutions, Electrodes and Electrical Setup

11.2.3 Evaluation of the Charge Due to Surface-Oxide Formation

11.2.4 Determination of Fractional Surface Coverage as a Function of Potential in Acid and Alkaline Solutions

11.2.5 Treatment of Results

11. ADSORPTION OF OXYGEN AND HYDROGEN ON NOBLE-METAL ELECTRODES

Introduction

In the study of electrode reactions at solid metals, noble metals have the advantage of being stable over a relatively large potential region. Thus a platinum electrode in purified deaerated 0.5 $\underline{M}$ H_2SO_4 solution will show practically zero faradaic current under steady-state conditions from 0.05 V to about 1.55 V vs RHE. Nevertheless, the system cannot be considered ideally polarized over this whole region, since hydrogen is adsorbed at cathodic potentials up to about +0.40 V and oxygen is adsorbed[*] at anodic potentials, starting at about 0.9 V. The existence of adsorbed hydrogen or oxygen can be detected by applying a potential transient of any form to the system. A transient faradaic current is then observed, which decays to zero some time after the potential transient has been applied.

When a galvanostatic (i.e. current-step) transient is applied, the formation or removal of a layer of adsorbed species is evidenced by a plateau in the observed potential/time plot, from which the surface concentration of adsorbed species can be calculated.[1] The system behaves as though it has acquired an

---

[*] An argument may be found even in fairly recent papers as to whether the species adsorbed at anodic potentials is "adsorbed oxygen" or a "surface oxide". The distinction between these seems to be a matter of semantics as long as it is borne in mind that a surface oxide which is one atomic layer thick is not expected to have the same properties as bulk oxide.

INTRODUCTION

additional large capacitance (in parallel with the double-layer capacitance) and this is termed the <u>adsorption pseudocapacitance</u> (cf. section II.8). If a linear potential sweep is applied, the regions of adsorption are characterized by peaks in the observed current/potential plot, which are due to the strong potential dependence of the adsorption pseudocapacitance.[2,3] The surface concentration in this case is calculated from the area under the peaks.

The potential-sweep method was applied to the study of oxygen and hydrogen adsorption on several noble metals by Will and Knorr.[4] The theory applicable to this technique was developed by Srinivasan and Gileadi[5] who showed that when adsorption and desorption are not reversible (which is the case for surface-oxide layers for all the metals studied even at the lowest sweep rate) the peak potential, and consequently the calculated dependence of surface concentration on potential, should depend on the sweep rate applied. Similarly, it was shown by Conway <u>et al</u>[6] that the surface-concentration/potential plot calculated from galvanostatic transients depended on the magnitude of the applied current step.

The inaccuracy inherent in the above techniques is avoided when potentiostatic (i.e. potential-step) transients are employed to determine the surface concentration of adsorbed species. Here the potential, which is initially held at a point where adsorption is absent, is stepped to a desired value and held constant until the current decays to its background value. The potential is then stepped back to its initial value and held there until the current decays once again to its background value at this potential. The charge measured during the transient yields the true dependence of surface concentration on potential.* The equality of the charge in the forward and reverse transients shows that the adsorbed species has been completely removed. The potential-step method has been applied recently by Icenhower, Urbach and Harrison[7] to the study of surface oxides on several noble metals at different pH values. The experiment given below follows in a large part the work of these authors.

11.1 Fast Linear Potential Sweep

11.1.1 General

The purpose of this experiment is to obtain a qualitative picture of the adsorption of oxygen and hydrogen on platinum and gold electrodes in acid and alkaline solutions. The sensitivity of the measurements to molecular oxygen and to surface-active impurities will also be tested.

11.1.2 The Cell, Solutions, Electrodes and Electrical Setup

The cell, electrodes and electrical setup required for this experiment are identical with those described in Experiment 9.2, except that an X-Y recorder should be used in place of the

---

* This is the "true dependence" in the sense that it corresponds to that which would have been obtained by the galvanostatic or linear potential-sweep techniques if the current and sweep rate, respectively, were kept sufficiently low so that the system did not depart significantly from equilibrium during the transient.

11.1 FAST LINEAR POTENTIAL SWEEP

oscilloscope. The reference electrode will be a palladium/hydrogen electrode in the form of a wire coated with insulating material (shrinkable teflon tubing) and bare only at its end. This is very suitable for the system studied here since it is a low-resistance indicator-type reference electrode reversible to hydrogen ions. Thus the potentials of hydrogen and oxygen evolution as well as the potentials for formation of adsorbed hydrogen and oxygen are independent of pH when measured against this reference electrode (cf. Experiment 2.2).

Platinum and gold working electrodes will be constructed by press-fitting into teflon. Both will be cleaned according to the procedure outlined in Experiment 6.1. Electrochemical pretreatment will be performed only with platinum electrodes.

The solutions used in this experiment will be 0.5 $\underline{M}$ H_2SO_4 and 1 $\underline{M}$ NaOH, made up of analytical-grade reagents and triply distilled water.

11.1.3 Determination of the Voltammogram in Pure Solutions

Add 0.5 $\underline{M}$ H_2SO_4 to the cell. Introduce the platinum working electrode and the counter and reference electrodes and deaerate by bubbling purified nitrogen for 10 minutes. Set the potentiostat to 0.0 Volt $\underline{vs}$ RHE[*] and adjust the amplitude of the

---

[*] Note that although the palladium/hydrogen electrode behaves reversibly with respect to H_3O^+ ions in solution, its potential differs by about 50 mV from that of a reversible hydrogen electrode in the same solution. This difference is caused by the lower fugacity of hydrogen dissolved in palladium.

potential sweep to 1.6 Volts. Set the sweep rate to 40 mV/sec, pretreat the working electrode electrochemically and immediately apply a single triangular wave starting from 0.0 Volt. Record the i/E transient on an X-Y recorder. Change the sweep rate to 100 mV/sec, decrease the sensitivity of the Y axis of the recorder by a factor of 2.5 and repeat the above procedure (including pretreatment). Record the i/E transient on the same sheet of paper. Stir the solution throughout the experiment and pass nitrogen over it.*

Repeat with a gold working electrode in a fresh 0.5 $\underline{M}$ H_2SO_4 solution and with both gold and platinum electrodes in 1 $\underline{M}$ NaOH. In the case of gold only the higher sweep rate will be used. The gold electrode will not be pretreated electrochemically. Instead it will be cycled several times until the i/E plot does not change further and only this last plot will be recorded. (The "steady-state" i/E curve should be reached in 5-10 cycles at most.)

11.1.4 The Effect of Impurities on the Voltammogram

A platinum working electrode and a 0.5 $\underline{M}$ H_2SO_4 solution will be used to test the effect of impurities on the current/potential plot. Repeat the first experiment of the previous section (at 40 mV/sec only) but saturate the solution with oxygen instead of nitrogen. Plot the steady-state voltammogram. Switch again to

---

* The reactions studied here are surface reactions, and are therefore not affected by stirring. The current is proportional to the sweep rate rather than to its square root, as is the case in diffusion-controlled reactions discussed in Experiment 9.

11.1   FAST LINEAR POTENTIAL SWEEP

nitrogen, continue cycling and plot the voltammograms after 1, 3, 6, 10, 15 and 20 minutes.

Prepare a solution with de-ionized or once-distilled water, deaerate with nitrogen and obtain the i/E plot as above.

Clean the cell and electrodes. Repeat the experiment in the pure system and record the i/E plot at 40 mV/sec. Introduce a small quantity of a dilute solution of a detergent. (Any detergent may be used and the final concentration should be about 5-10 mg/l.) Record several voltammograms, until a new steady state is reached.

11.1.5   Treatment of Results

Discuss the major features of the voltammograms observed for platinum and for gold. How does the pH affect the adsorption of oxygen and hydrogen on these metals?

What is the qualitative effect of sweep rate on the various peaks observed? Discuss.

Use the results of the measurements in 0.5 $\underline{M}$ H_2SO_4 to calculate the real area of the platinum and gold electrodes by determining the charge required to reduce the surface-oxide layers. Assume for the purpose of this calculation that a monolayer is formed under these conditions and that the charge required to form it is 0.43 mC/cm^2. Calculate the ratio of real to apparent surface area. (This quantity is called the roughness factor.)

Discuss the effect of impurities on the shape of the voltammogram. Estimate the lowest concentration of detergent of the type used in this experiment which could be detected by its effect on the shape of the steady-state voltammogram.

11.2 Potential-Step Transients

11.2.1 General

The purpose of this experiment is to measure the surface concentration or fractional surface coverage, θ, of adsorbed oxygen on platinum and gold electrodes in acid and alkaline solutions. The application of potential-step functions (potentiostatic transients) in electrochemistry requires great care in cell design and circuit elements, otherwise the shape of the resulting current transient may be determined not only by the electrode reaction being studied but also partly by these factors. This danger is avoided in the present experiment by the use of a fast electronic integrator. Since the charge is measured by integrating the i/t curve over a relatively long period (10-100 seconds), distortions in the shape of this curve at very short times do not affect the result. This technique, incidentally, allows much easier and more accurate correction for the background current to be made, as will be shown below.

11.2.2 The Cell, Solutions, Electrodes and Electrical Setup

The cell, solutions and electrodes are identical to those used in Experiment 11.1. The block diagram of the electrical circuit is shown in Fig. 11.1. The potential programmer serves as the source of potential for electrochemical pretreatment and for application of the potential step. In the absence of such a device, the potential can be stepped manually from one value to

11.2 POTENTIAL-STEP TRANSIENTS

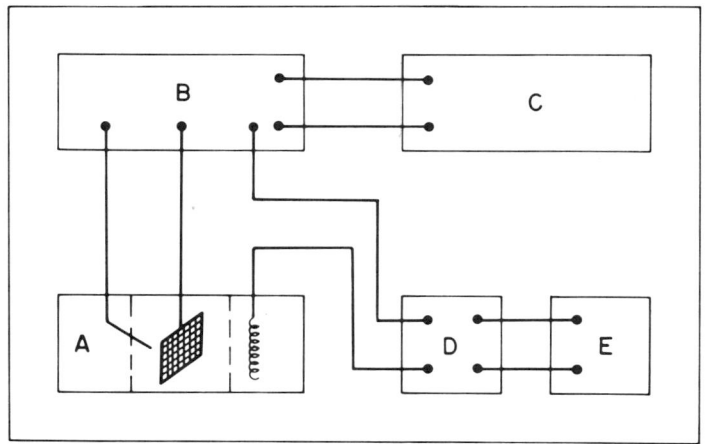

FIG. 11.1 BLOCK DIAGRAM FOR POTENTIAL-STEP MEASUREMENTS WITH CURRENT INTEGRATION
A - CELL;   B - POTENTIOSTAT;
C - PROGRAMMER;   D - INTEGRATOR;
E - STRIP-CHART RECORDER.

the next. The sequence of potentials applied to the platinum electrode is shown in Fig. 11.2a. The potential should be held at 0.45 Volt <u>vs</u> RHE (the last step in the pretreatment) until the current has declined to a small constant value. The gold electrode will be pretreated by cycling between 0.0 Volt and 1.6 Volt (RHE) until a steady-state voltammogram is observed. Cycling is terminated on the cathodic side, the potential is stepped to 0.45 Volt (RHE) for 30 seconds and the experiment continued as with the platinum electrode.

11.2.3 Evaluation of the Charge Due to Surface-Oxide Formation

A typical current/time transient is shown in Fig. 11.2b. A few seconds after the potential step is applied, the current decays to a small, constant background current. The transient

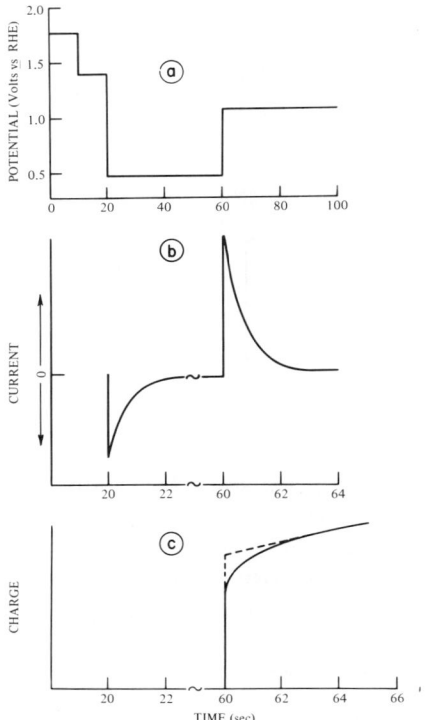

FIG. 11.2 ELECTRODE PRETREATMENT AND MEASUREMENT OF CHARGE DUE TO FORMATION OF SURFACE OXIDE LAYER
a - SEQUENCE OF POTENTIAL STEPS;
b - CURRENT/TIME TRANSIENTS;
c - CHARGE/TIME TRANSIENTS.

at the output of the integrator, giving the time variation of the total charge passed, is shown schematically in Fig. 11.2c. The linearly rising section of the curve corresponds to the constant background current. The true value of the charge associated with surface-oxide formation is obtained by extrapolating the linear section of the curve to zero time.

11.2 POTENTIAL-STEP TRANSIENTS

If a suitable integrator is not available, this experiment can still be performed by recording the current/time transients and integrating manually (e.g., by weighing the paper corresponding to the area under the curve). In this case, however, one should use an oscilloscope and camera to record the transient, since distortions due to the slow response of a pen recorder may cause major errors in the results. Correction for the background current is achieved by subtracting the product of residual current and integration time from the total charge measured.

11.2.4 Determination of Fractional Surface Coverage as a Function of Potential in Acid and Alkaline Solutions

Set up the cell with 0.5 $\underline{M}$ H_2SO_4 solution and use a platinum working electrode. Deaerate with purified nitrogen in the usual manner (cf. Experiment 11.1). Step the potential to 0.75 Volt $\underline{vs}$ RHE and record the charge at the output of the integrator. Adjust the sensitivity of the recorder to obtain highest accuracy and repeat the measurement several times to estimate the reproducibility of your measurements.

Determine the charge as a function of potential at 50 mV intervals up to 1.55 Volt $\underline{vs}$ RHE. Take at least two measurements at each potential. Repeat with a gold electrode in the same solution.

Repeat with both electrodes in 1 $\underline{M}$ NaOH solution.

11.2.5 Treatment of Results

Plot the charge against potential (on the RHE scale) for acid and alkaline solutions on the same sheet of graph paper for platinum and similarly on another sheet of graph paper for gold.

Calculate, in each case, the surface concentration and fractional coverage. For these calculations, use the real surface area of the electrode as measured in Experiment 11.1 and assume that the formation of a complete monolayer requires the passage of 0.43 mC/cm^2.

Correct the values of the charge obtained above for double-layer charging (assume an average value of $C_{dl} = 20$ $\mu F/cm^2$) and replot the corrected charge against potential for all the systems studied.

Calculate the reproducibility of your measurements and discuss the possible sources of error.

REFERENCES

1. E. Gileadi and B. E. Conway, "Modern Aspects of Electrochemistry", Vol. 3, Chap. 5, p. 351, Ed. J. O'M. Bockris and B. E. Conway, Butterworths, London, 1964.
2. E. Eucken and B. Weblus, Z. Electrochem. 55, 114 (1951).
3. B. E. Conway and E. Gileadi, Trans. Faraday Soc. 58, 2493 (1962).
4. F. G. Will and C. A. Knorr, Z. Elektrochem. 64, 258, 270 (1960).
5. S. Srinivasan and E. Gileadi, Electrochim. Acta 11, 321 (1966).
6. B. E. Conway, E. Gileadi and H. Angerstein-Kozlowska, J. Electrochem. Soc. 112, 341 (1965).
7. D. E. Icenhower, H. B. Urbach and J. H. Harrison, J. Electrochem. Soc. 117, 1500 (1970).

12. ELECTROSORPTION

Introduction

12.1 Linear Potential-Sweep Transients on Platinum

12.1.1 General
12.1.2 The Cell, Solutions, Electrodes and Electrical Setup
12.1.3 Background Current
12.1.4 Background Correction
12.1.5 Reproducibility of Charge Due to Benzene Oxidation
12.1.6 Treatment of Results

12.2 Determination of the Surface Concentration of Benzene as a Function of Sweep Rate and of Potential

12.2.1 General
12.2.2 The Cell, Electrodes and Electrical Setup
12.2.3 Distortion of the Transient at High Sweep Rate
12.2.4 Determination of Q as a Function of Sweep Rate
12.2.5 Determination of Q as a Function of Potential
12.2.6 Treatment of Results

12.3 Kinetics of Electrosorption of Benzene on Platinum

12.3.1 General
12.3.2 The Cell, Solutions, Electrodes and Electrical Setup
12.3.3 The Rate of Electrosorption in Unstirred Solution
12.3.4 The Rate of Electrosorption in Stirred Solution
12.3.5 Treatment of Results

12.4 The Isotherm for Electrosorption of Benzene

12.4.1 General
12.4.2 The Cell, Electrodes and Electrical Setup
12.4.3 Determination of the Adsorption Isotherm
12.4.4 Treatment of Results

12. ELECTROSORPTION

Introduction

Electrosorption is defined as the chemisorption of a species from solution on the surface of a conducting material (the electrode) the potential of which is controlled. Electrosorption differs from gas-to-solid chemisorption in two ways; the potential E is introduced as an added independent variable[*] and the adsorption process is in effect a replacement reaction in which a number of solvent molecules is desorbed from the surface for each solute molecule adsorbed. Electrosorption has been most widely studied on mercury. This is because the extent of adsorption (more precisely, the relative surface excess, cf. section II.4.2) can be evaluated by measurement of the variation of the interfacial tension with the chemical potential of the adsorbate in solution (cf. Experiment 3.3). Electrosorption on solid electrodes is more difficult to determine experimentally and the results obtained tend to be substantially less accurate. In view of the small quantities of material adsorbed on the surface (of the order of 10^{-10} mole cm^{-2}) an obvious approach to the experimental study of electrosorption is the use of radiotracer

---

[*] There is currently disagreement among authors as to whether the potential E or the surface-charge density q_M should be used as the independent variable. This argument is of little interest where electrosorption on solid electrodes is concerned, since for this case there is as yet no accurate method of determining q_M as a function of E.

techniques. This approach has been followed by the Frumkin school[1] mainly for the study of adsorption of ions, and by Bockris and coworkers[2] for the study of adsorption of neutral organic molecules. When the ratio between the volume of the solution and the area of the electrode in contact with it is sufficiently small* it is possible to measure adsorption by determining the change of concentration in solution with the use of a suitable analytical technique[3].

Electrochemical techniques for the study of adsorption of organic molecules on the surface of noble metals have been used by several authors[4-7]. In one of these, a short anodic pulse is applied to the electrode and the charge required to oxidize ("burn off" electrochemically) the adsorbate originally present on the surface is recorded. A blank experiment with a clean electrode must be performed in each case and the charge is obtained as the difference between the two runs. A rapid linear potential sweep has been most often used to measure electrosorption by the electrochemical technique, but a current- or potential-step function may also be used. The validity of the electrochemical techniques for the measurement of electrosorption is based on several assumptions. These are: (a) that oxidation is complete, leading to CO_2 as the only product; (b) that all the organic adsorbate is removed from the surface during the transient; (c) that no organic

---

* This is most easily achieved with electrodes of high roughness factor, e.g. platinized platinum, where the actual surface area can be greater than the geometrical area by a factor of a thousand or more.

species is desorbed during the transient without oxidation; (d) that there is no additional adsorption of solute on the surface during the transient; (e) that the difference between the background currents on the bare and on the covered electrode is negligible. When all these assumptions are valid, the electrochemical methods can yield accurate results at low surface concentration.

In a recent study by Gileadi, Duic and Bockris[8], the electrosorption of benzene on platinum was measured as a function of concentration and potential by the radiotracer and by the three electrochemical techniques. On the cathodic branch of the θ/E plot good agreement was found among results obtained by all techniques. On the anodic branch the three electrochemical techniques yielded consistently lower values of the surface coverage than did the radiotracer method. This is probably due to the fact that as the potential is made more anodic, an increasing proportion of the benzene is adsorbed in a partially dehydrogenated form, which requires fewer electrons for oxidation to CO_2. The radiotracer method measures the number of carbon atoms on the surface and is hence insensitive to the oxidation state of the adsorbed species.

It was concluded[8] that in a complete study of electrosorption of an organic species, the radiotracer technique and at least one electrochemical technique should be employed. The choice among the possible electrochemical methods is a matter of preference, or may depend on available instrumentation.

In the experiments described below, the technique of fast linear potential sweep has been chosen for the study of the electrosorption of benzene on platinum, under conditions similar to those reported in the original paper[8]. The radiotracer method is too involved to be proposed as an exercise.

12.1 Linear Potential-Sweep Transients on Platinum

12.1.1 General

The purpose of this experiment is to familiarize the student with the linear-potential-sweep technique and the general phenomena obtained with a platinum electrode in 0.5 $\underline{M}$ H_2SO_4 solution with and without benzene. Fig. 12.1 shows a typical current/potential plot for this system, in the absence of benzene, at a sweep rate of 40 mV/sec. On the left (at relatively cathodic potentials), two pairs of peaks are shown, corresponding to formation and ionization of adsorbed atomic hydrogen. The existence of two peaks indicates that hydrogen can be adsorbed on the surface in two modes[9,10]. The distance between the peaks (about 0.12 V) is a measure of the difference in energy of adsorption, which is about 2.5 kcal/mole. The peaks on the right are due to the formation and removal of a monolayer of oxide. Note that the hydrogen peaks on the anodic and cathodic branches of the sweep occur at almost the same potential and are very similar in shape. The oxide peaks have quite different shapes and occur at different potentials. The former represent a system which behaves essentially reversibly under the chosen experimental conditions while the latter are typical of an irreversible process.

The effect of addition of benzene to the solution is shown in Fig. 12.2. The current during the anodic sweep is higher, the difference being due to the oxidation of benzene adsorbed on

12.1 LINEAR POTENTIAL-SWEEP TRANSIENTS

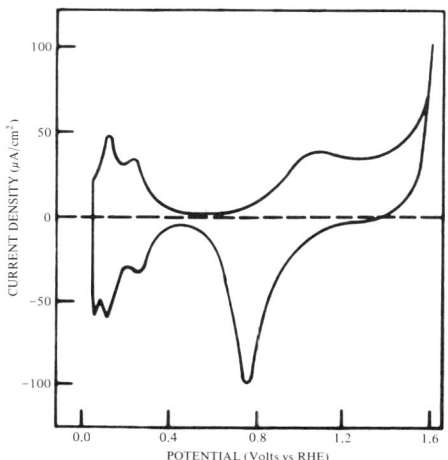

FIG. 12.1   POTENTIAL-SWEEP TRANSIENT FOR PLATINUM IN 0.5 $\underline{M}$ H_2SO_4.   v = 40 mV/sec.

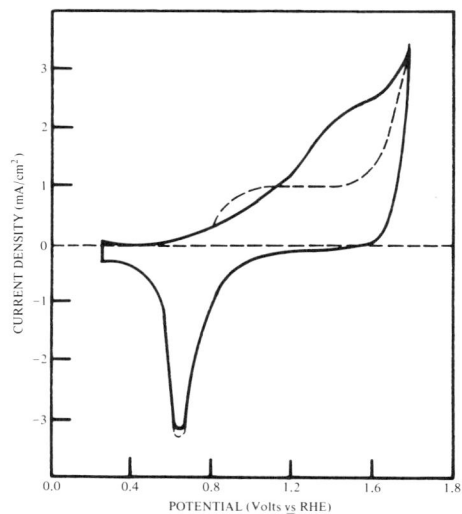

FIG. 12.2   POTENTIAL-SWEEP TRANSIENT FOR $2\mu\underline{M}$ BENZENE IN 0.5 $\underline{M}$ H_2SO_4 ON PLATINUM. (DOTTED LINE OBTAINED IN SECOND SWEEP) v = 1.0 Volt/sec.

the surface. If a second sweep is taken just after the first, the anodic current is identical to that observed in the absence of benzene (see dotted line in Fig. 12.2), showing that all of the adsorbate has been oxidized from the surface and that the time has been too short for readsorption to take place.* From the area between the dotted and the solid lines, one can calculate the charge due to oxidation of benzene adsorbed on the surface. Two small corrections may be applied in this measurement. It is noted in Fig. 12.2 that during the anodic sweep the dotted line is slightly higher than the full line over a narrow potential range. This is due to the fact that oxide formation is retarded and occurs at slightly higher potentials in the presence of adsorbed benzene. The small area between the dotted and solid lines due to this effect is usually subtracted from the overall area in order to obtain the charge due to the oxidation of benzene. On the cathodic branch of the sweep, the full line is seen to enclose a smaller area than the dotted line. This indicates that less oxide is formed in the presence of benzene than in its absence. The small area between the two lines should therefore be added to the total area representing the charge due to the oxidation of benzene on the surface.

---

* This is strictly true only when a proper sweep rate is chosen. If the sweep rate is too high, complete oxidation of adsorbate occurs only after several sweeps; if it is too low, readsorption from solution during the transient occurs.

12.1 LINEAR POTENTIAL-SWEEP TRANSIENTS

12.1.2 The Cell, Solutions, Electrodes and Electrical Setup

A cell such as shown in Fig. 8.4 will be used in these experiments. Since relatively high currents are to be expected during the transients, care must be taken to minimize the resistance in solution between the working and counter electrodes. A low-resistance reference electrode should preferably be used to eliminate distortions in the shape of the fast transient. A platinum electrode (about 0.2 cm^2) is suitable as a working electrode. A cylindrical gauze counter electrode surrounding the working electrode will provide adequately uniform distribution of the current density. A palladium wire loaded with hydrogen and placed close to the working electrode will serve as a low-resistance reference electrode (cf. Experiment 2.2). Its potential will be checked from time to time with respect to a commercial reference electrode placed in the auxiliary-electrode compartment of Fig. 8.4.

A stock solution of 10^{-3} $\underline{M}$ benzene in 0.5 $\underline{M}$ H_2SO_4 is used. The purity of the solution and particularly the absence of organic impurities is very important in these experiments.*

The electrical setup shown in Fig. 9.5 will be used. The current/potential transients will be recorded on a storage oscilloscope (the Tektronix type 564 oscilloscope with a 10 M Ω adaptor (replacing the impedance-matching unit) and a 3A3 plug-in unit is

---

* Further purification of sulfuric acid can be achieved by adding 5 percent by volume of concentrated H_2O_2, refluxing for half an hour and distilling off the water formed by decomposition of hydrogen peroxide.

suitable for this purpose). The oscilloscope is operated as an X-Y recorder. A suitable oscilloscope camera should be used to record the transients so that the areas between the curves can be measured with accuracy.

12.1.3 Background Current

Clean the cell carefully and add 0.5 $\underline{M}$ H_2SO_4. Introduce the electrodes. Deaerate by bubbling purified nitrogen for 10 minutes. Continue to pass the gas above the solution during the experiment.

Connect the working and reference electrodes to the horizontal input of the oscilloscope, by way of the 10 M Ω adaptor. Set the sensitivity on this scale to 0.2 V/cm. Measure the current on the vertical axis of the oscilloscope. Adjust the sensitivity to obtain highest accuracy. The setting will depend on the actual size of the working electrode and on the sweep rate.

Set the linear-sweep generator to 1 Volt/sec over the range 0.4-1.7 Volts vs RHE and set it on the single-pulse mode. Follow the electrochemical pretreatment procedure described in Experiment 6.1 and then apply a single triangular sweep (i.e. an anodic and cathodic linear potential sweep between the same potential limits and at the same sweep rate). Observe the shape of the current/potential transient on the storage oscilloscope (at this stage it is not necessary to take pictures unless a storage oscilloscope is unavailable). The transient should show the regions of oxide formation and reduction and the beginning of oxygen evolution. Increase the amplitude by about 0.4 Volt and shift in the cathodic direction, so that the range covered will now

12.1 LINEAR POTENTIAL-SWEEP TRANSIENTS

be 0.0-1.7 Volt (RHE). Observe the two peaks for ionization of adsorbed atomic hydrogen and the corresponding cathodic peaks for its deposition on the surface. A further increase of the potential range by about 0.1 Volt will clearly show a rise on the anodic side due to oxygen evolution and one on the cathodic side due to hydrogen evolution.

Decrease the potential span and adjust the range so that the hydrogen peaks will disappear and oxygen evolution will be barely detectable on the i/E plot. Take pictures at sweep rates of 0.1, 1 and 5 Volt/sec. Make sure to mark the exact experimental conditions on the back of each picture.

12.1.4 Background Correction

Add to the cell an accurately measured volume of 0.5 $\underline{M}$ H_2SO_4. Add enough of the stock solution of benzene to obtain a final concentration of 2×10^{-6} $\underline{M}$. Introduce the electrodes and deaerate by bubbling purified nitrogen for 10 minutes. Pretreat the electrode electrochemically, (cf. Experiment 6.1), allow the solution to rest for about 20 seconds (while nitrogen is passed above the solution) and apply a single triangular sweep in the range of 0.4-1.7 Volt (RHE). Apply a second and third sweep immediately after the first. Keep all three transients simultaneously on the oscilloscope memory. Note that the second and third transients practically coincide. Photograph these transients and compare quantitatively with the transient observed under identical conditions in the absence of benzene in the previous section. The limit of the sweep on the anodic side in this and all further experiments should be adjusted so that the currents in the presence and in the absence of benzene coincide, as seen in Fig. 12.2.

Repeat, for the same solution, at a sweep rate of 5 Volts/sec. Apply several sweeps until consecutive transients are identical.

These experiments show that most of the adsorbed benzene can be removed from the surface by oxidation during the first sweep and the surface is completely bare after a few more sweeps, the number depending on the experimental conditions. Readsorption does not occur to a significant extent during the short time of the sweep. Thus the background current can be measured in a solution containing benzene by simply measuring the second or consecutive transient (usually the third or fourth sweep will do for this purpose). The same electrode can be used throughout the experiment without further electrochemical pretreatment, as the sweep between the limits required for these experiments incidentally serves as a reproducible pretreatment procedure.

12.1.5 Reproducibility of Charge Due to Benzene Oxidation

The reproducibility of the charge measured due to the oxidation of benzene adsorbed on the surface depends on several factors, such as the use of a clean, reproducible surface, constant time and potential of adsorption, constant stirring conditions and measurement at the same sweep rate. In this section, the reproducibility of the results under identical experimental conditions will be tested.

Use the same solution and electrodes as in the previous section. Adjust the oscilloscope so that only the anodic part of the transient can be seen. This ensures increased sensitivity and

12.1 LINEAR POTENTIAL-SWEEP TRANSIENTS

accuracy in measurement of the charge.* At a sweep rate of 1.0 Volt/sec and a potential span as determined in the previous section, sweep several times without stirring until the electrode surface is free of benzene. Start the magnetic stirrer immediately after the last sweep and stir for exactly 120 seconds. Stop stirring, wait exactly 10 seconds, then apply a sweep. Photograph this transient but do not advance the film. Clean the surface by applying several sweeps. During this procedure the intensity of the beam on the oscilloscope should be decreased so that unwanted information is not recorded on the memory of the oscilloscope screen. Increase the intensity of the beam and obtain a background transient. Take a picture on the same film.

Repeat the experiment three more times under identical conditions.

12.1.6 Treatment of Results

Measure the area under the anodic and the cathodic oxide peaks in the background transient. Calculate the real surface area and the roughness factor assuming that about 0.43 mC/cm^2 are required to form a monolayer of adsorbed oxygen. Use the cathodic peak for this determination.

Copy the transients on paper, cut out the area between the curves and weigh. Subtract the weight of the small area

---

* The correction due to the difference between the amounts of surface oxide formed in the presence and absence of benzene (cf. the cathodic branch in Fig. 12.2) is small and may be neglected in these exercises.

where the background is above the curve with benzene. Calculate the charge per unit area of the electrode due to oxidation of benzene. Calculate the surface concentration of benzene as measured in this experiment, assuming complete oxidation to CO_2.

From the background transients at different potential spans, estimate the range of stability of water under these experimental conditions. Compare with the thermodynamic range of stability. Comment on the reversibility of the hydrogen- and oxygen-evolution reactions on platinum in 0.5 $\underline{M}$ H_2SO_4.

Which of the assumptions required for the validity of the electrochemical technique for the measurement of surface coverage have been tested in this experiment? Comment on the results.

What is the accuracy which can be expected in this kind of experiment? Discuss both the systematic and the random sources of error.

12.2 Determination of the Surface Concentration of Benzene as a Function of Sweep Rate and of Potential

12.2.1 General

Two of the necessary conditions for the correct determination of the surface concentration of organic adsorbates are: (i) that the transient is of short enough duration, that readsorption from solution during the sweep is insignificant and (ii) that the transient is of long enough duration to ensure complete oxidation of the adsorbate initially present on the surface. If either condition is not fulfilled, the charge (Q) measured (which should

12.2 DETERMINATION OF SURFACE CONCENTRATION

be proportional to the surface concentration of the adsorbed species) decreases with increasing sweep rate. Under favorable conditions a plateau may be observed in the plot of Q vs sweep rate as shown in Figs. 12.3 and 12.4 for benzene[8] and methyl alcohol[4] respectively. The limits of the plateau are determined at one end by the rate of adsorption and at the other by the rate of oxidation. The width of the plateau (i.e. the range of sweep rates over which it exists) tends to increase with decreasing concentration in solution, and the whole plateau may occur over different ranges of sweep rate depending on the compound being adsorbed and the experimental conditions, as seen by comparison of Figs. 12.3 and 12.4. Measurements of surface coverage must be performed at sweep rates in the region of the plateau and clearly, unless conditions can be found under which there is a plateau in the Q vs log v plot, the method cannot yield meaningful results for surface-coverage determination.

In this experiment, the charge Q due to benzene oxidation in a 2×10^{-6} M solution of benzene in 0.5 M H_2SO_4 will be determined as a function of sweep rate. The real charge is that measured in the plateau region. After the appropriate sweep rate has been determined in this manner, the dependence of Q (or surface coverage) on the applied potential will be studied.

12.2.2 The Cell, Electrodes and Electrical Setup

The cell, electrodes and electrical setup will be exactly as in Experiment 12.1.

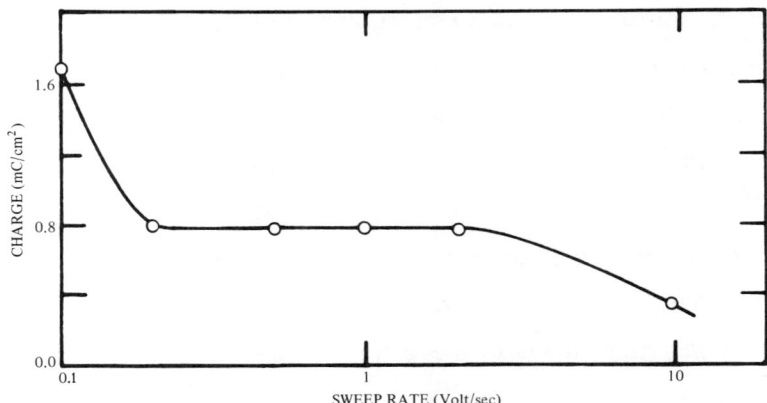

FIG. 12.3   DEPENDENCE OF THE APPARENT CHARGE DUE TO OXIDATION OF BENZENE ON THE SWEEP RATE.   2 $\mu\underline{M}$ BENZENE IN 0.5 $\underline{M}$ H_2SO_4.

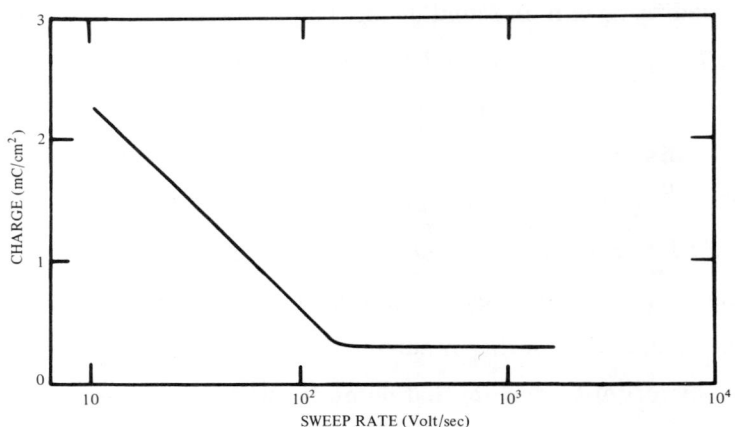

FIG. 12.4   DEPENDENCE OF THE APPARENT CHARGE DUE TO OXIDATION OF METHANOL ON THE SWEEP RATE.   1 $\underline{M}$ CH_3OH IN 1 $\underline{M}$ $HClO_4$.

12.2 DETERMINATION OF SURFACE CONCENTRATION

12.2.3 Distortion of the Transient at High Sweep Rate

The upper limit of sweep rates which can be used in this experiment depends on the intrinsic properties of the instruments used and on the configuration of the cell. The limiting factors are usually the slewing rate of the potentiostat and the resistance of the reference electrode.* If the sweep applied to the electrode is distorted, the results obtained may be grossly in error.

A simple way to detect distortions in the sweep is to measure the potential of the electrode as a function of time or of the output potential of the sweep generator. To do this, use a solution which is 2×10^{-6} $\underline{M}$ in benzene and 0.5 $\underline{M}$ in H_2SO_4. Connect the working and reference electrodes to the horizontal input of the oscilloscope as in Experiment 12.1 (using a $10^7 \Omega$ adaptor for the reference electrode). Connect the vertical input of the oscilloscope to the output terminals of the function generator. Set the scale on both axes of the oscilloscope to 0.2 Volt/cm and sweep once at 1.0 Volt/sec from 0.4 to 1.7 Volt (RHE). A straight line of unit slope and zero intercept should be observed on the oscilloscope. Repeat at increasing sweep rates up to the point where a marked shift from zero intercept or deviation from the straight line can be observed. Between sweeps allow two minutes with stirring to reach equilibrium coverage by benzene. This is preferable because the distortions in the

---

* A further error may arise from a substantial iR drop which will cause a nonlinear change of electrode potential with time. This, however, will not be detected in the experiment suggested in this section, where only changes in the shape of the transient applied to the electrode are seen.

applied sweep increase with increasing current and the above experiment should be done under the worst conditions encountered during the experiment.

12.2.4 Determination of Q as a Function of Sweep Rate

Use a solution of 2×10^{-6} $\underline{M}$ benzene in 0.5 $\underline{M}$ H_2SO_4.* Set the potential to 0.4 Volt $\underline{vs}$ RHE and the amplitude to 1.3 Volt. The function generator should be operated in the single-pulse mode. Stir the solution well for two minutes. Stop stirring, allow the solution to rest for 10 seconds and apply several pulses at 0.05 Volt/sec until the i/E transients are identical. Record only this transient, which is used as the background i/E curve on the memory of the oscilloscope. Now stir for two minutes, stop for 10 seconds and apply a triangular wave at the same sweep rate. Record on the memory of the oscilloscope and take a picture of both transients. Repeat the experiment at increasing sweep rates, in steps, up to 15 volt/sec or as fast as you can without causing a distortion in the shape of the transient (see previous section). Repeat in a more concentrated benzene solution (1×10^{-5} $\underline{M}$).

Copy the transients photographed in this experiment and determine Q for different sweep rates by weighing the paper between the two curves on each picture, as in Experiment 12.1. Plot Q as a function of log v for the two concentrations of benzene

---

* The limit of the sweep on the anodic side should be adjusted so that the currents in the presence and in the absence of benzene coincide, as seen in Fig. 12.2.

12.2 DETERMINATION OF SURFACE CONCENTRATION

tested. Choose a sweep rate most suitable for further determinations of the surface coverage by benzene.

12.2.5 Determination of Q as a Function of Potential

Use the same solution as in the previous section (2×10^{-6} $\underline{M}$ benzene, 0.5 $\underline{M}$ H_2SO_4). Set the potential of the electrode at 0.10 Volt $\underline{vs}$ RHE. Set the amplitude of the triangular sweep and the sweep rate at the values selected in the previous section and apply several pulses without stirring until all the benzene has been removed from the electrode surface (as seen by the coincidence of consecutive traces on the oscilloscope). Stir the solution for two minutes while the electrode is held at a potential of 0.10 Volt $\underline{vs}$ RHE. Stop stirring for 10 seconds and apply a sweep. Apply a second and third sweep immediately after the first. The coincidence of the traces in the second and third sweeps serves to test the assumption that all the benzene on the surface was removed during the first sweep. If this is not so, a slower sweep rate (but still in the plateau region found in the previous section) should be used. Take a picture of the transient and the background (i.e. second transient) displayed simultaneously, copy on paper and determine the charge Q by weighing the section of the paper between the curves as above (cf. Experiment 12.1). Repeat the experiment, starting at 0.30, 0.50 and 0.70 Volt $\underline{vs}$ RHE. The upper limit of the pulse should always be the same.

12.2.6 Treatment of Results

Discuss in detail the factors causing distortion in the shape of the transient at high sweep rates. How can these factors be reduced?

Discuss the factors which determine the length of the plateau in the plot of Q vs log v. How will the upper and the lower limits of the plateau be affected by (i) the concentration of benzene in solution, (ii) the equilibrium concentration of benzene on the surface (in the same solution and at the same potential), (iii) stirring, (iv) the viscosity of the solvent, (v) temperature?

Convert the values of Q found in this experiment to surface concentrations in units of mole/cm^2, assuming complete oxidation of benzene to CO_2.

12.3 Kinetics of Electrosorption of Benzene on Platinum

12.3.1 General

The rate of electrosorption depends on two factors. One is the kinetics of the adsorption process itself and the other is the rate of mass transport to the electrode. The kinetics of adsorption is expected to control the overall rate of electrosorption when the breaking of covalent bonds is involved in the process and/or when the concentration of adsorbate in solution is high. The electrosorption of saturated hydrocarbons may serve as an example[13]

$$[C_3H_8]_{soln} + n[H_2O]_{ads} \rightleftharpoons [C_3H_7^\bullet]_{ads} + n[H_2O]_{soln} + H^+ + e_M^- \quad [12.1]$$

Mass transport may be expected to be the rate-controlling step if the structure of the molecule does not change in the process of adsorption and/or its concentration in solution is very low, as is

12.3 KINETICS OF ELECTROSORPTION

the case, for example, with ethylene[14]

$$[C_2H_4]_{soln} + n[H_2O]_{ads} \rightleftharpoons [C_2H_4]_{ads} + n[H_2O]_{soln} \qquad [12.2]$$

The rate of electrosorption of benzene on platinum was shown to be diffusion-controlled[15] and the kinetic equations for this case have been evaluated.[12] Solving the equations for semi-infinite linear diffusion and assuming a linear adsorption isotherm one has:

$$\theta(t) = \frac{2 D^{1/2} \theta_{eq}}{\pi^{1/2} K \Gamma_m} t^{1/2} \qquad [12.3]$$

where θ_{eq} is the value of the fractional coverage at equilibrium with a given concentration in solution and $\theta(t)$ is the momentary value of the coverage. Γ_m is the maximum surface coverage (in mole/cm^2) and K is the equilibrium constant for the electrosorption process. Equation [12.3] is valid only for short times, when $\theta(t)/\theta_{eq} \ll 1$. In Fig. 12.5 the variation of $\theta(t)$ with $t^{1/2}$ is shown. From the initial slope, $[d\theta(t)/d(t^{1/2})]_{t \to 0}$, it is possible to evaluate K, provided that the diffusion coefficient, D, and the maximum surface coverage, Γ_m, are known independently.

In this experiment, the rate of electrosorption of benzene on platinum will be studied in unstirred and in stirred solutions. The hydrodynamic equations for the latter case have not been solved (except for the rotating-disc electrode; cf. Experiment 6.1) but the effect of stirring on the rate of electrosorption will be shown unambiguously, demonstrating that the process is not kinetically controlled.

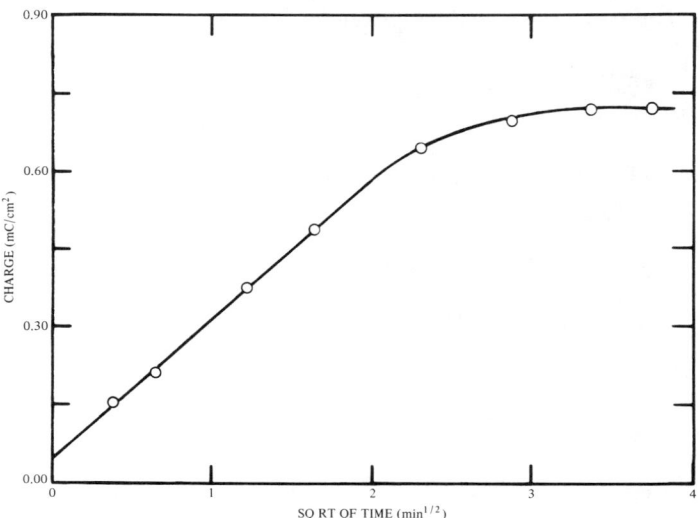

FIG. 12.5   VARIATION OF THE CHARGE (OR FRACTIONAL COVERAGE) WITH $t^{1/2}$ FOR ELECTROSORPTION OF BENZENE FROM A QUIESCENT SOLUTION. 2 $\mu\underline{M}$ BENZENE, 0.5 $\underline{M}$ H_2SO_4.

12.3.2 <u>The Cell, Solutions, Electrodes and Electrical Setup</u>

These will be as in Experiment 12.1. The solution will be 2×10^{-6} $\underline{M}$ benzene, 0.5 $\underline{M}$ H_2SO_4.

12.3.3 <u>The Rate of Electrosorption in Unstirred Solution</u>

Choose an appropriate sweep rate and potential span as determined in Experiment 12.2. (If this experiment has not been performed, the student may use v = 0.5-1 Volt/sec and potential limits of 0.4-1.7 Volt <u>vs</u> RHE.*) All experiments in this section

---

* The limit of the sweep on the anodic side should be adjusted so that the currents in the presence and in the absence of benzene coincide, as seen in Fig. 12.2.

12.3 KINETICS OF ELECTROSORPTION

are performed without stirring. Between measurements stir the solution for about 20 seconds to destroy any concentration gradients which may have formed near the electrode surface. Set the initial potential at 0.40 Volt vs RHE and clean the electrode surface by applying several pulses. When the surface is clean, wait exactly 60 seconds and then apply two consecutive pulses, the first to determine the current needed to oxidize the benzene on the surface and the second to obtain the background. Take a picture of these two transients and determine the charge due to the oxidation of benzene, as in Experiment 12.1. Stir the solution for 20 seconds and repeat the above experiment with an adsorption time of two minutes. Repeat with increasing adsorption times up to about 20 min. Some 8-10 points should be taken so that an accurate plot of Q vs $t^{1/2}$ can be made. Note that θ and hence Q should be proportional to $t^{1/2}$ at short times. It is best, therefore, to choose adsorption times whose square roots are fairly equally spaced.

12.3.4 The Rate of Electrosorption in Stirred Solution

Repeat the above experiment with stirring during the time of adsorption (stirring should be stopped after an exactly measured time and the pulse applied immediately). Much shorter adsorption times are required in this experiment. The first measurement should be taken after 5 seconds and the last after 120 seconds.

12.3.5 Treatment of Results

Plot Q as measured in the unstirred solution against $t^{1/2}$. Calculate the equilibrium constant for electrosorption assuming:

$$\Gamma_m = 2.5 \times 10^{-10} \text{ mole/cm}^2$$

$$D = 6.6 \times 10^{-6} \text{ cm}^2/\text{sec}$$

roughness factor = 2

If the initial rate of adsorption were maintained throughout, how long would it take for θ to reach θ_{eq}? Estimate the times required for the ratio θ/θ_{eq} to reach values of 0.5, 0.9 and 0.95. Plot Q as measured in the stirred solution against $t^{1/2}$. Compare the initial slope to that observed in the unstirred solution.

If benzene were oxidized to CO_2 instead of being adsorbed, what would be the current density corresponding to the initial rate of adsorption in unstirred and in stirred solutions?

12.4   The Isotherm for Electrosorption of Benzene

12.4.1 General

The fractional surface coverage of an electrode may be expressed as a function of four variables

$$\theta = f(E, C, T, P) \qquad [12.4]$$

At constant temperature and pressure, θ can be represented on a three-dimensional diagram as a function of potential, E, and concentration in solution, C. A cross section of this diagram at constant values of C yields plots of θ vs E as in Experiment 12.2. In the present experiment the usual isotherm of θ vs concentration at constant potential will be determined.

The Frumkin adsorption isotherm may be applied to the

12.4 ISOTHERM FOR ELECTROSORPTION

analysis of the results obtained in this experiment. Thus, at constant potential

$$\frac{\theta}{1-\theta}\exp(r\theta/RT) = KC \qquad [12.5]$$

This gives, at vanishingly low coverage

$$\left(\frac{d\theta}{dC}\right)_{\theta\to 0} = K \qquad [12.6]$$

or at slightly higher coverage

$$\left[d\left(\frac{\theta}{1-\theta}\right)/dC\right]_{\theta\to 0} = K \qquad [12.7]$$

Equation [12.6] is valid for values of θ smaller than 0.10. The range of applicability of equation [12.7] depends on the value of the parameter r in equation [12.5] which determines the rate of change with coverage of the apparent standard free energy of electrosorption (cf. section II.7.3). For the system studied here, equation [12.7] was found to apply up to $\theta \cong 0.30$. A first estimate of the value of the parameter r may be obtained by plotting θ vs log C at intermediate values of θ ($0.3 < \theta < 0.7$ in the present system). In this range equation [12.5] can be simplified to the form

$$\theta = \frac{2.3\,RT}{r}\log KC \qquad [12.8]$$

Equation [12.5] may be rearranged to the form

$$\theta = \frac{2.3\,RT}{r}\log K + \frac{2.3\,RT}{r}\log\frac{C(1-\theta)}{\theta} \qquad [12.9]$$

A better value of r can be obtained by plotting θ vs log[C(1-θ)/θ]

according to this equation (which is applicable at all values of θ, provided that the electrosorption obeys the Frumkin isotherm).

12.4.2 The Cell, Electrodes and Electrical Setup

As in the previous exercises in this group of experiments.

12.4.3 Determination of the Adsorption Isotherm

Prepare a solution 1×10^{-6} $\underline{M}$ in benzene in 0.5 $\underline{M}$ H_2SO_4, deaerate by bubbling purified nitrogen for 10 minutes, and measure the charge due to oxidation of benzene adsorbed on the surface as described in Experiment 12.3. Stir for three minutes to ensure that adsorption equilibrium has been attained. The potential at which adsorption is carried out is maintained at 0.40 Volt $\underline{vs}$ RHE. Increase the concentration of benzene in the same solution to 2×10^{-6} $\underline{M}$ by adding the appropriate volume of stock solution of benzene and measure Q. Repeat your measurement with increasing concentrations of benzene in the same solution until Q becomes constant. Take a sufficient number of experimental points to permit accurate construction of the isotherm. A rough estimate of Q should be made each time so that the best concentration values may be used. If necessary, prepare new solutions so that Q may be measured in regions where an insufficient number of points have been taken.

12.4.4 Treatment of Results

Plot Q $\underline{vs}$ concentration and determine its saturation value. Assuming that benzene is oxidized completely to CO_2 and taking

12.4 ISOTHERM FOR ELECTROSORPTION

Γ_m = 2.5 x 10^{-10} mole/cm^2, calculate the roughness factor. Replot the fractional coverage θ as a function of concentration, and determine the equilibrium constant K from the initial slope (cf. Equation [12.6]).

Plot $\theta/(1 - \theta)$ vs concentration and determine K from the initial slope. Compare the two values of K with each other and with the value obtained in Experiment 12.3, from measurement of the initial rate of adsorption.

Calculate the parameter r with the use of equations [12.8] and [12.9]. Compare and discuss.

REFERENCES

1. N. A. Balashova and V. E. Kazarinov, "Electroanalytical Chemistry - A Series of Advances", Vol. III, Chap. 3, A. Bard, Editor. Marcel Dekker, 1969. (Further references are listed in this review paper).
2. E. Gileadi, J. Electroanal. Chem. $\underline{11}$, 137 (1966).
3. R. G. Barradas and B. E. Conway, J. Electroanal. Chem. $\underline{6}$, 314 (1963).
4. M. W. Breiter and S. Gilman, J. Electrochem. Soc. $\underline{109}$, 622 (1962).
5. S. Gilman, J. Phys. Chem. $\underline{66}$, 2657 (1962).
6. S. Gilman, Trans. Faraday Soc. $\underline{61}$, 2546 (1965).
7. S. B. Brummer, J. I. Ford and M. J. Turner, J. Phys. Chem. $\underline{69}$, 3424 (1965).
8. E. Gileadi, Lj. Duic and J. O'M. Bockris, Electrochim. Acta $\underline{13}$, 1915 (1968).
9. F. G. Will and C. A. Knorr, Z. Elektrochem. $\underline{64}$, 258, 270 (1960).
10. F. G. Will, J. Electrochem. Soc. $\underline{112}$, 451 (1965).
11. S. Srinivasan and E. Gileadi, Electrochim. Acta $\underline{11}$, 321 (1966).
12. A. K. N. Reddy, "Electrosorption", Chap. 3, E. Gileadi, Editor, Plenum Press, 1967.
13. J. O'M. Bockris, G. E. Stoner and E. Gileadi, J. Phys. Chem. $\underline{73}$, 427 (1969).
14. E. Gileadi, B. T. Rubin and J. O'M. Bockris, J. Phys. Chem. $\underline{69}$, 3335 (1965).
15. W. Heiland, E. Gileadi and J. O'M. Bockris, J. Phys. Chem. $\underline{70}$, 1207 (1966).

13. ABSORPTION OF HYDROGEN IN PALLADIUM AND DIFFUSION THROUGH IT

Introduction

13.1 The Solubility and Diffusion Coefficient of Hydrogen in Palladium

13.1.1 General

13.1.2 The Cell, Solutions, Electrodes and Electrical Setup

13.1.3 Determination of D and C^o

13.1.4 Treatment of Results

13.2 Equilibrium Between Adsorbed and Absorbed Hydrogen in Palladium

13.2.1 General

13.2.2 The Cell, Solutions and Electrodes

13.2.3 Electrical Circuit

13.2.4 Determination of D and C^o

13.2.5 Choice of Suitable Current Density for the Determination of Q_H

13.2.6 The Variation of C^o and θ with i_{pol} or E_{pol}

13.2.7 The Real Surface Area

13.2.8 Treatment of Results

Appendix 13.1

13. ABSORPTION OF HYDROGEN IN PALLADIUM AND DIFFUSION THROUGH IT

Introduction

The absorption of hydrogen by palladium was first observed by Graham[1] as far back as 1866. Since then it has been found that most metals absorb hydrogen, some with an endothermic heat of absorption and others with an exothermic heat of absorption. Among the former, one finds metals such as Fe, Cu, Ni with relatively low solubilities for hydrogen. Among the latter are metals such as Pd, Ti, Zr which can absorb hydrogen to such an extent that they swell and lose their mechanical strength and other metallic properties.

Two metals have attracted particular interest in this respect, mainly for technological reasons. Iron, and particularly high strength steels, absorb very minute amounts of hydrogen. When the solubility limit is exceeded by forcing too much hydrogen into the metal, the phenomenon of "hydrogen embrittlement" occurs, which leads to a drastic decrease in the tensile strength of the metal and often to sudden (catastrophic) failure in service. Hydrogen is, of course, forced into the metal inadvertently during processes of chemical cleaning of the surface in acid to remove scale (pickling), electroplating and as a result of corrosion when the metal part is in service. In all these cases, hydrogen evolution may be a side reaction that leads to absorption of hydrogen as shown by the equations below for the case of spontaneous corrosion of iron.

INTRODUCTION

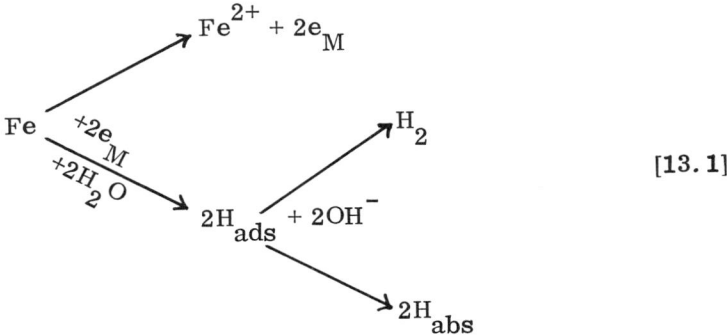

[13.1]

The problem of absorption of hydrogen in iron and its alloys has been studied in detail by Smialowski[2] and by others.[3]

The hydrogen/palladium system has attracted considerable interest because of its use as a permselective membrane for the generation of pure hydrogen. Since the solubility and diffusion coefficients of hydrogen in palladium are relatively large, kinetic and thermodynamic measurements can be made conveniently. The publications dealing with various aspects of the H/Pd system are numerous. Much of the available knowledge is summarized in a book[4] and a volume of review papers[5] published recently.

Thermodynamic studies of the H/Pd system have shown that two phases can co-exist. The α phase is the only stable phase when the concentration of hydrogen is below 3 atom percent. When more hydrogen is absorbed, another phase is formed (the β phase) which corresponds to an atomic ratio of H/Pd = 0.60. The two phases can be distinguished by X-ray diffraction since the lattice parameters of the α phase are identical to those of the pure metal while in the β phase the lattice is expanded by about

3% in each direction and this leads to a volume increase of about 10%. Both phases can be found in a palladium sample when the average concentration is in the range of 3 - 60 atom percent. In this range, the equilibrium hydrogen pressure is independent of concentration.

Palladium can be "charged" with hydrogen (i.e., it can be made to absorb hydrogen) by placing it in contact with molecular hydrogen at a given partial pressure p or by applying to it a suitable potential in an aqueous solution. That the two techniques are thermodynamically equivalent can be seen if one considers the equilibria in equation [13.2].

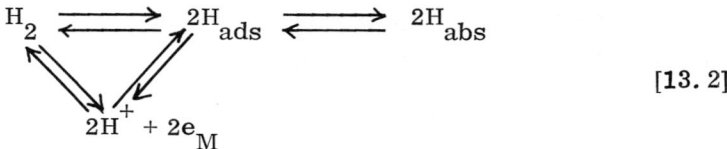

[13.2]

The partial pressure is related to the applied potential through the Nernst equation. Thus, a potential equal to that of a reversible hydrogen electrode (RHE) in the same solution corresponds to $p = 1.0$ atm and a potential change of about 30 mV ($2.3RT/2F$) is equivalent to an order of magnitude change in pressure. The power of electrochemical methods of measurement of hydrogen diffusion in metals is readily seen from this relationship. Thus, application of a potential of, say, $E = -0.60$ V (RHE) is equivalent to a hydrogen partial pressure of $p = 10^{20}$ atm., while $E = +0.60$ V (RHE) corresponds to $p = 10^{-20}$ atm. It should be noted that the above correspondence between applied potential and the equivalent pressure exists even if the equilibria shown in equation [13.2] are not maintained. Thus, the activity of H_{ads} is a function of the applied potential in a given solution, and this can be related thermodynamically (through the Nernst equation) to an equivalent pressure. Experimentally one finds, for example, that pure iron does not absorb hydrogen from the gas phase since dissociation of H_2 to $2\ H_{ads}$ does not occur on its surface. Absorption does occur when the sample of Fe is placed in solution and a cathodic potential of $E = 0.5$ Volt (RHE), equivalent to a pressure of $p \cong 10^{17}$ atm, is applied.

13.1 The Solubility and Diffusion Coefficient of Hydrogen in Palladium

13.1.1 General

In this experiment, an electrochemical technique[6-8] will be employed to measure the diffusion coefficient D and the solubility C^o of hydrogen in palladium, under a variety of experimental conditions. The advantages of the electrochemical technique are twofold. First, as noted above, by adjusting the potential one can apply a wide range of equivalent hydrogen pressures without resort to high-pressure or high-vacuum techniques. Secondly, the hydrogen forced through a thin palladium membrane can be transformed to H^+ ions when it reaches the solution on the other side of the membrane, and thus the permeation rate will be proportional to the current required to maintain a high anodic potential ($p \rightarrow 0$) on one side of the membrane. That this allows very accurate and sensitive measurements of the permeation rate to be made, can be seen from the fact that a current of $1\ \mu A\ cm^{-2}$

corresponds to a permeation rate of H_2 of about 10^{-7} ml cm^{-2} sec^{-1} at STP.

The diffusion equation

$$\frac{\partial C(x,t)}{\partial t} = D\frac{\partial^2 C(x,t)}{\partial x^2} \qquad [13.3]$$

has been solved[9] for the initial condition

$$C(x,0) = 0 \qquad [13.4]$$

and for the boundary conditions

$$C(0,t) = C^o \qquad [13.5]$$

$$C(L,t) = 0 \qquad [13.6]$$

Put into words, the conditions are that before the experiment is started the membrane contains no hydrogen and, at any time after the start of the experiment, the concentration on one side (x = 0) is a constant, C^o, and on the other side (x = L) it is equal to zero.

The solution of equation [13.3] with the above initial and boundary conditions yields a series which, to a good approximation,* can be given by its first term

$$\frac{i(t)}{i(\infty)} = \left(\frac{2}{\pi^{1/2}}\right)\left(\frac{1}{\tau^{1/2}}\right)\exp\left(-\frac{1}{4\tau}\right) \qquad [13.7]$$

in which $i(t)$ and $i(\infty)$ are the measured currents (proportional to

---

* This approximation is valid for $i(t)/i(\infty) \leq 0.96$.

13.1 SOLUBILITY AND DIFFUSION COEFFICIENT

the permeation rate of hydrogen) at a time t and at steady state, respectively, and τ is a dimensionless parameter given by

$$\tau = \frac{Dt}{L^2} \qquad [13.8]$$

where L is the thickness of the membrane. A plot of $\gamma \equiv i(t)/i(\infty)$ vs τ, calculated according to equation [13.7] is shown in Fig. 13.1. The values of τ corresponding to some selected values of γ are given in Table 13.1 below.

Table 13.1 Selected Values of γ and τ

$\gamma = i(t)/i(\infty)$	$\tau = Dt/L^2$
0.1	0.066
0.3	0.101
0.5	0.138
0.7	0.192
0.9	0.304

Equation [13.8] may be rewritten in the form

$$D = \tau_\gamma L^2 / t_\gamma \qquad [13.9]$$

where t_γ is the time required to reach a current $i(t) = \gamma i(\infty)$ and τ_γ is the corresponding value of τ. A test of the validity of the assumptions leading to equation [13.7] is to calculate D from equation [13.9] for different values of γ and show that it is independent of γ. An alternative method of testing equation [13.7] is to use membranes of different thickness and plot t_γ vs L^2.

The shapes of the concentration profiles of hydrogen

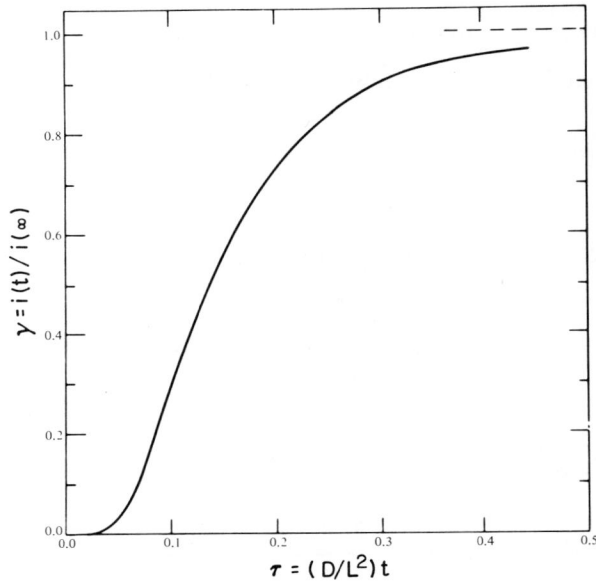

FIG. 13.1   CALCULATED PLOT OF THE RATIO $i(t)/i(\infty)$ AS A FUNCTION OF THE DIMENSIONLESS PARAMETER τ.

inside the membrane for different values of γ have been calculated,[10] and are shown in Fig. 13.2. The measured current is proportional to $-\left(\frac{\partial C}{\partial x}\right)_{x=L}$ and this varies from zero to a steady-state value of C^o/L. Thus, for steady-state conditions we have

$$i(\infty) = \frac{nFDC^o}{L} \qquad [13.10]$$

and knowing the value of D from the transient, we can calculate the concentration $C^o \equiv C(0,t)$.

In certain cases, when D is small or when relatively thick membranes are to be used, it is not convenient to wait until the current has reached its steady-state value. In such an event, the

13.1   SOLUBILITY AND DIFFUSION COEFFICIENT   479

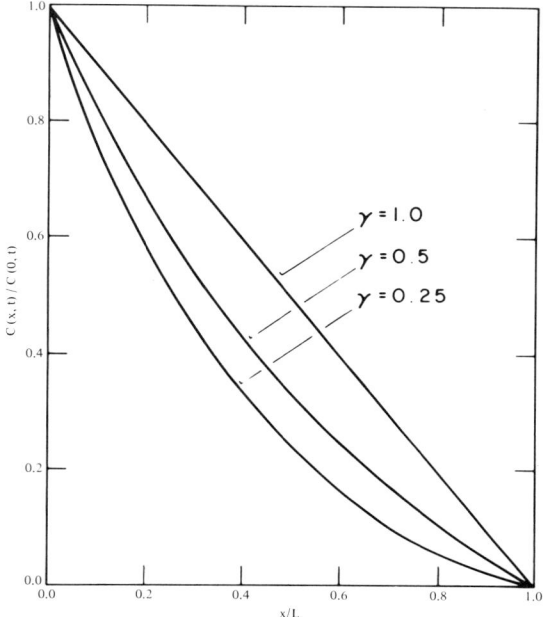

FIG. 13.2   CONCENTRATION PROFILES OF HYDROGEN INSIDE A PALLADIUM MEMBRANE.

diffusion coefficient may be calculated from the initial part of the transient. For two values of the current we may write, by combining equations [13.7] and [13.8]

$$\frac{i_2(t)}{i_1(t)} = \left(\frac{t_1}{t_2}\right)^{1/2} \exp\frac{L^2}{4D}\left(\frac{1}{t_2} - \frac{1}{t_1}\right) \qquad [13.11]$$

Thus, the diffusion coefficient can be calculated by measuring the current at two points during the transient. Alternatively, one may choose a fixed value of the ratio $i_2(t)/i_1(t)$ at different points along the transient and plot log (t_1/t_2) vs $[(1/t_2)-(1/t_1)]$. A straight line with a slope of $L^2/2D$ should result, from which D can be evaluated.

13.1.2 The Cell, Solutions, Electrodes and Electrical Setup

The cell, shown in Fig. 13.3, is similar to cells described in the literature,[6-8] but is simpler to assemble and requires only a small volume of solution. A palladium membrane (0.01 cm thickness) divides the two parts of the cell (which act as two separate cells) and serves as the working electrode for both. On the "polarization side" (high equivalent pressure) small cathodic currents, i_{pol}, are applied. These currents are chosen so that the potential is maintained in the range E_{pol} = +0.06 Volt to +0.24 Volt (RHE), corresponding to a range of equivalent pressures of $10^{-2} - 10^{-8}$ atm.* On the "diffusion side", the potential is controlled by means of a potentiostat and set to a value of $E_d \cong$ +0.60 Volt (RHE), corresponding to an effectively zero equivalent pressure (p = 10^{-20} atm.)

The solution is 1 $\underline{M}$ H_2SO_4 made up of analytical-grade acid and triply distilled water. It is deaerated before introduction into the cell by having purified nitrogen bubbled through it for 10 minutes. The cell is enclosed in a plastic glove-bag which is continuously purged with nitrogen.

Platinum counter electrodes and commercial reference electrodes (Hg/Hg_2SO_4 is preferable, if available) are to be used on both sides of the membrane.

---

* Alternatively a potentiostat can be used and the potential controlled in the same range.

13.1   SOLUBILITY AND DIFFUSION COEFFICIENT

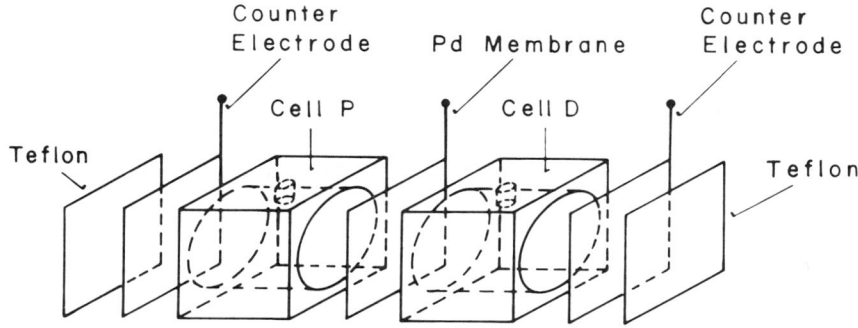

FIG. 13.3   CELL FOR MEASUREMENT OF HYDROGEN PERMEATION THROUGH METALLIC MEMBRANES.

The electrical circuit is shown in Fig. 13.4. The current $i(t)$ on the "diffusion side" is measured with a strip-chart recorder.

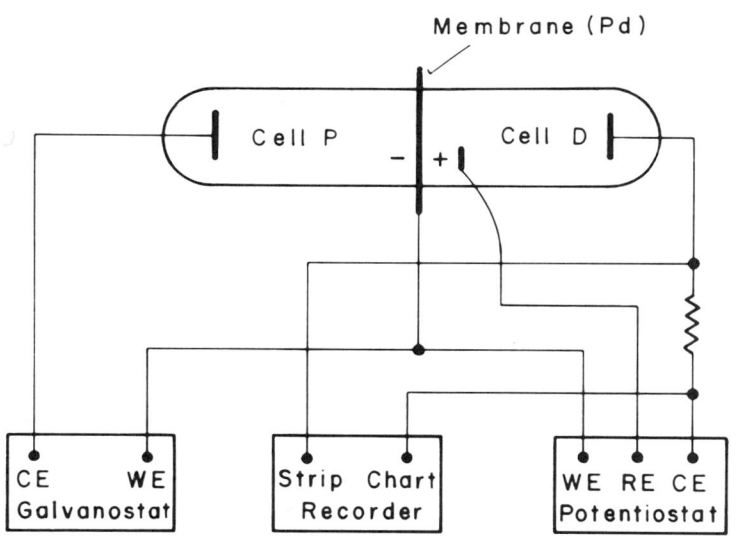

FIG. 13.4   BLOCK DIAGRAM OF THE ELECTRICAL CIRCUIT EMPLOYED FOR THE MEASUREMENT OF HYDROGEN PERMEATION.

13.1.3 __Determination of D and C^o__

Clean the palladium membrane by refluxing with benzene, rinsing with alcohol and drying at reduced pressure. Anneal for four hours at 900°C in an atmosphere of argon. Cool slowly (over an hour or two) maintaining the inert-gas atmosphere until the membrane has reached room temperature.* Set up the cell, fill the plastic glove-bag with nitrogen and introduce the deaerated solution into the cell by means of a syringe. To remove hydrogen from the system, apply a small anodic current, $i_{pol} = 5 \mu A/cm^2$, on the "polarization side" and an anodic potential, $E_d = +0.60$ Volt (RHE) on the "diffusion side". Maintain these conditions until the potential on the polarization side reaches about +0.60 Volt (RHE) and until the current on the diffusion side has reached a small, nearly constant value, not exceeding 2-3 $\mu A/cm^2$.**

Set the recorder chart speed and sensitivity to convenient values, considering that the diffusion coefficient is of the order of 2×10^{-7} cm^2/sec and noting that i(t) is very close to the cathodic polarizing current i_{pol}, as has been observed in the literature.[11]

---

* The experiment can be performed successfully without annealing but the values of D and C^o measured in this case may differ markedly from the best literature values.

** This stage of removing the residual hydrogen from the membrane can be performed more conveniently if a potentiostat is used on both sides. If a galvanostat is used on the polarization side, as suggested above, the current may have to be adjusted during this pretreatment stage so that a positive potential of about +0.60 Volt (RHE) is maintained until just before the start of the transient.

13.1 SOLUBILITY AND DIFFUSION COEFFICIENT

Set the recorder in motion and apply a cathodic polarization current of $i_{pol} = 0.1$ mA/cm^2. Measure the polarization potential, E_{pol}, and record the diffusion current until i(t) reaches its steady-state value i(∞). Repeat the experiment several times. Between runs remove hydrogen from the membrane electrochemically as described above. In one of the runs increase the sensitivity of the recorder and the chart speed and record only that part of the transient for which $\gamma = i(t)/i(\infty) \leq 0.3$.

In one of the above runs measure i(∞) as a function of E_d (in the range of $E_d = 0.30 - 0.90$ Volt (RHE)) after steady state has been reached and while i_{pol} is maintained constant.

Obtain diffusion-current transients for three additional values of i_{pol} (0.05, 0.15, 0.3 mA/cm^2). Measure the potential E_{pol} on the polarization side. Note that only the sensitivity of the recorder needs to be adjusted before each run.

Use membranes of varying thickness, as available, in the range of $5 \times 10^{-3} - 5 \times 10^{-2}$ cm and obtain a transient for each, using a polarizing current density of $i_{pol} = 0.1$ mA/cm^2. Measure the potential on the polarization side in each case.

13.1.4 Treatment of Results

For each experiment, calculate the diffusion coefficient D from $t_{1/2}$ (i.e., from the value of t for which $\gamma = 0.50$ and $\tau = 0.138$). Does D depend on i_{pol} or on the thickness of the membrane?

Choose one transient and with it calculate D from the time corresponding to each of the values of γ given in Table 13.1. What can be concluded from the values obtained in this way?

Use the data obtained in the first part of the transient (where measurement was conducted only to $\gamma \leq 0.3$) to calculate D from equation [13.11].

Discuss the effect of the potential E_d applied to the diffusion side of the membrane on the steady-state diffusion-current density $i(\infty)$.

Calculate the concentration of hydrogen C^o for all measurements at constant membrane thickness. Plot log C^o vs E_{pol} (the potential measured on the polarization side).

Plot $t_{1/2}$ vs L^2 from measurements with membranes of different thicknesses.

Plot $i(\infty)$ vs $1/L$. Compare this with equation [13.10] and explain any discrepancies. What would you expect this plot to be like if the polarization side of the membrane were controlled potentiostatically rather than galvanostatically?

Calculate the ratio $i(\infty)/i_{pol}$ in each experiment and discuss.

13.2 Equilibrium Between Adsorbed and Absorbed Hydrogen in Palladium

13.2.1 General

In the following experiment we will determine the ratio of concentrations of hydrogen on the surface of palladium and in the bulk just below the surface. The procedure will largely be that of Breger and Gileadi.[11] The surface concentration Γ (mole/cm^2) will be determined by application of galvanostatic transients, as discussed below. The bulk concentration C^o (mole/cm^3) will be evaluated from diffusion transients taken in the same system, as

13.2 ADSORBED AND ABSORBED HYDROGEN

in Experiment 13.1.

For the equilibrium

$$H_{ads} \rightleftarrows H_{abs} \qquad [13.12]$$

we can write a dimensionless equilibrium constant if we express both surface and bulk concentrations in terms of the respective mole fractions. Thus, on the surface

$$X_s = \frac{\Gamma}{\Gamma_{max}} = \theta \qquad [13.13]$$

and in the bulk

$$X_b \cong \frac{c^o}{\rho/M} \qquad [13.14]$$

where M and ρ are the atomic weight and density of palladium, respectively. The equilibrium constant

$$K = \frac{X_b}{X_s} \qquad [13.15]$$

was found[11] to be in the range of $(1.5 - 15) \times 10^{-4}$, indicating that hydrogen is much more stable on the surface than in the bulk of palladium.

Measurement of the surface concentration of hydrogen on palladium by fast galvanostatic transients will follow the procedure first developed by Pearson and Butler,[12] (who used platinum electrodes). A constant anodic current is applied and the variation of potential with time is followed on an oscilloscope. The charge required to ionize all the hydrogen atoms initially adsorbed on the surface is determined from the length of the plateau, as shown schematically in Fig. 13.5. The section AC corresponds to the total charge passed during ionization of hydrogen, while the section AB corresponds to the true charge, corrected for the process of double-layer charging, which takes place simultaneously. A detailed discussion of this and other methods for the determination of surface coverage by hydrogen has been given in a review[13] in which further references may be found. A difficulty exists in the case of palladium, since some of the hydrogen absorbed in the metal may reach the surface during the transient and be ionized, thus giving erroneously high values for the apparent surface coverage. To eliminate this source of error, the experiment is repeated with increasing current densities (shorter transients) until the charge due to ionization of adsorbed hydrogen no longer depends on the duration of the transient. At this point, the contribution due to absorbed hydrogen may be considered negligible.

A further difficulty, which is common to all measurements on solid electrodes, is the uncertainty concerning the true surface area (or the roughness factor) of the electrode. Different methods have been proposed to measure this quantity but common to all is the involvement of one or more assumptions which cannot be independently verified. In the present experiment, it will be assumed that, at $E = 0.0$ Volt (RHE), the surface is completely covered by adsorbed hydrogen ($\theta = 1$) and the charge corresponding to this layer, per cm^2 of real surface area, is 0.23 mC. The charge Q measured at this starting potential will serve in this way to determine the real surface area of the electrode used. While the assumption of $\theta = 1$ at $E = 0.0$ Volt (RHE) may be

13.2   ADSORBED AND ABSORBED HYDROGEN

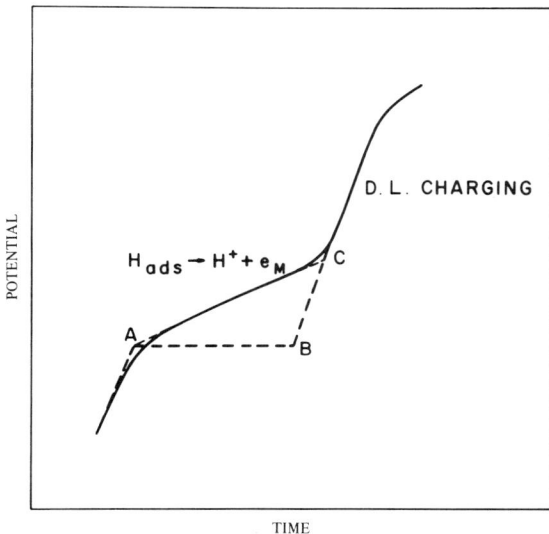

FIG. 13.5   SHAPE OF THE E/t CURVE FOR AN ANODIC GALVANOSTATIC PULSE ON PLATINUM IN ACID SOLUTION.

criticized, it can hardly be questioned that $\theta \geqslant 0.8$ under these conditions, so that the area calculated by the above procedure is probably close to the true value though it may in fact be slightly low.

13.2.2   The Cell, Solutions and Electrodes

The cell and solutions are as in Experiment 13.1. The reference electrode should be positioned as close as possible to the membrane working electrode (use a Luggin capillary, if possible) to minimize iR drop. Care should also be taken to position the counter electrode so that the current density on the working electrode will be uniform.

13.2.3 Electrical Circuit

In the first part of this experiment the values of D and C^o are determined and the electrical circuit is identical to that in Experiment 13.1. In the second part, another galvanostat is connected on the polarization side and it is set to deliver currents in the range of 0.05 - 0.5A, to oxidize the hydrogen atoms adsorbed on the surface. The galvanostat used to deliver the small polarizing current i_{pol} (in the range of 10 - 500 $\mu A/cm^2$) may be replaced by a battery and suitable set of resistors. The duration of the anodic current pulse should be limited so that extensive oxidation of the surface is avoided. This is achieved by means of the circuit shown in Fig. 13.6 which will turn off the current when the potential has reached a certain preset value.[*] The variation of potential with time during the transient is followed on an oscilloscope which is triggered by the applied current pulse. The circuit providing the polarizing current i_{pol} need not be disconnected during the transient, since its contribution to the total current during the transient is negligible.

13.2.4 Determination of D and C^o

Obtain a diffusion transient (i.e., the variation of i(t) with t) and determine D and C^o following the procedure outlined in Experiment 13.1. Use a membrane thickness of 0.01 cm and apply a polarizing current density of i_{pol} = 0.10 mA/cm^2 and

---

[*] The mode of operation of this circuit is discussed in the appendix to this experiment.

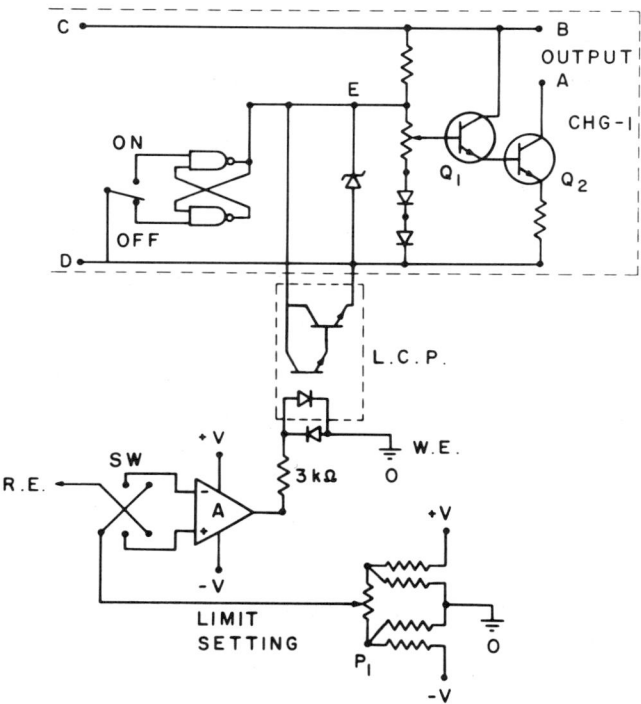

FIG. 13.6 CIRCUIT FOR GALVANOSTATIC STRIPPING WITH PROVISION FOR DISCONNECTING THE CURRENT AUTOMATICALLY AT A PRESET POTENTIAL.

L.C.P. - LIGHT-COUPLED PAIR SUCH AS GE, HBI OR EQUIVALENT; A - OPERATIONAL AMPLIFIER SUCH AS 741 OR EQUIVALENT.

measure the potential E_{pol} on the polarization side. Maintain steady-state diffusion and determine θ and C^o as functions of i_{pol} as outlined below.

13.2.5 Choice of Suitable Current Density for the Determination of Q_H

Apply an anodic pulse of 50 mA/cm^2 and, from section AB (Fig. 13.5), determine Q_H, the charge required to ionize all the hydrogen initially adsorbed on the surface. Wait about one minute to ensure that steady-state diffusion has been reestablished and apply a pulse of 100 mA/cm^2. Repeat with increasing current densities (in steps of 50 mA/cm^2) up to 500 mA/cm^2. Determine Q_H in each experiment and define the region where Q_H no longer varies systematically with applied current density. Use a current density in this range for all further experiments.

13.2.6 The Variation of C^o and θ with i_{pol} or E_{pol}

Use a range of values of i_{pol} from 10 μA/cm^2 to 500 μA/cm^2. Allow steady state to be reached in each case and measure E_{pol} and $i(\infty)$. Apply a galvanostatic pulse of suitable current density, as determined in the previous section, and obtain Q_H for each value of i_{pol}.

13.2.7 The Real Surface Area

Adjust the value of i_{pol} until the potential on the polarization side is E_{pol} = 0.0 Volt (RHE). Allow sufficient time for steady-state diffusion to be reached and apply an anodic pulse to determine Q_H.

13.2.8 Treatment of Results

Plot Q_H as a function of applied current density and

13.2 ADSORBED AND ABSORBED HYDROGEN

determine the region of invariance of Q_H. Discuss the various causes of variation of Q_H with current density.

Make a plot of E_{pol} vs log i_{pol} as observed under steady-state diffusion conditions.

Determine the real surface area and the roughness factor from the value of Q_H at $E_{pol} = 0.0$ Volt (RHE). Assume that $Q_H = 0.23$ mC/cm^2 at this potential.

Calculate C^o and θ for all values of i_{pol} employed and plot each quantity against E_{pol}.

Calculate the equilibrium constant K defined by equation [13.15] and plot log K against both E and θ. Discuss these plots.

How is it possible that on the two sides of the same metallic membrane (which must have the same electrical potential in any point on it, being a very good conductor) two opposite electrochemical reactions take place, namely the deposition and ionization of atomic hydrogen?

REFERENCES

1. T. Graham, Phil. Trans. Roy. Soc. (London), 136, 349, 573, (1866).

2. M. Smialowski, "Hydrogen in Steel", Pergamon Press, 1962.

3. W. Beck, E. J. Jankowsky and P. Fischer, "Hydrogen-Stress Cracking of High-Strength Steel", Report No. NADC-MA-7140, Naval Air Material Center, Warminster, Pennsylvania 18974, U. S. A. (1971).

4. F. A. Lewis, "The Hydrogen-Palladium System", Academic Press, 1967.

5. Engelhard Industries, Inc., Technical Bulletin Vol. VII, No. 1/2 (1966).

6. A. N. Frumkin and N. A. Aladzhalova, Acta Phys. Chim. U. S. S. R. 19, 1 (1944).

7. S. Schuldiner and J. P. Hoare, J. Electrochem. Soc. 103, 178 (1956).

8. M. A. V. Devanathan and Z. Stachursky, Proc. Roy. Soc. (London) 270A, 90 (1962).

9. J. McBreen, L. Nanis and W. Beck, J. Electrochem. Soc. 113, 1218 (1966).

10. E. Gileadi, M. A. Fullenwider and J. O'M. Bockris, J. Electrochem. Soc. 113, 926 (1966).

11. V. Breger and E. Gileadi, Electrochim. Acta 16, 177 (1971).

12. J. D. Pearson and J. A. V. Butler, Trans. Faraday Soc. 34, 1163 (1938).

13. E. Gileadi and B. E. Conway, "Modern Aspects of Electrochemistry", Ed. J. O'M. Bockris and B. E. Conway, Vol. 3, p. 348, Butterworths 1964.

Appendix 13.1

The circuit shown in Fig. 13.6 may be employed to turn off a galvanostat when the potential of the working electrode has reached a certain preset value, thus avoiding excessive oxidation of the surface during application of an anodic pulse. This circuit can be coupled directly to a type CHG-1 galvanostat produced by ELRON or similar units which have an electronic on/off switch. For use with other galvanostats certain minor modifications may have to be introduced.

The amplifier A is employed as a comparator. One input is connected to the reference electrode while the other input is connected to a potentiometer P_1 the setting of which determines the potential at which the galvanostat will be turned off. The "common" output connection of the power supply ($\pm$ 15 volt) is connected to the working electrode.

When the potential of the reference electrode (with respect to the working electrode) reaches a value equal to that set on the potentiometer P_1, the output of amplifier A changes from $-V$ to $+V$. This causes the LCP (light-coupled pair) to turn off the output current of the galvanostat. The reversing switch SW changes the mode of operation, allowing the limiting potential to be reached either from the anodic or the cathodic direction.

14. KINETIC STUDIES OF THE Ni(OH)$_2$/NiOOH SYSTEM

Introduction

14.1 The Charge Capacity of NiO$_x$ Electrodes

14.1.1 General

14.1.2 The Cell, Solutions and Electrical Setup

14.1.3 Preparation of Electrodes

14.1.4 Determination of the Charge Capacity

14.1.5 Variation of the Charge Capacity with Charging Rate

14.1.6 Treatment of Results

14.2 Open-Circuit-Decay Measurements on NiO$_x$ Electrodes

14.2.1 General

14.2.2 The Cell, Solutions, Electrodes, and Electrical Setup

14.2.3 Measurement of the Reversible Potential for the Ni(OH)$_2$/NiOOH System

14.2.4 Polarization Curves for Oxygen Evolution on NiO$_x$

14.2.5 Open-Circuit Decay of Overcharged NiO$_x$ Electrodes

14.2.6 Treatment of Results

14. KINETIC STUDIES OF THE Ni(OH)$_2$/NiOOH SYSTEM

Introduction

A very short discussion of the fundamental properties of batteries and fuel cells has been presented in section II.12. A detailed account of primary and secondary cells may be found in the books of Vinal[1,2] and systems suitable for high-energy-density batteries have been discussed by Jasinski[3]. The electrochemical reactions taking place at the two electrodes constitute only one of several technological problems involved in the development of batteries. Finding a suitable separator, which is compatible with the solvent, has sufficiently high mechanical strength and also low electrical resistance is often a major problem. Shelf life of primary batteries may depend on the resistance of the casing to corrosion and on the rate of chemical side-reactions causing self discharge. The useful service life of secondary batteries (i.e. the number of charge/discharge cycles which can be performed before breakdown) depends on the rate of recrystallization of the electrode material. This process may cause a drastic decrease in available surface area. Service life also depends on the tendency to form dendrites which often pierce the separator and cause internal short circuits in the battery.

The nickel oxide electrode which will be studied in this experiment is part of the nickel/cadmium battery first built by Jungner and the nickel/iron battery developed by Edison. The modern version of the nickel oxide electrode in this battery is a plaque produced by sintering fine-grain nickel particles on a nickel or stainless-steel grid, which serves both as a mechanical support and as the current collector. The electrode is impregnated with $Ni(NO_3)_2$, which is then converted <u>in situ</u> to $Ni(OH)_2$ and oxidized to NiOOH[4]. The electrode is operated in concentrated KOH solution (20-30% by weight) and its potential varies only slightly with the concentration of the electrolyte, although the rate of self discharge was found to decrease with increasing concentration of KOH.

A detailed review covering various aspects of the nickel/cadmium battery has been published recently[5]. The behavior of the nickel oxide electrode has been studied by Wynne-Jones and co-workers[6-9] and by Conway and co-workers[10-13]. The charging process at the solid/liquid interface may be described by the simple equation:

$$Ni(OH)_2 + OH^- \longrightarrow Ni(O)(OH) + H_2O + e_M \qquad [14.1]$$

The charge is propagated in the solid phase by a proton-transfer mechanism[14] (cf. section II.12.2) which can be represented schematically by the equation

$$\underset{OH \quad HO}{Ni \overset{O----HO}{\diagup} Ni} \longrightarrow \underset{OH \quad HO}{Ni \overset{OH----O}{\diagup} Ni} \qquad [14.2]$$

The partially charged nickel oxide is a nonstoichiometric compound the composition of which may be represented as NiO_x where $1 \leqslant x \leqslant 1.7$. The upper limit of x depends on the concentration of KOH in solution. Values of x greater than 1.5 have been interpreted alternatively as indicative of the existence of a Ni^{4+} species or as adsorption of oxygen on the surface of NiOOH. The reversible potential of the $Ni(OH)_2$/NiOOH couple was found to be independent of the degree of charge of the electrode above 10% charge (i.e. in the range $0.05 \leqslant x \leqslant 0.5$)[13]. Studies by X-ray crystallography[9,10] showed a continuous variation of the lattice parameters with the degree of charge of the electrode.

14.1 The Charge Capacity of NiO_x Electrodes

14.1.1 General

In this experiment the most fundamental characteristics of a battery electrode will be studied, namely the variation of potential with time during charge and discharge and the dependence of the charge capacity on the charging rate. Two types of nickel oxide electrodes will be used for this purpose. The commercial nickel plaque type[4,5] and a thin home-made type as prepared by Wynne-Jones et al.[6] A distinct difference will be observed between the behavior of the two types of electrodes, due to the fact that in the former most of the active material is in the bulk of the solid phase, while in the latter essentially all of it is on the surface.

14.1.2 The Cell, Solutions and Electrical Setup

The cell shown in Fig. 14.1 will be used in this experiment. Careful fabrication of the teflon sleeve of the working electrode

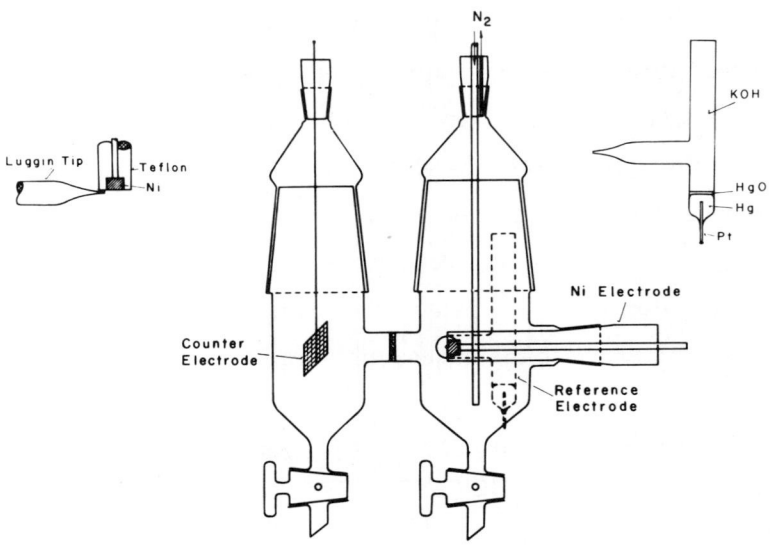

FIG. 14.1 H-CELL FOR STUDYING REACTIONS INVOLVING GAS EVOLUTION.

will eliminate leakage of the solution. There are two reference-electrode compartments on opposite sides of the cell and the ends of the Luggin capillaries reach to the edge of the working electrode. The counter electrode, (platinum wire or gauze), is separated from the working electrode by means of a medium-pore sinter.

Measurements will be performed in 1 $\underline{M}$ KOH solutions. For preparation of the layer of nickel hydroxide, a solution of 0.1 $\underline{M}$ $NiSO_4$, 0.1 $\underline{M}$ NaAc (sodium acetate) and 0.001 $\underline{M}$ KOH will be employed. All solutions will be made up of analytical-grade reagents and triply distilled water.

The simple galvanostatic circuit shown in Fig. 2.4 will be employed, with a strip-chart recorder to measure the variation of potential with time.

14.1 CHARGE CAPACITY

14.1.3 Preparation of Electrodes

Home-made working electrode

This electrode is constructed from a nickel rod* 0.5 - 1.0 cm in diameter, press-fitted into a teflon rod in a configuration similar to a rotating-disc electrode. Preparation of the nickel hydroxide surface will follow one of the procedures described by Wynne-Jones et al.[6] Degrease the electrode, then polish with α-alumina powder (particle size 0.3 μ), immerse for a few seconds in a 20% HCl solution and wash thoroughly with triply distilled water. Place the electrode in a simple single-compartment cell (a beaker with a suitable cover will do) which contains a solution of 0.1 $\underline{M}$ $NiSO_4$, 0.1 $\underline{M}$ NaAc and 1×10^{-3} $\underline{M}$ KOH at room temperature. The electrode should be positioned at an angle to allow easy escape of gas bubbles. Arrange the platinum-wire counter electrode to face the working electrode, thus providing relatively even current distribution. Polarize the electrode galvanostatically at a current density of 1 mA/cm^2 for one-minute periods in the anodic and cathodic directions. A satisfactory $Ni(OH)_2$ layer is generated after 4-6 such anodic/cathodic cycles.

Transfer the working electrode to the cell shown in Fig. 14.1, prepare two reference electrodes in situ as described below and add to the cell a 1 $\underline{M}$ KOH solution. Apply an anodic

---

* It is not necessary to use very high-purity metal for this purpose. Commercial nickel having a purity of 99.4% was found to be suitable for these experiments.

current density of 2 mA/cm^2 galvanostatically (cf. Fig. 2.4) and follow the potential on a strip-chart recorder (sensitivity 25 mV/cm) until oxygen evolution starts. At this point reverse the current and follow the variation of potential with time until the onset of hydrogen evolution. The beginning of gas evolution at each end of the charging cycle is accompanied by a sharp change of potential with time. Repeat about 4-6 times. Avoid prolonged gas evolution at either end of the cycle. This cycling procedure causes aging of the electrode and makes its further behavior more reproducible.

Commercial working electrode

A commercial working electrode should be immersed in 1 M KOH and aged by anodic and cathodic cycling as described above. Since such electrodes have a large charge capacity, a small section, about the size of a single square on the metal grid (ca 2 x 2 mm) should be used. The current used for aging an electrode this size should be about 5 mA*.

Reference electrodes

The Hg/HgO reference electrodes are prepared in situ as follows: Introduce into the reference-electrode compartment (Fig. 14.1), enough mercury to cover the platinum wire completely. Spread red mercuric oxide in a thin layer (about 1 mm.) on the surface of the mercury with a glass rod and mix it slightly with

---

* If manufacturer's specifications are available, the electrode should be aged by cycling at a current density corresponding to 1-3 minutes charging rate.

the mercury. Fill the cell and the reference-electrode compartments with KOH solution. The reference electrode may also be prepared after the solution has been introduced into the cell. In this case, however, care should be taken that HgO does not accidentally enter the working-electrode compartment. An excessive amount of HgO should be avoided since this material is not conducting and might insulate the mercury surface from the solution. It is best to have a cell with two reference electrodes. Equality of the potentials of the two (to within a few millivolts) is an indication of proper operation of both.

14.1.4 <u>Determination of the Charge Capacity</u>

Set up the cell, (Fig. 14.1), with the home-made working electrode. Connect it for galvanostatic operation (Fig. 2.4), with a strip-chart recorder for following the potential during the charging and discharging of the electrode.

Charge the working electrode at a current density of 0.5 mA/cm^2 until oxygen evolution starts. Allow about one minute on open circuit, then apply the same current density in the cathodic direction and continue until hydrogen is evolved for about two minutes to ensure complete reduction of NiOOH to $Ni(OH)_2$. Estimate the charge capacity of the electrode from the length of the plateau in the potential/time plot during discharge.

14.1.5 Variation of the Charge Capacity with Charging Rate*

In this experiment, charging rates of about 5 minutes and about 30 seconds will be used. From the value of the charge capacity obtained in the previous section, calculate the current required in each case.

Stir the solution and bubble purified nitrogen through it during the whole experiment.

Charge the home-made working electrode at the calculated current density until oxygen is evolved. Allow one minute at open circuit (in order to remove oxygen from the vicinity of the electrode) and then discharge at the same current density. At the end of the discharge step, continue to evolve hydrogen for two minutes to ensure complete reduction of NiOOH. If, for any reason, the cycle is to be repeated, allow one minute at open circuit for removal of hydrogen from the vicinity of the electrode.

In this manner, obtain the charge and discharge E/t curves for both charging rates.

Repeat this section with the small commercial working electrode. The current density required for the above charging rates may be obtained from manufacturer's data or estimated from the cathodic branch of the last aging cycle.

---

* The charge or discharge rate of a battery is often defined in terms of the time that would be required to charge or discharge it completely, assuming that all the charge is available.

14.1.6 Treatment of Results

Determine the anodic and cathodic charge from the length of the plateau at each of the two charging rates for the two types of electrodes employed in this experiment. Comment on the difference in the behavior of the two electrodes.

In each case, calculate the charging efficiency (defined as the ratio of cathodic to anodic charge).

Replot the E/t curves in terms of potential vs charge. Note the loss in energy (increase in overpotential) with increase in charging rate.

14.2 Open-Circuit-Decay Measurements of NiO_x Electrodes

14.2.1 General

Open-circuit-decay measurements are based on the idea that the electrode/solution interface may be represented by an equivalent circuit comprising a capacitor (the ionic-double-layer capacitance and the adsorption pseudocapacitance combined in parallel) and a complex resistor (the faradaic resistance) in parallel with it. When a steady-state potential or current is applied, the capacitor is charged to a fixed value and a current corresponding to the value of the potential applied flows across the interface. When the polarizing circuit is disconnected, (i.e. the electrode is placed on open circuit), the capacitor is discharged across the complex resistor and the potential decays with time. It should be borne in mind that the values of both the capacitor and the resistor may vary with potential (and time) in a manner characteristic of the system studied. To describe the situation at the instant the polarizing current or potential is turned off, we may write (cf. section II.8.4)

$$i(t=0) = -C\left(\frac{\partial \eta}{\partial t}\right)_{t \to 0} \qquad [14.3]$$

where $i(t=0)$ is the steady-state polarization current and C is the value of the capacitance at the steady-state overpotential, $\eta(t=0)$, corresponding to the current density $i(t=0)$. By polarizing at different current densities it is possible to use equation [14.3] to evaluate C as a function of η.

Equation [14.3] may be extended and written in the form

$$i = -C\left(\frac{\partial \eta}{\partial t}\right) \qquad [14.4]$$

where i is the current which would flow under steady-state polarization at an overpotential η equal to the value of the overpotential reached at time t and $\partial \eta/\partial t$ is the slope measured at that time*. Thus, equation [14.4] can be used to obtain the total capacitance of the interphase as a function of overpotential from analysis of a single open-circuit-decay curve[15].

If we write the current i in the form

$$i = i_o \exp\left(\frac{\eta}{b_a^t}\right) \qquad [14.5]$$

---

* It is tacitly assumed in deriving this equation that the mechanism during polarization and during open-circuit decay is the same, so that for a given value of the overpotential the same current flows across the interface.

14.2 OPEN-CIRCUIT-DECAY MEASUREMENTS

and combine the last two equations, we have

$$-\frac{i_o}{C} dt = \exp\left(-\frac{\eta}{b_a'}\right) d\eta \qquad [14.6]$$

where $b_a' = b_a/2.3$. Upon integration and rearrangement this yields

$$\eta = a - b_a \log(t + \tau) \qquad [14.7]$$

in which τ is a constant of integration given by

$$\tau = \frac{b_a C}{i(t=0)} \qquad [14.8]$$

Equation [14.7] is an approximate result, based on the assumption that the capacitance is not a function of potential over the potential region of interest. If C is not constant but varies exponentially with potential, a linear relationship between η and $\log(t+\tau)$ would still be obtained, but the slope will no longer be equal to the Tafel slope[12]. The value of C in equation [14.8] will in any case correspond to the initial overpotential at which the circuit is polarized. The parameter τ can be evaluated from the open-circuit-decay transient in the following manner. Differentiating equation [14.7] and rearranging, we have

$$-\frac{dt}{d\eta} = \frac{\tau}{b_a} + \frac{t}{b_a} \qquad [14.9]$$

Thus, a plot of $dt/d\eta$ vs t yields $1/b_a$ as the slope and τ/b_a as the intercept. If the correct value of τ is used to plot η vs $\log(t+\tau)$, a straight line should be obtained, extending to very short times. Moreover, it has been shown that all points obtained by starting open-circuit decay from different polarization current densities or overpotentials fall on the same line if the proper value of τ is used for each set of data.

In the experiments below, open-circuit-decay measurements will be taken in two potential regions. First, open-circuit decay and buildup of potential (following anodic and cathodic polarization, respectively) will be studied with the commercial working electrode half-charged, (formal composition $NiO_{1.25}$). Then the home-made electrode will be overcharged and open-circuit decay will be measured following oxygen evolution at different current densities. The results of the latter experiments will be compared with polarization curves measured on the same electrodes.

14.2.2 <u>The Cell, Solutions, Electrodes and Electrical Setup</u>

These will be exactly as in Experiment 14.1. The commercial working electrode will be used to study the behavior of the partially charged system ($NiO_{1.25}$) and the home-made electrodes will be employed for studies of oxygen evolution and open-circuit decay on overcharged electrodes (NiO_x with $x > 1.5$). The latter should be prepared just before each experiment.

14.2.3 <u>Measurement of the Reversible Potential for the Ni(OH)$_2$/NiOOH System</u>

The procedure suggested by Conway <u>et al</u>[10,13] will be followed.

Determine the charge capacity of the electrode as in

14.2 OPEN-CIRCUIT-DECAY MEASUREMENTS

Experiment 14.1. Choose a current density corresponding to a charging rate of about ten minutes. Charge the electrode to half its full charge (Q = 50%), stop charging and follow the decay of potential on open circuit. After the potential has become steady, continue charging to Q = 75%, then reverse the current and discharge at the same rate back to Q = 50%. Stop the current and follow the buildup of potential with time on open circuit, until a constant potential is reached.

Increase the current density by a factor of two. Discharge to Q = 25%, reverse the current and charge to Q = 50%, then follow open-circuit decay. Charge to Q = 75%, discharge back to Q = 50% and follow the buildup of potential at open circuit.

14.2.4 Polarization Curves for Oxygen Evolution on NiO_x

Charge the home-made electrode at a current density of 10 mA/cm^2 and continue steady-state oxygen evolution for two minutes, or until the potential has reached a constant value. Decrease the current density in about ten steps to 0.1 mA/cm^2 and measure the steady-state potential at each current density. Repeat the measurement with the current density increasing by the same steps.

14.2.5 Open-Circuit Decay of Overcharged NiO_x Electrodes

Charge the home-made electrode at a current density of 5 mA/cm^2 until the potential reaches a constant value. Switch off the current and follow the decay of potential on open circuit. Repeat the measurement at a current density of 0.5 mA/cm^2.

The strip-chart recorder should be adjusted to a speed of about 1 cm/sec and set in motion just before the current is switched off. It will be necessary to change the recorder speed and perhaps its sensitivity at longer times, when the rate of change of potential with time has decreased.

14.2.6 Treatment of Results

Determine the reversible potential for the $Ni(OH)_2/NiOOH$ system by extrapolating the E vs $\log(t+\tau)$ plots obtained for the half-charged electrodes to their point of intersection. Compare with the results given in the literature.

Calculate the parameter τ for this system making use of equation [14.9]. From it calculate the electrode capacitance.

Plot η vs $\log i$ for the oxygen-evolution reaction on the overcharged NiO_x electrode, and determine the Tafel parameters b_a and i_o.

Find the value of τ for open-circuit decay following oxygen evolution at 5 and 0.5 mA/cm^2 and plot η vs $\log(t+\tau)$ on a single graph paper. Determine the slope of this plot and compare to the Tafel slope b_a for the system. Discuss the possible reasons for the difference between these slopes.

Assuming that the capacitance changes with potential according to the equation

$$C = A \exp(-B\eta) \qquad [14.10]$$

obtain the relationship between the Tafel slope and the open-circuit-decay slope, $\partial\eta/\partial \log(t+\tau)$. (Hint: integrate equation [14.6]).

14.2 OPEN-CIRCUIT-DECAY MEASUREMENTS

Estimate the iR correction (cf. Introduction section to Experiment 2) from the difference between the value of the overpotential at steady state and its extrapolated value obtained from the $\eta/\log(t+\tau)$ plot at $t = 0$.

REFERENCES

1. G. W. Vinal, "Primary Batteries", J. Wiley, 1950.
2. G. W. Vinal, "Storage Batteries", J. Wiley, New York, 1967.
3. R. Jasinski, "High Energy Batteries", Plenum Press, N. Y. 1967.
4. E. J. Casey, P. L. Bourgault and P. E. Lake, Can. J. Technol. 34, 95 (1956).
5. P. C. Milner and U. B. Thomas, "Advances in Electrochemistry and Electrochemical Engineering", Vol. 5, Chap. 1, ed. P. Delahay and C. W. Tobias, Interscience Publishers N. Y. (1967).
6. G. W. D. Briggs, E. Jones and W. F. K. Wynne-Jones, Trans. Faraday Soc. 51, 1433 (1955).
7. E. Jones and W. F. K. Wynne-Jones, Trans. Faraday Soc. 52, 1260 (1956).
8. G. W. Briggs and W. F. K. Wynne-Jones, Electrochim. Acta, 7, 241 (1962).
9. G. W. D. Briggs, G. W. Stott and W. F. K. Wynne-Jones, Electrochim. Acta, 7, 249 (1962).
10. P. L. Bourgault and B. E. Conway, Can. J. Chem. 38, 1557 (1960).
11. B. E. Conway and P. L. Bourgault, Can. J. Chem. 37, 292 (1959).
12. B. E. Conway and P. L. Bourgault, Trans. Faraday Soc. 58, 593 (1962).
13. B. E. Conway and E. Gileadi, Can. J. Chem. 40, 1933 (1962).
14. F. P. Kober, J. Electrochem. Soc. 112, 1064 (1965).
15. B. E. Conway, E. Gileadi and M. Dzieciuch, Electrochim. Acta 8, 143 (1963).

ϕ_2	potential of outer Helmholtz plane
$\phi_{\neq}$	potential at site of activated complex
$\Delta\phi$	inner potential difference across interphase
$\Delta\phi_r$	absolute metal/solution potential difference at equilibrium
$\Delta\phi_z$	potential difference across interphase at $q_M = 0$
χ	surface potential
$\Delta\chi$	surface potential difference
χ	$\left(\dfrac{12}{7}\right)^{1/2} \lambda \left[= \left(\dfrac{12}{7}\right)^{1/2} \dfrac{k_c t^{1/2}}{D^{1/2}} \text{ or } \left(\dfrac{12}{7} \cdot Kk_f t\right)^{1/2} \text{ according to reaction conditions} \right]$
$\Delta\psi$	outer potential difference
ω	angular velocity

LIST OF SYMBOLS

ν	kinematic viscosity
ξ	$\left(= \dfrac{D_{Ox}}{D_{Red}}\right)^{1/2}$
ρ_x	charge density per unit volume at a distance x from the surface of the electrode
$\rho_1(k)$, $\rho_2(k)$	reaction-order parameters
$\Delta\rho$	difference in density (between aqueous solution and mercury)
τ	transition time (chronopotentiometry)
τ_f, (τ_r)	forward (reverse) transition time (chronopotentiometry with current reversal)
τ	drop time (of mercury)
τ	time of measurement of capacitance in double-layer experiment
τ	mean lifetime of species in solution
τ	constant of integration in open-circuit-decay measurement $\left(= \dfrac{b_a C}{i(t=0)}\right)$
τ	dimensionless parameter $\left(= \dfrac{Dt}{L^2}\right)$ in hydrogen-permeation experiment
ϕ	inner potential of a phase
ϕ_M	inner potential of metal
ϕ_S	inner potential of the solution
ϕ_x	inner potential at a distance x from the electrode surface
ϕ_o	inner potential at the limit of the diffuse double layer

LIST OF SYMBOLS

Γ	surface concentration
Γ_i	surface excess of component i
Γ_i'	relative surface excess of component i
Γ_m	maximum surface concentration
δ	Nernst diffusion-layer thickness
ε	dielectric constant
η	overpotential
η_a	activation overpotential
η_d	diffusion overpotential
η_R	resistance overpotential
θ	fractional surface coverage
θ_{eq}	fractional coverage at equilibrium
$\theta(t)$	momentary value of fractional coverage
θ	ratio of concentrations of reactants and products of reduction reaction at electrode surface $\left(\dfrac{C_{Ox}}{C_{Red}}\right)_{x=0}$
κ	Debye-Huckel reciprocal length
λ	$(k_c t^{1/2}/D^{1/2})$ or $(Kk_f t)^{1/2}$ according to reaction conditions
μ	chemical potential
$\bar{\mu}$	electrochemical potential
μ^o	standard chemical potential
μ	thickness of reaction layer
$\vec{\mu}$	dipole moment
ν	stoichiometric number

LIST OF SYMBOLS

V_{sq}	voltage of square-wave signal
$V_{out}(max)$	maximum output voltage of galvanostat
V_{RE}, (V_{WE}), (V_{CE})	potential of reference, (working), (counter) electrode
V_{RE-WE}	voltage difference between reference and working electrodes
W	weight of mercury drop
x_i	generalized coordinate
X_i	mole fraction of species i
z_i	valency of ion i (including sign)
z(a, Red)	electrochemical reaction order for species Red in oxidation reaction
z(a, Ox)	electrochemical reaction order for species Ox in oxidation reaction
z(c, Red)	electrochemical reaction order for species Red in reduction reaction
z(c, Ox)	electrochemical reaction order for species Ox in reduction reaction
Z	impedance
α	transfer coefficient
β	symmetry factor
γ	interfacial tension
γ_{max}	interfacial tension at electrocapillary maximum
γ	in hydrogen-permeation experiment $(= \frac{i(t)}{i(\infty)})$
γ	in potentiostat with positive feedback, the fraction of V_{out} at amplifier 3 which is fed back to amplifier 1

LIST OF SYMBOLS

r_1, r_2	radii of curvature of mercury in capillary
R	gas constant
R_{in}	input resistance of operational amplifier
R_f	feedback resistance of operational amplifier
R_F	average faradaic resistance of an electrochemical reaction at equilibrium
R_p	internal resistance of glass electrode
S_D	separation factor for deuterium
S_T	separation factor for tritium
t_f	time of current reversal in chronopotentiometry
T	temperature (°K)
U	energy of lateral interaction between water molecules on the surface
v	sweep rate dE/dt
v	rate of chemical reaction
v_i^e	exchange rate of reaction
V	volume of solution
V	volume of mercury delivered per unit time
V_d	volume of mercury drop
V_{in}	input voltage of operational amplifier
V_{out}	output voltage of operational amplifier
$V_+, (V_-)$	potential at positive (negative) input of operational amplifier
V_{CM}	voltage with respect to ground of each input to operational amplifier
ΔV	input bias voltage ($= V_{in}/V_{RE}$)

LIST OF SYMBOLS

n_a	number of electrons taking part in the rate-determining step
$\vec{n}, (\overleftarrow{n})$	number of electrons passed before (after) the rate-determining step per act of the overall reaction
$n_i(x)$	number of ions per unit volume at a distance x from the electrode
n_i^o	number of ions per unit volume in the bulk of the solution
N	Avogadro's number
$N\uparrow, (N\downarrow)$	number of molecules of water in the "up" ("down") position per cm^2
ΔP	pressure difference required to bring mercury to a certain point in the capillary
q_M	surface charge density on metal
q_-	excess of surface charge due to anions
q_-^{M-2}	part of excess surface charge due to specific adsorption
q_-^{2-S}	part of excess surface charge due to adsorption in the diffuse double layer
q_F	faradaic charge during transient, related to change of θ
Q	electrical charge
Q_H	charge required to ionize all the hydrogen initially adsorbed on the surface
r	interaction parameter in Frumkin isotherm
r	radius of capillary

LIST OF SYMBOLS

k_f, (k_b)	specific rate constant for forward (backward) reaction in an equilibrium
k_f^o	specific rate constant for forward reaction under zero field ($\Delta\phi$ = o) conditions
$k_{f,s}$	specific rate constant for forward reaction at the standard potential E^o
k_1, k_{-1}, k_2	electrochemical rate constants (potential dependent)
k_s	standard rate constant (η = o)
k'	charge needed to form a monolayer of adsorbed intermediates
K	integral capacity
K	equilibrium constant of chemical reaction $(= \dfrac{k_f}{k_b}$ or $\dfrac{k_1}{k_{-1}})$
K	$(= R_f/R_{in})$ under certain conditions equal to gain of operational amplifier
K	$nFD^{1/2}C^o\pi^{-1/2}(1 + \xi\theta)^{-1}$
L	distance between electrodes in a thin-layer cell
L	thickness of palladium membrane
m	flow rate of mercury from capillary
m	number of solvent molecules replaced by each molecule of adsorbed species
M	site on metal surface
n	number of electrons passed per act of the overall reaction

LIST OF SYMBOLS

i_{corr}	corrosion current density
i_d	diffusion current (polarographic)
$\vec{i}$, $\overleftarrow{i}$	partial current densities for forward and backward reactions
i_p	peak current density (potential sweep)
i_{pol}	current density on the polarization side of a metal membrane
i_{cell}	current flowing through electrochemical cell
i_{ac}	activation-controlled current density
i^o	initial current density (constant-potential electrolysis)
$i(\infty)$	steady-state current (in hydrogen permeation experiments)
i_m	amplitude of sinusoidal current wave
i_{in}	input current to operational amplifier
i_b	input bias current to operational amplifier
Δi_b	input offset current to operational amplifier
$i_{out}(max)$	maximum output current of galvanostat
I	ionic strength
k	Boltzmann constant
k_c, (k_a)	specific rate constant for cathodic (anodic) reaction
k_c^r, (k_a^r)	specific rate constant for cathodic (anodic) reaction at the reversible potential
k_c^o, (k_a^o)	specific rate constant for cathodic (anodic) reaction under zero-field ($\Delta\phi = 0$) conditions

E_p	peak potential (potential sweep)
E_{pol}	potential applied on the polarization side of a metal membrane
$E_{1/4}$	value of E for which $t = \tau/4$ (chronopotentiometry)
E_{corr}	corrosion (mixed) potential
$E^{\neq}$	apparent energy of activation
f	activity coefficient
f	frequency of sine wave
F	Faraday constant
$\vec{F}$	electric field in double layer
g	number of nearest neighbors of a water molecule on a surface
$\Delta G^{o\neq}$	standard free energy of activation
$\Delta \bar{G}^{o\neq}$	standard electrochemical free energy of activation
ΔG^o	standard electrochemical free energy of reaction
ΔG_θ^o	apparent standard free energy of adsorption
h	Planck's constant
i_c, (i_a)	cathodic (anodic) current
i_o	exchange-current density
$i_{o,c}$ $(i_{o,a})$	exchange-current density for cathodic (anodic) process-spontaneous corrosion
i_L	limiting current density
$i_{L,c}$, $(i_{L,a})$	limiting current density for cathodic (anodic) process

LIST OF SYMBOLS

CMRR	common-mode rejection ratio of operational amplifier
C_i	concentration of ith species
$C(s)$	concentration at ϕ_2 (OHP)
C^o	concentration in bulk of solution
C^o	initial concentration (constant potential electrolysis)
d	diameter of an oriented layer of water molecules on the surface
D	diffusion coefficient
e_o	electronic charge
e_M	electron in the metal
E	potential with respect to some reference electrode
E_z	potential of zero charge
$\overline{E}$	potential on the rational scale ($\overline{E} = E - E_z$)
E_N	shift in E_z as θ goes from zero to unity
E_+, (E_-)	potential measured with respect to reference electrode reversible to some cation (anion) in the solution
E_r	reversible potential
E^o	reversible standard potential
E_{RE}, (E_{WE})	metal/solution potential difference at the reference (working) electrode
E_{rc}, (E_{ra})	reversible potential for a cathodic (anodic) process
$E_{1/2}$	polarographic half-wave potential

LIST OF SYMBOLS

a	Tafel parameter ($b \log i_o$)
a	distance of closest approach of ion
a_i	activity of species i
A	area of electrode
A_o	open-loop gain of operational amplifier
A_{CM}	common-mode gain of operational amplifier
A_i	affinity of ith step in a reaction sequence
b	Tafel slope ($\pm \partial\eta/\partial \log i$)
b'	Tafel slope ($\pm \partial\eta/\partial \ln i$)
b_c, (b_a)	Tafel slope for reduction (oxidation) reaction
B	hydrodynamic parameter for rotating-disc electrode ($i_L/\omega^{1/2}$)
B	excess of water molecules oriented in one direction at the surface of an electrode ($\frac{N\uparrow - N\downarrow}{N\uparrow + N\downarrow}$)
C	differential capacity
C_{dl}	double-layer capacity
C_o	double-layer capacity at $\theta = 0$
C_1	double-layer capacity at $\theta = 1$
C_H, C_{M-2}	capacity of Helmholtz layer
C_G, C_{2-S}	capacity of Gouy–Chapman layer (diffuse double layer)
C_ϕ	adsorption pseudocapacity
C_{in}	input capacitance to operational amplifier
C_f	feedback capacitance in operational amplifier

REFERENCES (cont'd)

16. E. Gileadi and B. E. Conway in "Modern Aspects of Electrochemistry", ed. J. O'M. Bockris and B. E. Conway, Vol. 3, Butterworths, London, 1964.

INDEX

Activation energy (see energy)
adsorption, contact, 14
    of neutral species, 96
    specific, effect on electro-
        capillary maximum, 36, 241, 251
adsorption-desorption peaks, 241
affinity of reaction, 77
amalgam preparation, 354
amplifier follower, 156
anodic protection, 124

Batteries, 127

Capacitance, diffuse-double-layer, 238
    double-layer, integral, 2, 23
        differential, 2, 4, 6, 30, 185, 234, 237, 241, 252
    Helmholtz-layer, 238
capacitance meter, 241
cathodic discharge of metal ions, 138
cathodic protection, 124
charge density, excess surface, 4, 28, 31, 255
charge transfer, 51

chronopotentiometry, 63, 399
    thin-layer, 64
    with current reversal, 402, 411
common-mode rejection ratio, 169
constant-current electrolysis
    (see electrolysis)
constant-potential electrolysis
    (see electrolysis)
contact adsorption (see adsorption)
corrosion inhibitors, 126
corrosion potential (see potential)
current density, activation-
    controlled, 310, 323
        anodic, 45
        cathodic, 45
        corrosion, 113
        diffusion-limited, 359
        exchange, 43, 46, 68, 321
            variation with reversible
            potential and with
            concentration, 335
        limiting, 59, 310
        net, 45
current-to-voltage converter, 163
current variation with time, at
    constant potential, 106
    during growth of mercury
        drop, 287
cyclic voltammetry, 370

523

Debye reciprocal length, 5
dielectric constant, in bulk of
  solution, 15
  in double layer, 15
differentiator (electronic), 161
diffuse-double-layer theory (see
  double layer, Gouy-Chapman
  model)
diffusion coefficient of hydrogen
  in palladium, 475
diffusion current, polarographic,
  60
discharge rate of a battery, 130
double layer, capacitance (see
  capacitance)
  effect of solvent on, 13
  Gouy-Chapman model, 3, 237
  Helmholtz model, 1
  Stern model 6, 237
drop time, 34
dummy cell for testing potentiostat, 185, 195

Electrocapillary electrometer, 32
electrocapillary maximum, 31, 234
  shift in, as result of
    specific adsorption, 36, 241
electrocrystallization, 140
electrode, dynamic hydrogen, 214
  indicator, 207
  palladium/hydrogen, 220
    preparation, 224
  parallel-plane, 210
  platinized-platinum, preparation, 216
  reference, 208
  silver/silver perchlorate,
    in propylene carbonate, 226
  streaming-mercury, 247
electrodics, vii
electrolysis, constant-current,
  383
  constant-potential, 61, 388
electroplating, 138, 140
electrosorption (see adsorption)
electrowinning, 147
energy of activation, apparent,
  74
equivalent circuit, 89
Esin-Markov effect, 239, 251
exchange current density (see
  current density)
exchange rate of reaction, 77

Faradaic efficiency, 66, 142
faradaic resistance (see
  resistance)
feedback, 153
field-effect transistor, 156
free-energy, of activation,
  standard, 44
    standard electrochemical,
      45, 59
  of adsorption, apparent
    standard, 82, 86

of reaction, standard
  electrochemical, 49
frequency response of equivalent
  circuit, 89
Frumkin correction for double-
  layer effects, 56
Frumkin isotherm, 83
  electrosorption of benzene,
    466
fuel cells, 127, 131

Gain, 165
galvanostat, 179
galvanostatic-transient technique,
  350, 484

Hanging mercury drop, preparation,
  353
Helmholtz plane, inner and outer,
  7, 14
    concentration of ionic
      species at outer, 263
hydrogen-evolution reaction, 294
hydrogen/oxygen fuel cell, 132

Ilkovic equation, 60
immunity, 121
input bias current, 167, 177, 188
input bias voltage, 188
input impedance, 170
input offset current, 169
input offset voltage, 166, 178
integrator (electronic), 161
interfacial tension (see tension)
interphase, vii
  polarizable and non-
    polarizable, 21
inverting-voltage amplifier, 157
ionics, vii
irreversible thermodynamics, 76
isotope effects on reaction
  mechanism, 73, 295

Kinematic viscosity, 309

Langmuir adsorption isotherm, 79
limiting current density (see
  current density)
Lippman equation, 30
Luggin capillary, 184, 192, 210

Maximum-bubble-pressure method,
  34
maximum input voltage, 170
mixed potential (see potential)
multistep processes, 52

Nickel oxide electrode, 128, 495
number of electrons transferred
  per act of overall reaction,
  60
    determination by chrono-
      potentiometry, 63, 401

Index

determination by constant-potential electrolysis, 61, 388
determination by linear potential sweep, 65, 383
determination by polarography, 60
determination by potential-step method, 62
determination by thin-layer chronopotentiometry, 64

Open-circuit decay, 92, 503
open-loop gain, 165, 172
operational amplifier, 152
output current, 170
output swing voltage, 170
overpotential, 43, 46
    activation, 58
    diffusion, 58
    resistance, 59
oxygen-evolution reaction, 55

Passivity, 115
pH, effect on reaction order, 72
polarographic diffusion current (see diffusion current)
polarography, 60
potential, chemical, 16
    corrosion, 112
    difference across interphase, 16, 46
    difference between two identical phases, 19
    difference between two phases at equilibrium, 20
    diffuse-double-layer, 4
    electrochemical, 17
    inner, 17, 26
    inner Helmholtz plane, 7
    metal-surface, 4
    mixed (see potential, corrosion)
    outer, 18
    outer Helmholtz plane, 7, 238
        dependence on applied potential, 264
    quarter-wave, 401, 404
    rational, 103, 237, 265
    variation with distance, 5
    zero-charge, 6, 31, 234, 237
        determination with streaming-mercury electrode, 247
potential/pH (Pourbaix) diagrams, 117
potential-sweep, linear, 65, 370, 448
potentiostat, 181
    with positive feedback, 191, 195

potentiostatic-transient technique, 357, 433
pseudocapacitance, adsorption, 86, 433

Quarter-wave potential (see potential)
quasi-equilibrium, 75, 78

Rate constant, specific, 45, 372, 380, 404
reaction layer, 271
reaction order, 69, 80, 329
reaction rate, 44
resistance, faradaic, 23, 47, 144, 185
    solution, 144
rise time of potentiostat, 190
rotating-disc (or-cone) electrode, 299, 309
roughness factor, 133, 437, 486

Slewing rate, 171, 176, 190
solubility of hydrogen in palladium, 475
solvent effect on double layer (see double layer)
specific adsorption (see adsorption)
Stern model of double layer (see double layer)
stoichiometric number, 53, 68
streaming-mercury electrode (see electrode)
summing inverting amplifier, 159
surface excess, 28
    relative 29, 235, 255
surface tension (see tension)
symmetry factor, $\overline{45,}$ 49

Tafel equation, 43
Tafel slope, 43, 67, 373
Temkin isotherm, 83
tension, interfacial, 27, 32, 36, 234, 253
thin-layer chronopotentiometry (see chronopotentiometry)
throwing power, 143
transfer coefficient, 54, 109, 315, 322
transition time, 63, 400

Voltage follower, 154
voltage subtractor, 159
voltammetry at constant current (see chronopotentiometry)
voltammetry at constant potential, 418
    two-step, 421

/541.37246471I>C1/